Materials Horizons: From Nature to Nanomaterials

Series Editor

Vijay Kumar Thakur, School of Aerospace, Transport and Manufacturing, Cranfield University, Cranfield, UK

Materials are an indispensable part of human civilization since the inception of life on earth. With the passage of time, innumerable new materials have been explored as well as developed and the search for new innovative materials continues briskly. Keeping in mind the immense perspectives of various classes of materials, this series aims at providing a comprehensive collection of works across the breadth of materials research at cutting-edge interface of materials science with physics, chemistry, biology and engineering.

This series covers a galaxy of materials ranging from natural materials to nanomaterials. Some of the topics include but not limited to: biological materials, biomimetic materials, ceramics, composites, coatings, functional materials, glasses, inorganic materials, inorganic-organic hybrids, metals, membranes, magnetic materials, manufacturing of materials, nanomaterials, organic materials and pigments to name a few. The series provides most timely and comprehensive information on advanced synthesis, processing, characterization, manufacturing and applications in a broad range of interdisciplinary fields in science, engineering and technology.

This series accepts both authored and edited works, including textbooks, monographs, reference works, and professional books. The books in this series will provide a deep insight into the state-of-art of Materials Horizons and serve students, academic, government and industrial scientists involved in all aspects of materials research.

Review Process

The proposal for each volume is reviewed by the following:

1. Responsible (in-house) editor
2. One external subject expert
3. One of the editorial board members.

The chapters in each volume are individually reviewed single blind by expert reviewers and the volume editor.

Arbind Prasad · Vimal Katiyar
Editors

Resorbable Polymers for Bioimplants and Fixation Devices

Springer

Editors
Arbind Prasad
Department of Mechanical Engineering
Katihar Engineering College
Katihar, Bihar, India

Vimal Katiyar
Department of Chemical Engineering
Indian Institute of Technology Guwahati
Guwahati, Assam, India

ISSN 2524-5384 ISSN 2524-5392 (electronic)
Materials Horizons: From Nature to Nanomaterials
ISBN 978-981-96-9305-4 ISBN 978-981-96-9306-1 (eBook)
https://doi.org/10.1007/978-981-96-9306-1

This Springer imprint is published by the registered company Springer Nature Singapore Pte Ltd.
The registered company address is: 152 Beach Road, #21-01/04 Gateway East, Singapore 189721, Singapore

This book is dedicated to all engineers, scientists, doctors, health professionals, biotechnologists, researchers, and academicians…

Preface

The proposed book titled *Resorbable Polymers for Bioimplants and Fixation Devices* will be based on the latest research outcomes, perspective research, and current research ongoing in the areas of internal fixation devices, scaffolds, bioimplants, emerging resorbable polymeric nanomaterials, and its composites. There is a huge demand for resorbable composites especially biocompatible in nature for application in bone implants, drug delivery systems, wound healing, hydrogels, biomaterials for bone fracture fixations, and other medical implants. Resorbable composites from natural resources have significant and versatile roles for their worthwhile applications in almost all domains of biomedical applications. The self-healing and bioinspired materials in the area of biomedical application play a significant role in fixing the problems associated. This proposed book covers exhaustive studies and relevant case studies about emerging resorbable materials, their synthesis, characterization, and application in various domains of biomedical areas. The advances in resorbable polymeric material, implantable devices, drug delivery, bioimaging and scaffolds, natural and polymeric material for implants and drug delivery application has established their importance for the development and utilization of implants, fixation devices, dental implants, wound healing, and artificial organ developments. Traditional procedures, implants, and other biomedical devices based on metals have many limitations such as stress shielding, and stress underlying bone damage. The emergence of bioresorbable composites based on naturally sourced biomaterials has always greater impact on their application.

The advantages associated with bioresorbable composites such as ease of modification of the chemical, physical, surface, and biomimetic properties of polymers have made them very dear to the researchers working in resorbable implants, drug delivery, biomaterials, bioimplants, fixation devices, and other biomedical field. A variety of research results are published almost daily on polymers and composites by enhancing their properties as biomaterials to meet the ongoing and evolving challenges in biomedical applications. This book *Resorbable Polymers for Bioimplants and Fixation Devices* deals with the latest bio-based resorbable composites, processing, and characterization for implants, fixation devices, drug delivery, and introduction of multifunctional nanomaterials, composites, biological, and toxicological studies which will have very much impact in biomedical industries and academics. There will be a total of 16 chapters introduced in this book which would be beneficial for budding researchers, scientists, material scientists, academicians, industry professionals, clinical professionals, Engineers, pharma industries, Health professionals, and allied science professionals.

Katihar, India Dr. Arbind Prasad
Guwahati, India Dr. Vimal Katiyar

Acknowledgments We express our gratitude to Springer Publishing House (Springer Nature Singapore Pte Ltd.) and the editorial team for their suggestions and support during the completion of this book. We are grateful to all contributors and reviewers for their illuminating views on each book chapter in Resorbable Polymers for Bioimplants and Fixation Devices.

Contents

Editors and Contributors

About the Editors

Dr. Arbind Prasad has completed his Ph.D. (Mechanical Engineering) from the Indian Institute of Technology Guwahati, Assam, India. He has earned a gold medal in his M.Tech. He is currently working as an assistant professor and head of mechanical engineering in the Department of Science, Technology and Technical Education, Government of Bihar, at Katihar Engineering College, Katihar, Bihar, India. He has four granted patents out of his research work. He has 16 international journal papers, edited 18 books (Wiley Publishing, CRC Press (Taylor & Francis) USA, De Gruyter, Elsevier, Apple Academic, etc.), 48 book chapters, and 16 reputed international conference papers to his credit. He has completed numerous projects funded by the state government. He is an associate editor for the *International Journal of Materials, Manufacturing and Sustainable Technologies* (IJMMST) and an early career editor for the *International Journal of Mathematical, Engineering and Management Sciences* (IJMEMS).

Prof. Vimal Katiyar is currently serving as a professor in the Chemical Engineering Department at the Indian Institute of Technology Guwahati (IIT Guwahati), Assam, India. He has vast experience in teaching, research, and administrative programs. He has published more than 150 articles in various reputed international journals. He holds about 29 granted/published patents to his credit. He has edited six books. His research interests include biodegradable polymers, resorbable scaffolds, plastic and its processing, biomaterials, resorbable polymers, processing, nanomaterials, drug delivery, 3D bioprinting, internal fixation devices, sustainable polymer synthesis and scale-up activities, polymer processing, rheology, structure property relationship, development of polymer-based nanomaterials and its technologies polymer degradation and migration toxicological studies etc.

Contributors

Sruthi Ann Alex Department of Biomedical Engineering, SRM Institute of Science and Technology, Kattankulathur, Tamil Nadu, India

Amit Arora Galgotias University, Greater Noida, Uttar Pradesh, India

Ranjit Barua Centre for Healthcare Science and Technology, Indian Institute of Engineering Science and Technology, Howrah, West Bengal, India;
Omdayal Group of Institutions, Uluberia, Harinarayan Chak, West Bengal, India

Asim Kumar Basak Department of Allied Health Sciences, Brainware University, Kolkata, West Bengal, India

Meghali Bhattacharjee Department of Chemical Engineering, National Institute of Technology Andhra Pradesh, Tadepalligudem, Andhra Pradesh, India

Gourhari Chakraborty Department of Chemical Engineering, National Institute of Technology Andhra Pradesh, Tadepalligudem, Andhra Pradesh, India

Dipankar Das Department of Allied Health Sciences, Brainware University, Kolkata, West Bengal, India

Sudipto Datta Department of Materials Engineering, Indian Institute of Science, Bangalore, Karnataka, India

Sriparna De Department of Allied Health Sciences, Brainware University, Kolkata, West Bengal, India

Revathi Devulapalli Functional Nanomaterials and Polymer Nanocomposite Laboratory, Department of Physics, VIT-AP University, Amaravati, Guntur, Andhra Pradesh, India

Sunkari Dinesh Functional Nanomaterials and Polymer Nanocomposite Laboratory, Department of Physics, VIT-AP University, Amaravati, Guntur, Andhra Pradesh, India

Maneesh Dubey Department of Mechanical Engineering, IIT Madras, Chennai, India

S. Gnanavel Department of Biomedical Engineering, SRM Institute of Science and Technology, Kattankulathur, Tamil Nadu, India

Ruchin Kacker SISTec GN, Bhopal, Madhya Pradesh, India

Subhasish Kamilya Department of Allied Health Sciences, Brainware University, Kolkata, West Bengal, India

S. K. Khadheer Pasha Functional Nanomaterials and Polymer Nanocomposite Laboratory, Department of Physics, VIT-AP University, Amaravati, Guntur, Andhra Pradesh, India

Hemnath Anandan Kumar Department of Mechanical Engineering, Amrita School of Engineering, Amrita Vishwa Vidyapeetham, Chennai, India

Medi Harshith Kumar Department of Chemical Engineering, National Institute of Technology Andhra Pradesh, Tadepalligudem, Andhra Pradesh, India

Bidyanand Mahto Government Engineering College, Vaishali (Under Department of Science, Technology and Technical Education, Government of Bihar), Vaishali, Bihar, India

Syed Naveed Ul Meiraj Indian Institute of Information Technology, Design and Manufacturing, Kancheepuram, Chennai, Tamilnadu, India

Agnish Pahari Department of Allied Health Sciences, Brainware University, Kolkata, West Bengal, India

Subhasree Panda Functional Nanomaterials and Polymer Nanocomposite Laboratory, Department of Physics, VIT-AP University, Amaravati, Guntur, Andhra Pradesh, India;
Department of Chemical Engineering, Yeungnam University, Gyeongsan, Republic of Korea

Atanu Kumar Paul Department of Chemical Engineering, National Institute of Technology Andhra Pradesh, Tadepalligudem, Andhra Pradesh, India

Arbind Prasad Department of Mechanical Engineering, Katihar Engineering College (Under Department of Science, Technology and Technical Education, Government of Bihar), Katihar, Bihar, India

Varatharajan Prasannavenkadesan Trinity College Dublin, Dublin, Ireland

A. Rajeshwari Department of Biotechnology, The American College, Madurai, Tamil Nadu, India

S. Ram Prasanth Indian Institute of Science, Bangalore, Karnataka, India

Rambha Meghana Sai Department of Chemical Engineering, National Institute of Technology Andhra Pradesh, Tadepalligudem, Andhra Pradesh, India

Nisha Shankhwar Special Centre for Nanoscience, Jawaharlal Nehru University, New Delhi, India

P. J. Shiny Division of Medical Research, SRM Medical College Hospital & Research Centre, SRM Institute of Science and Technology, Kattankulathur, Tamil Nadu, India

Sanjay Kumar Singh SISTec GN, Bhopal, Madhya Pradesh, India

V. Subha Department of Chemical and Biological Engineering, University of Sheffield, Sheffield, UK

M. Vishnuvarthanan Department of Mechanical Engineering, Kalasalingam Academy of Research and Education, Srivilliputhur, Tamil Nadu, India

Chapter 1
Background, History, and Introduction to Resorbable Composites for Implants and Fixation Devices

Nisha Shankhwar and Arbind Prasad

1 Introduction

Resorbable composites, also referred to as biodegradable or bioabsorbable materials, represent a class of cutting-edge biomaterials engineered for temporary structural and functional support in biomedical applications [1]. These materials are specifically designed to degrade within the body after fulfilling their purpose, avoiding the need for surgical removal and reducing associated risks and costs [2]. Typically composed of polymeric matrices reinforced with bioactive fillers such as ceramics, glasses, or nanostructured components, resorbable composites offer a highly customizable solution that combines mechanical strength, bioactivity, and controlled degradation kinetics to meet diverse clinical needs [3].

At the heart of resorbable composites is the synergy between the polymer matrix and reinforcing agents. The polymer matrix, often derived from materials such as polylactic acid (PLA), polyglycolic acid (PGA), or polycaprolactone (PCL), serves as the primary structural component. These polymers are chosen for their biocompatibility, processability, and predictable hydrolytic degradation behavior. Reinforcing agents like hydroxyapatite (HA), β-tricalcium phosphate (β-TCP), or bioactive glass are incorporated to enhance mechanical properties, promote cellular adhesion, and stimulate natural tissue regeneration processes. Advances in nanotechnology have

Supplementary Information The online version contains supplementary material available at https://doi.org/10.1007/978-981-96-9306-1_1.

N. Shankhwar (✉)
Special Centre for Nanoscience, Jawaharlal Nehru University, New Delhi, India
e-mail: shankhwarnisha3july@gmail.com; nisha2023@jnu.ac.in

A. Prasad
Mechanical Engineering Department, Katihar Engineering College (Under Department of Science, Technology and Technical Education, Government of Bihar), Katihar, Bihar, India

A. Prasad and V. Katiyar (eds.), *Resorbable Polymers for Bioimplants and Fixation Devices*, Materials Horizons: From Nature to Nanomaterials,
https://doi.org/10.1007/978-981-96-9306-1_1

further improved the precision of material properties, enabled fine-tuned degradation rates and enhanced biological performance [4]. The development of resorbable composites has been transformative for medical implants and fixation devices. Early medical implants relied on non-degradable metals such as stainless steel and titanium, which, while strong and durable, often necessitated secondary surgeries for removal. In the 1960s, the introduction of biodegradable polymers like PLA and PGA marked a significant shift. These polymers were initially employed in surgical sutures and later adapted for other medical uses. The integration of bioactive ceramics in the 1980s further enhanced their utility, particularly in bone fixation and tissue engineering applications. The twenty-first century has seen the advent of smart resorbable materials, integrating nanostructures and tailored degradation mechanisms for specific clinical scenarios [5]. Resorbable composites have revolutionized various fields of medicine due to their versatile properties. In orthopedics, they are extensively used for bone fixation devices such as plates, screws, and pins. These devices provide temporary support while gradually transferring load to healing tissues as they degrade. In maxillofacial and dental surgery, resorbable scaffolds and membranes support bone and periodontal regeneration. In cardiovascular interventions, bioresorbable stents have emerged as a viable alternative to permanent implants, minimizing risks of late-stage complications. Soft tissue engineering has also benefited, with applications in wound healing, tendon repair, and neural scaffolds [6]. Despite their immense potential, resorbable composites face several challenges. Achieving precise control over degradation rates to match tissue healing dynamics remains a critical concern. Mechanical strength must be balanced with bioactivity, as excessive reinforcement can hinder degradation, while insufficient strength may compromise functionality. The development of smart materials capable of responding to physiological conditions and providing feedback for dynamic adjustments represents an exciting frontier. Emerging technologies, such as 3D printing and machine learning, are expected to play a pivotal role in the next generation of resorbable composites, enabling the design of patient-specific solutions and accelerating their clinical adoption [7].

This chapter offers a detailed exploration of resorbable composites, covering their scientific principles, historical advancements, and current applications in bioimplants and fixation devices. It begins with an overview of the fundamental properties and design principles, followed by a discussion of their evolution and integration into modern medicine. The transformative role of these materials in various biomedical fields is examined, alongside a critical analysis of the challenges and emerging innovations shaping their future. By synthesizing current knowledge and projecting future trends, the chapter aims to provide a comprehensive understanding of how resorbable composites are driving advancements in regenerative medicine and personalized healthcare.

2 Historical Background

The development of resorbable composites reflects decades of innovation, bridging the gap between temporary structural support and regenerative medicine. These materials, known for their ability to degrade safely within the body, have evolved from rudimentary natural materials to sophisticated engineered solutions tailored for modern biomedical challenges. This historical perspective underscores the transformative impact of resorbable composites in medical science, setting the stage for future breakthroughs in their design and application.

3 Evolution of Biomaterials

The concept of biomaterials dates back to ancient times when natural substances like bone, ivory, and wood were used in prosthetics and surgical tools. While biocompatible, these materials were inconsistent in mechanical properties and prone to degradation in biological environments. The industrial revolution introduced durable metals like stainless steel and titanium, which offered mechanical strength but lacked bioactivity and required surgical removal. The mid-twentieth century marked a paradigm shift with the introduction of synthetic polymers such as polylactic acid (PLA) and polyglycolic acid (PGA), which were biodegradable and metabolized into non-toxic by-products, laying the groundwork for modern resorbable implants [8]. Incorporating biodegradability into medical design addressed a critical limitation of traditional implants: their permanence. PLA and PGA initially gained prominence in sutures, where their controlled degradation eliminated the need for removal (Fig. 1).

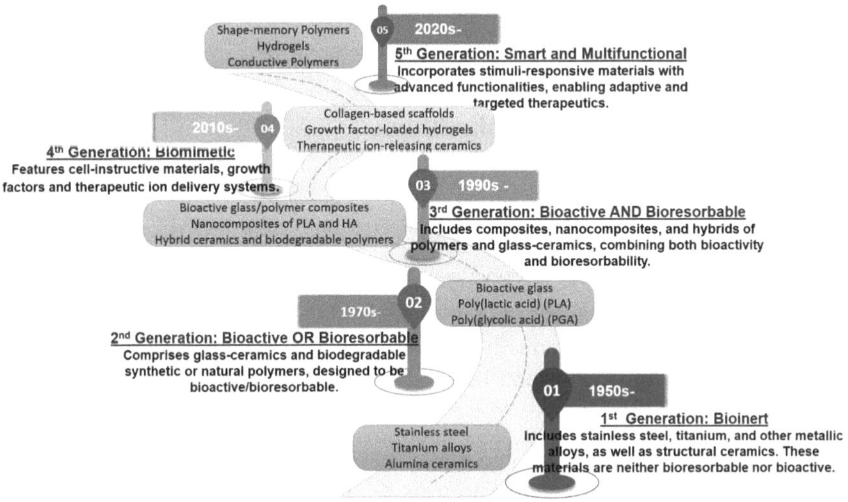

Fig. 1 Historical progression from Bioinert to smart and multifunctional biomaterials

Researchers quickly recognized their broader potential in tissue engineering and regenerative medicine, inspiring efforts to optimize these polymers for a range of medical applications [9].

4 Milestones in Resorbable Composite Development

1. ***Early Adoption of PLA and PGA (1960s–1980s)***: The first major milestone in resorbable materials was the use of PLA and PGA for sutures like Dexon™ and Vicryl™. These materials were praised for their predictable degradation profiles and mechanical integrity. However, limitations such as brittleness and lack of bioactivity spurred further research into blending and copolymerization, leading to the development of polylactic-co-glycolic acid (PLGA) as a versatile alternative [10].
2. ***Introduction of Copolymers and Bioactive Fillers***: By the 1980s, the incorporation of bioactive ceramics, such as hydroxyapatite (HA) and β-tricalcium phosphate (β-TCP), transformed resorbable composites. These additions not only enhanced mechanical properties but also promoted osteoconduction, making them suitable for bone regeneration. Hybrid materials combining polymers with nanostructured fillers emerged, allowing precise tailoring of degradation rates and bioactivity for specific clinical scenarios [11].
3. ***Commercialization of Resorbable Devices***: The late 1990s saw the commercialization of resorbable fixation devices like BioSorb® screws and LactoSorb® plates. These products demonstrated the feasibility of resorbable implants in orthopedic and craniofacial surgery, offering reliable support during healing and eliminating the need for removal. Their success catalyzed further development, leading to a broader range of applications including cardiovascular stents and drug delivery systems [12].

5 Pioneering Clinical Applications

The first clinical applications of resorbable fixation devices were groundbreaking in proving their potential.

Orthopedics: In the 1980s, PLA screws were used in anterior cruciate ligament (ACL) reconstruction. Clinical studies highlighted their ability to provide sufficient mechanical support while gradually transferring load to healing tissues. These findings were pivotal in establishing resorbable composites as viable alternatives to metallic implants [13].

Craniofacial Surgery: Resorbable plates made of PLA-PGA blends were introduced for treating pediatric craniosynostosis, a condition requiring corrective surgery. A seminal 1990 study demonstrated successful outcomes, with patients benefiting from reduced complications and elimination of growth interference compared to permanent metal implants [14].

Dental Applications: Bioactive composites integrating β-TCP gained traction in dental implants, where their osteoconductive properties accelerated bone regeneration around implant sites. These materials proved particularly useful in alveolar ridge augmentation and periodontal defect repair [15].

Building on these historical achievements, research continues to push the boundaries of resorbable composites. Emerging trends include the integration of nanotechnology, 3D printing for patient-specific implants, and smart materials capable of responding to environmental cues. These innovations promise to enhance the precision and efficacy of resorbable devices, solidifying their role in personalized medicine and regenerative therapies.

6 Principles of Resorbable Composites

Resorbable materials play a pivotal role in modern biomedical applications, particularly in implants and fixation devices. These materials are designed to degrade naturally over time in the body, supporting healing processes and gradually transferring load to regenerating tissue. Below is a detailed discussion of the various resorbable materials commonly used in composites, including synthetic and natural polymers, inorganic fillers, and nanomaterials. The continuous evolution of resorbable composites, driven by the integration of synthetic and natural polymers, inorganic fillers, and nanomaterials, has revolutionized biomedical applications, particularly in tissue engineering and regenerative medicine. By understanding the properties, degradation mechanisms, and applications of each component, researchers can design highly effective and personalized solutions for a wide range of clinical needs.

6.1 Synthetic Polymers

Polylactic Acid (PLA): PLA is one of the most widely used synthetic biodegradable polymers. It is derived from renewable resources like cornstarch or sugarcane, making it an eco-friendly choice. PLA is known for its strength, mechanical integrity, and slow degradation rate, which is advantageous for load-bearing applications. It undergoes hydrolytic degradation, where water breaks down the ester bonds in the polymer chain. PLA has been successfully used in bone fixation devices, sutures, and drug delivery systems [16].

Recent Developments: Recent studies have focused on improving PLA's mechanical properties and degradation rate by incorporating copolymers like polylactic-co-glycolic acid (PLGA) and blending it with bioactive fillers to enhance its bioactivity and osteoconductivity for bone regeneration applications [17].

Polyglycolic Acid (PGA): PGA, another aliphatic polyester, is known for its faster degradation compared to PLA. This is due to its more hydrophilic nature, which accelerates the hydrolytic breakdown of its polymer chains. PGA has been extensively used in sutures and scaffolds for tissue engineering, as its rapid degradation can be advantageous in applications requiring faster resorption. However, its brittleness limits its use in load-bearing applications [18].

Recent Developments: Researchers are focused on improving the material's mechanical properties by copolymerizing it with PLA and other polymers, resulting in a more versatile material with a controlled degradation profile [19].

Polycaprolactone (PCL): PCL is a synthetic polymer with a slower degradation rate than PLA and PGA, making it suitable for applications that require long-term mechanical support. It has been used in large bone defect repairs, nerve regeneration, and drug delivery systems. PCL is often blended with other polymers to modulate its degradation rate and mechanical properties. It undergoes hydrolysis, breaking down into non-toxic by-products like caproic acid [20].

Recent Developments: Studies have explored PCL-based scaffolds for nerve tissue engineering, highlighting its potential for use in chronic conditions that require prolonged healing times [21].

Polyhydroxybutyrate (PHB) and Polyhydroxyvalerate (PHBV): PHB and PHBV are biopolyesters produced through bacterial fermentation of sugars. PHB is biodegradable, biocompatible, and non-toxic, making it a promising material for medical applications. PHBV, a copolymer of PHB and hydroxyvalerate, is more flexible than PHB, which improves its processing and mechanical properties. Both materials are used in tissue engineering, drug delivery systems, and wound healing applications [22]. Researchers are focusing on the biocompatibility and controlled degradation of PHB and PHBV in tissue engineering, particularly for soft tissue regeneration [23].

6.2 Natural Polymers

Collagen

Collagen, the most abundant protein in the human body, is widely used in tissue engineering and wound healing due to its excellent biocompatibility and ability to promote cell adhesion and growth. Collagen-based scaffolds are often used in soft tissue repair, skin regeneration, and bone healing. It can be derived from various sources, including animal tissues, and has a natural affinity for promoting tissue integration and reducing inflammation [24].

Recent Developments: Collagen is being increasingly used in combination with synthetic polymers to improve the mechanical properties of scaffolds, creating hybrid materials that can support a wider range of applications [25].

Chitosan

Chitosan is a polysaccharide derived from the shells of crustaceans and is used in biomedical applications due to its antimicrobial properties, biodegradability, and ability to promote tissue growth. Chitosan is used in wound healing, drug delivery systems, and scaffolds for tissue regeneration, particularly for soft tissues. It also serves as a hemostatic agent in various surgical applications [26].

Recent Developments: Recent studies focus on modifying chitosan to enhance its mechanical strength and degradation rate, especially for bone tissue engineering applications [27].

Alginate

Alginate is a natural polysaccharide obtained from brown algae, and it is widely used in biomedical applications due to its biocompatibility, gel-forming ability, and ease of processing. It is commonly used in drug delivery systems, wound healing, and as a scaffold for tissue regeneration. Alginate's ability to form hydrogels allows it to hold water and maintain tissue viability [28].

Recent Developments: Researchers are exploring alginate-based hydrogels for 3D cell cultures and personalized medicine applications, providing a conducive environment for cellular growth and differentiation [29].

Gelatin

Gelatin, derived from collagen, is a biodegradable material often used in tissue engineering and drug delivery. It is particularly advantageous in creating scaffolds for soft tissues due to its ability to mimic the extracellular matrix. Gelatin-based composites promote cellular attachment and proliferation, making it an effective choice for wound healing and regenerative therapies [30]. Gelatin is being combined with synthetic polymers to create composite materials that improve mechanical strength while maintaining bioactivity and biodegradability [31].

6.3 Inorganic Fillers

Bioactive Glasses (*e.g.,* 45S5 Bioglass): Bioactive glasses, particularly 45S5 Bioglass, are used in resorbable composites for bone repair. These materials are known for their ability to bond with bone tissue and release ions that stimulate cellular activity and promote osteogenesis. Bioglass has been successfully used in bone grafts, dental implants, and coatings for metal implants [32].

Recent Developments: Research is focused on improving the ion release profiles of bioactive glasses, which can further enhance their osteoconductivity and integration with bone tissue [33].

Hydroxyapatite (HA): HA is a naturally occurring mineral form of calcium apatite, and it is one of the most common inorganic fillers in bone repair composites. It is highly osteoconductive, providing a scaffold for new bone growth. HA is often combined with polymers to improve the mechanical properties and bioactivity of the composite material [34].

Recent Developments: The use of nano-sized HA (nHA) has gained attention due to its increased surface area and enhanced interactions with biological tissues, improving bone integration [35].

Tricalcium Phosphate (TCP): TCP, particularly β-TCP, is a resorbable ceramic widely used in bone tissue engineering. It supports osteoconduction and provides a scaffold for new bone formation. Its resorption rate can be tuned to match bone healing, making it an ideal filler in composites for bone regeneration [36].

6.4 Nanomaterials

Nanohydroxyapatite (nHA): nHA is an enhanced form of hydroxyapatite that mimics bone mineral more closely. It is used to improve the mechanical properties and bioactivity of bone-regenerative composites. nHA has shown significant promise in improving the osteoconductivity and cellular responses of bone scaffolds [37].

Carbon-Based Nanomaterials: Carbon nanotubes (CNTs) and graphene oxide are used to reinforce resorbable composites due to their superior mechanical strength and conductivity. These nanomaterials also have the potential to serve as carriers for drug delivery, providing multifunctional composites that promote healing and regeneration [38].

7 Material Classes in Resorbable Composites

Resorbable composites are engineered materials designed to degrade within the human body in a controlled manner, promoting tissue healing while eliminating the need for surgical removal. These composites typically combine various material classes—polymers, ceramics, metals, and natural materials—to achieve the desired mechanical properties, degradation rates, and biocompatibility. The integration of these components facilitates the development of medical devices such as bone implants, fixation devices, and tissue scaffolds.

Table 1 offers an overview of the primary types of resorbable composites—spanning polymer-based, ceramic-based, metallic, and natural material composites—along with their key components and typical applications across various medical fields. This concise summary will enable readers to better understand the diversity of

Table 1 Overview of resorbable composite materials and their clinical applications

Composite type	Key materials	Common applications	References
Polymer-based	PLA, PGA, PCL, PLGA, PLLA	Orthopedic fixation, bone regeneration, sutures, meshes	[39–41]
Ceramic-based	HA, β-TCP, Bioactive Glass, Magnesium Alloys	Bone substitutes, drug delivery, craniofacial reconstruction	[42–44]
Metallic-based	Magnesium, Zinc Alloys, Polymer Coatings	Load-bearing implants, bone fixation devices	[45, 46]
Natural material-based	Collagen, Chitosan, Alginate, Gelatin	Soft tissue repair, drug delivery scaffolds	[47, 48]

materials and their clinical uses, enhancing their overall comprehension of resorbable composites in biomedical applications.

Below is an overview of the various material classes used in resorbable composites, along with insights into recent research and developments.

7.1 Polymer-Based Resorbable Composites

PLA/PGA Blends: Polylactic acid (PLA) and polyglycolic acid (PGA) are two of the most commonly used synthetic polymers in resorbable composites. Both are aliphatic polyesters, but they differ in their degradation rates—PLA degrades slower than PGA, making them complementary when blended. These blends are utilized in applications such as bone fixation devices, sutures, and scaffolds for tissue engineering. The combination of PLA and PGA offers a balance between mechanical strength and controlled degradation, making them ideal for applications where both structural integrity and predictable resorption are essential.

Recent Developments: Recent studies have focused on optimizing PLA/PGA blends by modifying their molecular weight and incorporating bioactive fillers to enhance their osteoconductivity. Additionally, the copolymerization of PLA and PGA has been explored to adjust the degradation rate further, allowing for more tailored solutions in regenerative medicine [39].

PLLA-TCP and PCL-HA Systems: Poly-L-lactic acid (PLLA) is another popular polymer used in resorbable composites. When combined with bioactive fillers like tricalcium phosphate (TCP), PLLA forms a composite material suitable for bone repair. TCP, known for its osteoconductivity, enhances the bone healing properties of PLLA, making these composites useful in load-bearing bone applications. Similarly, polycaprolactone (PCL) is often combined with hydroxyapatite (HA), another material with excellent bone-regenerative properties. PCL-HA composites are used in bone tissue engineering due to their slow degradation rates and ability to provide long-term mechanical support during the healing process. PLLA-TCP and PCL-HA systems have been advanced with the incorporation of nanomaterials and improved processing techniques to optimize their mechanical properties and degradation profiles. Recent research has also focused on using these composites in combination with growth factors and other bioactive agents to promote faster and more efficient bone regeneration [40].

Multi-Component Hybrid Polymers: Hybrid polymers combine two or more polymeric materials, often synthetic and natural polymers, to combine the best attributes of each. In bone repair, e.g., combining PLA with collagen or gelatin can provide both the strength of synthetic polymers and the biological activity of natural materials, promoting tissue integration. These hybrids offer flexibility, biocompatibility, and tunable mechanical properties, making them ideal candidates for scaffolds in bone, cartilage, and soft tissue repair. Hybrid systems are being increasingly explored for their versatility, where the incorporation of natural polymers like collagen or chitosan in synthetic polymer matrices can provide improved mechanical and biological responses. The recent trend is to develop multi-component scaffolds with enhanced drug delivery capabilities and a controlled release of growth factors to facilitate tissue regeneration [41].

7.2 Ceramic-Based Composites

HA/β-TCP Ceramics with Polymers: Hydroxyapatite (HA) and β-tricalcium phosphate (β-TCP) are ceramics known for their excellent osteoconductivity, making them ideal for use in bone regeneration. When combined with polymers such as PLA or PCL, these ceramics form composites that can mimic the mechanical and biological properties of bone. The addition of polymers improves the processability and mechanical properties of HA/β-TCP ceramics while maintaining their bioactivity and providing controlled degradation rates. Recent studies focus on the development of HA/β-TCP composites that incorporate bioactive agents, such as growth factors or stem cells, to enhance bone healing. Additionally, 3D printing technologies are being used to create these composites with intricate microstructures that can better mimic natural bone tissue [42].

Bioactive Glass-Polymer Systems: Bioactive glasses, such as 45S5 Bioglass, are materials that can bond with bone tissue, promoting new bone growth through the release of ions like calcium, phosphorus, and silicon. These glasses are combined with polymers to create composites that are used in bone regeneration. The polymers enhance the ease of fabrication, mechanical properties, and processability, while the bioactive glass enhances the material's ability to support bone growth. Recent innovations in bioactive glass-polymer composites include the optimization of glass composition to increase the release of beneficial ions and the use of nanostructured bioactive glasses for better tissue integration. The combination of bioactive glasses with biopolymers such as collagen is gaining attention for improving the biological performance of the composite materials [43].

Magnesium-Based Biodegradable Ceramics: Magnesium and its alloys are promising biodegradable materials with potential applications in bone implants and fixation devices. Magnesium-based composites offer the advantage of being lightweight and biodegradable, with mechanical properties similar to natural bone. These composites are often combined with polymers to improve their mechanical stability and control their degradation rates.

Recent Developments: Recent research has focused on improving the corrosion resistance and biocompatibility of magnesium alloys. One approach involves coating the magnesium alloys with bioactive polymers or incorporating nanoparticles to regulate the degradation process and prevent premature failure [44].

7.3 Metallic Resorbable Composites

Magnesium and Zinc Alloys with Polymer Coatings: Magnesium and zinc alloys are gaining attention as biodegradable metals for use in orthopedic implants. Their advantage lies in their mechanical strength and biocompatibility, but their rapid degradation in the body can lead to adverse effects such as hydrogen gas formation. To address this, these alloys are often coated with resorbable polymers, which control their degradation rate and improve their biocompatibility. The latest developments involve the design of polymer coatings that can regulate the degradation process of magnesium and zinc alloys to ensure their controlled resorption over time. Coatings of PLA or PCL, e.g., have been found to significantly improve the performance of magnesium-based composites in bone fixation applications [45].

Hybrid Metal-Polymer Scaffolds: Hybrid metal-polymer scaffolds are gaining traction in resorbable composite technology due to their ability to combine the mechanical strength of metals with the biological properties of polymers. These scaffolds are particularly useful in applications where both structural support and tissue regeneration are required, such as in bone and vascular tissue engineering. Recent studies are focusing on creating hybrid scaffolds with finely tuned properties, such as enhancing the mechanical properties of metals like magnesium with the degradation control offered by polymers. These hybrids are being used in complex, high-load-bearing applications, with some systems incorporating growth factors or nanoparticles to further enhance tissue integration [46].

7.4 Natural Material Composites

Collagen-Bioactive Glass Systems: Collagen, as a natural protein, is widely used in tissue engineering due to its biocompatibility and ability to promote cellular growth. When combined with bioactive glasses, these composites enhance both the mechanical properties and bioactivity, providing a scaffold for bone regeneration. Collagen provides a natural environment for cell attachment, while bioactive glasses promote mineralization and accelerate bone healing. Recent studies have focused on improving the fabrication techniques for collagen-bioactive glass composites, especially in the creation of 3D scaffolds that can better mimic the natural bone structure [47].

Chitosan-HA Systems: Chitosan, a natural polymer derived from chitin, is another widely studied material for use in resorbable composites. When combined with hydroxyapatite, it forms a composite material that is both biodegradable and osteoconductive, making it suitable for bone tissue engineering applications. Chitosan also has inherent antimicrobial properties, which enhance the material's potential for use in infection-prone areas. Recent advances in chitosan-HA composites focus on enhancing their mechanical strength and degradation rate. Additionally, the incorporation of nanoparticles into chitosan-HA composites has been explored to improve their bioactivity and stimulate faster bone regeneration [48].

The application of the above composite types can be seen in Fig. 2.

8 Requirements for Implants and Fixation Devices

Resorbable composites are essential components in modern biomedical applications, particularly in the development of implants and fixation devices. For these materials to be effective in supporting tissue regeneration while eventually degrading within the body, they must meet stringent requirements related to biological compatibility, mechanical properties, degradation kinetics, and regulatory standards. Below is a detailed exploration of these crucial requirements, accompanied by recent research insights.

Table 2 will summarize the critical mechanical, biological, and regulatory properties required for resorbable composites to be effective in clinical applications. It will be a concise reference point for understanding the essential criteria for material selection.

8.1 Biological Compatibility

Biocompatibility Standards: For any implant or fixation device, biological compatibility is paramount. Biocompatibility refers to the ability of a material to perform

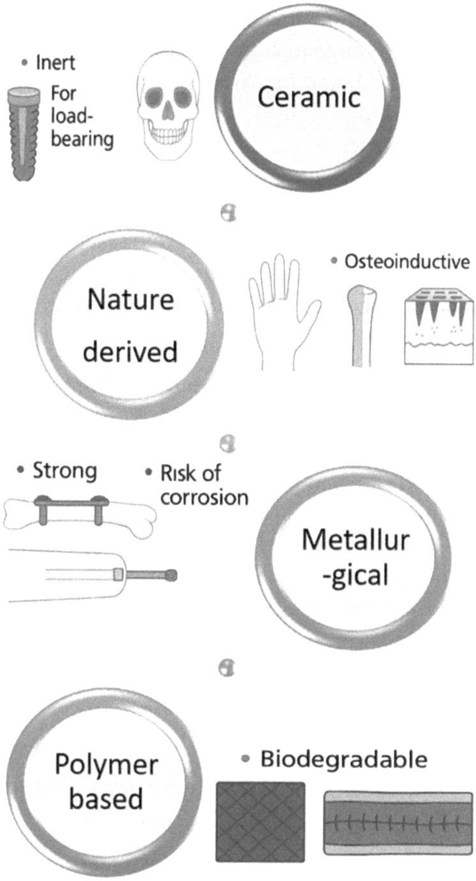

Fig. 2 Applications of general types of resorbable composites used in biomedical treatment

Table 2 Key properties of resorbable composites for biomedical applications

Property	Key requirements	Relevant materials	References
Biocompatibility	Minimal inflammatory response and tissue integration	PCL, HA, Collagen-based systems	[49, 50]
Mechanical strength	Adequate strength to support bone healing or soft tissue repair	PLA, PCL, HA/β-TCP composites	[51, 52]
Degradation rate	Synchronization with tissue healing rate	PLGA, PLA-PGA blends, Magnesium alloys	[53, 54]
Regulatory compliance	Adherence to ISO, ASTM standards for safety and efficacy	Materials across all classes, especially for clinical applications	[55, 56]

its intended function without causing adverse biological reactions, such as inflammation, immune rejection, or toxicity. The International Organization for Standardization (ISO) and the American Society for Testing and Materials (ASTM) provide global standards (ISO 10993, ASTM F-751) for evaluating the biocompatibility of materials used in medical devices. Recent studies have focused on the development of materials that not only comply with biocompatibility standards but also enhance biological activity. For example, incorporating bioactive ceramics like hydroxyapatite (HA) or bioactive glasses into resorbable composites has been shown to improve osteointegration and promote tissue regeneration. These materials foster cellular adhesion and growth, reducing the risk of complications such as implant rejection or infection [49].

Immune Response Management: Managing the immune response is another critical aspect of implant success. Resorbable materials must degrade in a way that avoids triggering an excessive immune reaction. Research has shown that synthetic polymers such as poly(lactic-co-glycolic acid) (PLGA) and polycaprolactone (PCL) exhibit relatively low immunogenicity, but their degradation by-products can still cause inflammation. Recent innovations in designing polymers with tailored degradation profiles have allowed for better control of immune responses. Additionally, bioactive ceramics incorporated into polymer matrices can help modulate the immune response by promoting tissue healing and minimizing inflammatory reactions during resorption [50].

8.2 Mechanical and Physical Properties

Strength, Toughness, and Elastic Modulus: Implants and fixation devices must possess appropriate mechanical properties to withstand physiological stresses without compromising the surrounding tissue. The strength, toughness, and elastic modulus of resorbable composites are crucial for ensuring they provide adequate mechanical support during the healing process. For example, composites made from poly-L-lactic acid (PLLA) combined with ceramics like tricalcium phosphate (TCP) demonstrate enhanced mechanical properties suitable for bone fixation. These composites must maintain structural integrity long enough to support the tissue while the bone heals and regains strength.

Recent research in polymer-ceramic composites has focused on enhancing the toughness and elasticity of resorbable materials. Advanced polymer blends, such as PLLA and PCL, are being optimized for better toughness without sacrificing the material's resorbability. The inclusion of nanomaterials such as nanohydroxyapatite (nHA) has been shown to improve both the strength and elasticity of resorbable composites, ensuring that they perform well under load-bearing conditions [51].

Load-Bearing Capabilities and Degradation Synchronization: One of the primary challenges in implant and fixation device design is ensuring that the material's mechanical properties are synchronized with its degradation rate. For example, during the healing process, the mechanical load on the device should gradually shift from the implant material to the regenerating tissue. If the material degrades too quickly, it may not provide sufficient support; if it degrades too slowly, it may impede tissue healing.

Recent research has introduced strategies for tuning the degradation kinetics of resorbable composites to match the mechanical load-bearing requirements at different stages of healing. This has led to the development of hybrid systems, where different materials degrade at different rates to provide optimal support throughout the healing process. For instance, PLLA and PCL composites are being used in combination with magnesium-based alloys to create materials that provide initial mechanical support, followed by controlled degradation once tissue healing progresses [52].

8.3 Degradation Kinetics

Tailoring for Specific Applications: The degradation kinetics of resorbable composites are critical for their performance in specific applications. While a material must degrade over time, the rate at which it does so must be precisely controlled to avoid premature failure or prolonged interference with the healing process. Resorbable materials for bone implants, e.g., need to degrade at a rate that matches bone healing, which typically takes several months to years.

Recent developments in the field include the design of composite materials with engineered degradation profiles. This is achieved by modifying the molecular structure of polymers or introducing specific additives that slow or accelerate degradation. For example, research has shown that adding bioactive glass to PLA/PGA composites can regulate the degradation rate by influencing the hydrolysis and mineralization processes, providing a tailored solution for bone regeneration [53].

Another advancement in degradation control involves the use of nanomaterials to influence degradation rates. Nanostructured materials have been shown to accelerate the resorption process by increasing the surface area and altering the hydrolytic breakdown of the polymer matrix. This capability allows for even more precise control over the resorption process to match the biological healing timeline [54].

8.4 Regulatory Requirements

ISO and ASTM Standards: To ensure the safety and effectiveness of resorbable materials, adherence to established regulatory standards is essential. The ISO 10993 series provides guidelines for the biological evaluation of medical devices, which includes tests for cytotoxicity, sensitization, and implantation response. Similarly, ASTM standards, such as ASTM F-981 for testing the biocompatibility of resorbable materials, provide detailed protocols for evaluating the performance of materials in vivo. Recent advancements have also highlighted the need for more comprehensive testing to account for the complex nature of resorbable composites. For example, studies suggest that while existing standards are effective for individual materials, new composite systems may require additional testing for their combined biological and mechanical properties, as well as their long-term degradation effects [55].

Manufacturing and Sterilization Constraints: Manufacturing and sterilization of resorbable composites are critical factors in the development of implantable medical devices. The materials used must be compatible with current manufacturing techniques, such as injection molding, 3D printing, or electrospinning, while maintaining their biological and mechanical properties. Additionally, the sterilization process must not compromise the integrity of the composite.

Recent developments in manufacturing techniques, such as additive manufacturing (3D printing), are enabling more precise control over the scaffold architecture, allowing for better performance in tissue engineering applications. Sterilization techniques, such as gamma radiation and ethylene oxide gas, are being carefully optimized to ensure they do not interfere with the functionality of bioactive composites. Some studies have explored the use of electron beam irradiation as a more efficient alternative to conventional sterilization methods for resorbable composites [56].

9 Applications and Clinical Relevance

Resorbable composites play a crucial role in advancing medical treatments by offering safe, effective, and biofunctional materials for various implant applications. Their biodegradability allows for temporary mechanical support during the healing process, followed by complete resorption, eliminating the need for secondary surgical removal. The unique properties of resorbable composites make them especially valuable in orthopedic, soft tissue, craniofacial, dental, cardiovascular, neural applications, and emerging fields like drug delivery and wound healing. Below is a detailed exploration of these applications, supported by recent research.

9.1 Orthopedic Fixation Devices

Orthopedic implants are one of the most significant applications of resorbable composites. Materials such as PLA/PGA and PLA-TCP composites are commonly used in the development of orthopedic fixation devices, including screws, plates, pins, and intramedullary rods. These devices provide mechanical stability to fractured bones during the healing process, with the added benefit of resorption over time, eliminating the need for secondary removal surgery. Recent advancements have focused on improving the mechanical strength and degradation rates of these composites. For example, studies have demonstrated that PGA/PLA-TCP composites exhibit superior osteoconductivity and promote bone healing by facilitating the formation of new bone tissue. The combination of bioceramics like hydroxyapatite (HA) with resorbable polymers not only enhances the mechanical properties of these composites but also supports osteointegration, accelerating the healing process [57]. In addition to these conventional materials, hybrid systems combining resorbable polymers with magnesium-based alloys are emerging. These systems offer the advantage of controlled degradation rates that match the bone healing process more effectively than traditional composites.

9.2 *Soft Tissue Implants*

Soft tissue repair is another area where resorbable composites show significant promise. PCL-based systems are often employed in sutures, meshes, and scaffolds used in soft tissue repair. These materials are advantageous because of their ability to provide structural support to healing tissues, such as in hernia repairs or abdominal wall reconstructions, while resorbing in sync with tissue regeneration.

Tendon and ligament repair is also an area where resorbable composites have demonstrated clinical relevance. Studies have investigated PCL, collagen, and collagen-PCL blends to develop scaffolds that promote cellular infiltration, tissue regeneration, and biomechanical integration with the surrounding tissue. For example, PCL-based scaffolds have been shown to provide excellent structural support while allowing the gradual integration of new tissue, particularly in tendon and ligament healing [58].

9.3 *Craniofacial and Dental Applications*

In craniofacial and dental applications, resorbable composites are primarily used for bone substitutes and fixation devices. HA-PLA composites are one of the most widely used materials for bone substitutes due to their excellent osteoconductivity, biocompatibility, and resorbability. These materials are used for bone regeneration following trauma, tumor resection, or congenital deformities. For dental applications, resorbable composites are used in the form of membranes and fixation devices for guided bone regeneration (GBR) and dental implant procedures. These composites help in the regeneration of lost bone tissue by acting as a scaffold that supports the growth of osteoblasts while gradually degrading as new bone forms. Research into HA-PLA composites for dental applications has shown that these materials are effective in enhancing bone healing and reducing the risk of implant failure [59].

9.4 Cardiovascular and Neural Applications

Resorbable composites are also gaining traction in the cardiovascular and neural fields. In cardiovascular applications, resorbable materials are used for the development of stents and drug-eluting scaffolds. Resorbable stents are designed to provide temporary mechanical support to blood vessels after angioplasty, with the material dissolving over time to avoid long-term complications associated with metallic stents, such as thrombosis or restenosis. PCL and PLLA are often employed in these devices due to their controlled degradation rates and biocompatibility [60].

Neural regeneration is another promising area for resorbable composites, particularly in nerve injury treatment. PCL-based systems, often combined with chitosan, have been studied for their ability to promote nerve regeneration and guide axonal growth following peripheral nerve injury. These composites provide a conducive environment for neural cells to grow, while the resorption process helps avoid long-term foreign body presence in the neural tissue. Studies have demonstrated that PCL-chitosan scaffolds enhance neuronal survival and promote functional recovery after nerve injury [61].

9.5 Emerging Applications

Drug Delivery Vehicles: Recent advancements in drug delivery systems have highlighted the potential of resorbable composites as carriers for controlled drug release. PCL-HA microspheres, e.g., have been explored for their ability to deliver therapeutic agents at controlled rates, enhancing the bioavailability of drugs while supporting tissue regeneration. These microspheres are particularly useful for localized treatments, where a gradual release of drugs, such as growth factors or antibiotics, is required to promote healing at the site of injury or infection [62].

Wound Healing and Regenerative Medicine: In the field of regenerative medicine, resorbable composites are being developed for wound healing applications. The ability to combine resorbable polymers with bioactive ceramics, such as HA or bioactive glasses, has shown promise in creating wound dressings and scaffolds that not only provide mechanical support but also actively promote tissue regeneration. PCL and its copolymers have been used in creating scaffolds that allow for cell migration and tissue regeneration in chronic wounds or burn injuries. These composites can be tailored for various clinical conditions, providing a customizable solution for wound care and tissue repair [63].

The applications are summarized in Fig. 3.

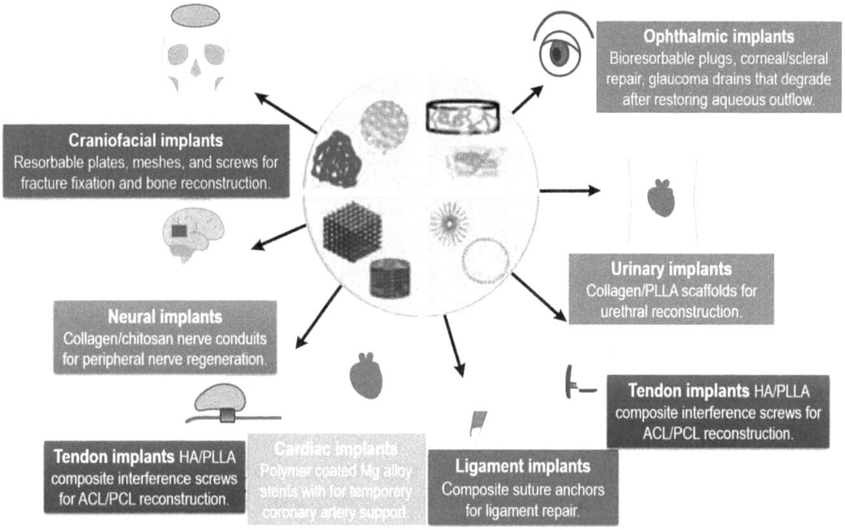

Fig. 3 Diagram showing applications of resorbable composites used as implants in human body parts

10 Challenges and Limitations

Despite the substantial progress in the development and application of resorbable composites for medical implants and fixation devices, several challenges remain that hinder their broader adoption and long-term clinical success. These challenges span material limitations, biological responses, and commercialization barriers, each of which must be addressed to improve the performance, safety, and scalability of resorbable composites.

10.1 Material Challenges

One of the major hurdles in the development of resorbable composites lies in balancing the mechanical strength with the degradation rates of the materials. While resorbable polymers such as PLA and PGA provide adequate mechanical support during the healing process, their degradation rate may not always align with the tissue regeneration process. If a composite degrades too quickly, it may fail to provide the necessary mechanical stability, whereas slower degradation can lead to prolonged foreign body presence, increasing the risk of infection or inflammation. Recent studies have sought to optimize the degradation rates of these materials, with the incorporation of bioactive fillers such as hydroxyapatite (HA) or bioactive glass showing promise in modulating degradation dynamics while enhancing mechanical properties [64]. Another material challenge is the brittle nature of ceramic-based composites, such as HA or tricalcium phosphate (TCP), which, while excellent for promoting bone growth, can be prone to cracking or breakage under load. This brittleness is particularly problematic in load-bearing applications, such as in orthopedic implants. Recent advancements in hybrid composites, combining ceramics with more ductile polymers like polycaprolactone (PCL), have been shown to improve the toughness of these materials without compromising their bioactivity [42]. However, achieving a balance between enhanced toughness and maintaining the desired resorbable properties remains a key challenge in composite design.

10.2 Biological and Clinical Challenges

Biological and clinical challenges are significant considerations in the use of resorbable composites for medical implants. Long-term biocompatibility remains a concern, particularly in the case of composite materials that degrade over time. While polymers such as PLA, PCL, and PGA are generally considered biocompatible, their degradation products may lead to local inflammation or immune responses in certain patients. In some cases, the acidic degradation products of PLA and PGA can lower the local pH, potentially leading to delayed wound healing or bone resorption. Addressing these concerns requires a detailed understanding of the long-term interaction between degradation products and biological tissues, as well as the development of new polymer systems that degrade in a more controlled and tissue-friendly manner [65]. Unpredictable inflammatory responses to resorbable implants are also a major clinical challenge. Studies have shown that although most resorbable materials elicit minimal immune responses, certain additives or fillers in composites can cause unexpected inflammation or even chronic foreign body reactions. Recent work has focused on improving the immunomodulatory properties of resorbable composites by selecting biocompatible bioactive fillers, optimizing material surface properties, and utilizing advanced polymer blends. These modifications help mitigate the risks of adverse reactions while promoting better integration with surrounding tissues [66].

10.3 Commercialization Barriers

While the potential of resorbable composites is clear, commercialization of these materials remains a significant barrier. Scaling production to meet the demand for clinical-grade implants presents several technical challenges. The synthesis of high-quality, reproducible resorbable composites often requires complex processing techniques and specialized equipment, which can increase manufacturing costs. This is particularly true for hybrid materials that combine polymers, ceramics, and metals, as these require precise control over each phase of production. Advances in manufacturing technologies, such as 3D printing and automated molding techniques, may help reduce costs and improve scalability, but these innovations are still in the early stages [67].

Regulatory and approval delays are another critical factor that impedes the commercialization of resorbable composites. The regulatory pathways for medical devices, particularly those involving novel composite materials, can be long and complex. Regulatory bodies such as the U.S. Food and Drug Administration (FDA) or the European Medicines Agency (EMA) require extensive preclinical and clinical testing to ensure the safety and efficacy of new resorbable materials. This process can delay the introduction of promising materials to the market and increase the overall time to commercialization. Recent efforts to streamline regulatory processes for innovative biomaterials are underway, but navigating the regulatory landscape remains one of the key barriers to market entry [68].

11 Future Trends in Resorbable Composites

The landscape of resorbable composites is evolving rapidly, with innovations in material design, manufacturing techniques, and multifunctional applications poised to enhance their role in biomedical implants and fixation devices. These advances will enable more personalized, efficient, and biocompatible solutions, meeting the growing demands of various clinical applications.

11.1 Advanced Designs

Nanocomposites with Enhanced Properties: Nanotechnology has become a key tool in improving the mechanical strength, bioactivity, and degradation behavior of resorbable composites. A notable example is the integration of nanohydroxyapatite (nHA) into polymer matrices, such as poly(lactic-co-glycolic acid) (PLGA) and polycaprolactone (PCL). These composites exhibit improved mechanical properties like increased tensile strength and stiffness compared to traditional composites, making them more suitable for load-bearing applications like bone regeneration. Moreover, the incorporation of nanoparticles facilitates better bonding with biological tissues, thereby accelerating bone healing. Recent studies by Chen et al. (2022) on the use of carbon-based nanocomposites have shown promise in orthopedic applications, particularly where electrical conductivity is essential, such as in neural regeneration [69].

Smart polymers that respond to external stimuli, like pH and temperature, are another exciting area of research. These polymers can be designed to degrade at controlled rates in response to environmental changes, allowing for more tailored healing processes. The use of thermoresponsive polymers, such as poly(N-isopropylacrylamide) (PNIPAM), in resorbable composites for drug release has been demonstrated by Xu et al. [70], who explored their potential in local growth factor delivery for accelerated tissue repair.

11.2 Innovative Manufacturing Approaches

3D Printing and Biofabrication of Resorbable Scaffolds: Additive manufacturing, particularly 3D printing, is transforming the production of resorbable scaffolds. This technique allows for precise control over the architecture of scaffolds, creating highly complex, patient-specific designs that better mimic the native tissue structure. Using materials like PLA, PCL, and PLA-PGA blends, 3D printing can produce scaffolds with tailored porosity, pore size, and mechanical properties, which are essential for effective tissue regeneration. Wang et al. (2024) demonstrated that 3D-printed PCL-based scaffolds could be used in bone regeneration, offering a customizable solution to meet the specific needs of patients [71]. Bioprinting, which integrates living cells into scaffolds, is pushing the boundaries of tissue engineering, as evidenced by the work of Wistner et al. (2023), who explored multi-material bioprinting to create scaffolds capable of promoting cellular infiltration and tissue growth [72].

11.3 Multifunctional Systems

Composites for Drug Delivery and Regeneration: The integration of resorbable composites with drug delivery capabilities is a major trend in the field. Composites that can release drugs locally over time not only support tissue regeneration but also reduce complications such as infection or inflammation. Research by Guo et al. [73] focused on the development of bioactive composite materials for orthopedic applications, demonstrating how resorbable devices like bone grafts could elute antibiotics or growth factors to enhance bone healing. Similarly, Skwira et al. [74] examined the use of bioactive glass-polymer composites for drug-eluting bone implants, showing how these systems improve recovery times and reduce the risk of post-surgical infection.

Stimuli-Responsive and Self-Healing Materials: Stimuli-responsive materials are also gaining attention in resorbable composite development. These materials can adapt to environmental stimuli such as mechanical stress or temperature changes, making them ideal for applications that require self-repair. Self-healing composites, like those being developed by Low et al. [75], employ healing agents embedded within the material to repair microcracks or damage, a crucial feature for implants that must maintain mechanical integrity over time. These self-healing properties could significantly enhance the longevity of implants and reduce the need for revision surgeries [75].

11.4 Integration with AI and Computational Models

Predictive Modeling for Material Design: Artificial intelligence (AI) and machine learning (ML) are becoming indispensable tools in the design of resorbable composites. These technologies enable researchers to predict material behaviors, such as degradation rates and biological responses, based on vast datasets. By applying AI to the material selection process, designers can optimize composite formulations, ensuring that they meet specific requirements for mechanical performance and biocompatibility. Kibrete et al. [76] demonstrated the use of AI in designing resorbable composites by predicting the interactions between different material components and their effect on the final performance in a biomedical context. This can be seen in Fig. 4.

Optimization of Mechanical and Biological Performance: Another promising trend is the use of computational models to simulate the mechanical and biological interactions of resorbable composites. These models, particularly finite element modeling (FEM), help researchers understand how composite materials will behave under physiological loading conditions, which is essential for applications like bone

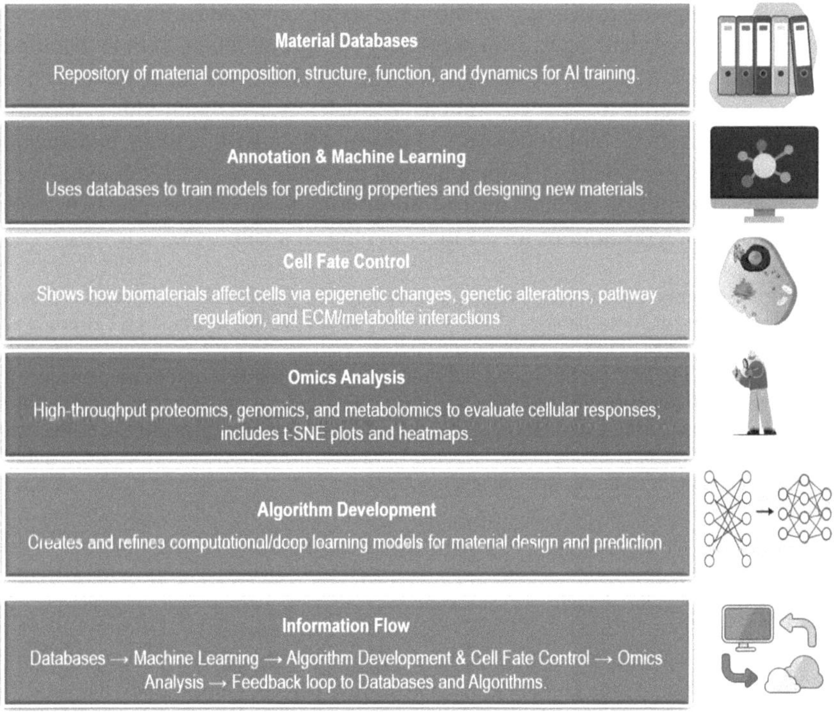

Fig. 4 Diagram illustrating AI-driven strategies for designing and developing resorbable biomaterials

fixation devices. Alamene et al. [77] used FEM to predict how different material compositions would impact the load-bearing capacity of implants, which is crucial for ensuring their effectiveness and safety.

12 Conclusion

The development of resorbable composites for biomedical applications has seen remarkable progress, transitioning from early uses of synthetic polymers like PLA and PGA to advanced materials integrating bioactive ceramics, nanocomposites, and hybrid systems. These composites now play a crucial role in orthopedic, craniofacial, cardiovascular, and neural applications, with enhanced mechanical strength and bioactivity. Innovations such as 3D printing and stimuli-responsive materials have further advanced the field, enabling patient-specific scaffolds with controlled degradation and integrated drug delivery. Despite these advancements, challenges remain, particularly in balancing mechanical strength with degradation rates, addressing biocompatibility issues, and overcoming production and regulatory barriers. The brittle nature of some ceramic composites and unpredictable immune responses are areas of ongoing research. Furthermore, scaling up production for clinical use presents logistical challenges, with regulatory delays continuing to hinder commercialization. Looking ahead, the future of resorbable composites lies in multifunctional systems that integrate tissue regeneration, drug delivery, and monitoring capabilities. The integration of 3D bioprinting and AI-driven design will facilitate more personalized and effective solutions. With continued innovation, resorbable composites are poised to revolutionize medical treatments, offering more effective, biocompatible, and patient-tailored solutions for a variety of clinical needs.

Acknowledgements The authors sincerely acknowledge the generous financial support provided by the Department of Biotechnology (DBT), Ministry of Science and Technology, Government of India, under Project No. DBT/2022-23/July/N/2240 through the DBT-RA Program. We extend our heartfelt gratitude to Prof. A. Srinivasan and Prof. Aynur Unal for their unwavering guidance and encouragement throughout this research. We also express our sincere appreciation to Dr. Mohit Kumar for his invaluable contributions, which significantly enhanced the success of this work. Lastly, a special note of gratitude to my daughter, Vanshika, whose unwavering love and inspiration have been a source of enduring strength and motivation.

References

1. Hwangbo J, Seo H, Sim G et al (2024) Bioresorbable polymers for electronic medicine. Cell Rep Phys Sci 5(8):102099. https://doi.org/10.1016/j.xcrp.2024.102099
2. Naser MA, Sayed AM, Abdelmoez W, El-Wakad MT, Abdo MS (2024) Biodegradable suture development-based albumin composites for tissue engineering applications. Sci Rep 14(1):7912. https://doi.org/10.1038/s41598-024-58194-5

3. García-Sobrino R, Muñoz M, Rodríguez-Jara E, Rams J, Torres B, Cifuentes SC (2023) Bioabsorbable composites based on polymeric matrix (PLA and PCL) reinforced with magnesium (Mg) for use in bone regeneration therapy: physicochemical properties and biological evaluation. Polymers 15(24):4667. https://doi.org/10.3390/polym15244667
4. Mohamed Jalaludeen A, Ramakrishnan R, Sundari Gunasekaran S et al (2024) Advancements in hydroxyapatite synthesis and surface modifications for emerging biomedical applications. Inorg Chem Commun 170:113414. https://doi.org/10.1016/j.inoche.2024.113414
5. Brochu BM, Sturm SR, Kawase De Queiroz Goncalves JA et al (2024) Advances in bioceramics for bone regeneration: a narrative review. Biomimetics 9(11):690. https://doi.org/10.3390/biomimetics9110690
6. Zong J, He Q, Liu Y, Qiu M, Wu J, Hu B (2022) Advances in the development of biodegradable coronary stents: a translational perspective. Mater Today Bio 16:100368. https://doi.org/10.1016/j.mtbio.2022.100368
7. Bandyopadhyay A, Mitra I, Goodman SB, Kumar M, Bose S (2023) Improving biocompatibility for next generation of metallic implants. Prog Mater Sci 133:101053. https://doi.org/10.1016/j.pmatsci.2022.101053
8. Reddy MSB, Ponnamma D, Choudhary R, Sadasivuni KK (2021) A comparative review of natural and synthetic biopolymer composite scaffolds. Polymers 13(7):1105. https://doi.org/10.3390/polym13071105
9. Samir A, Ashour FH, Hakim AAA, Bassyouni M (2022) Recent advances in biodegradable polymers for sustainable applications. NPJ Mater Degr 6(1):68. https://doi.org/10.1038/s41529-022-00277-7
10. Al-Shalawi FD, Mohamed Ariff AH, Jung DW et al (2023) Biomaterials as implants in the orthopedic field for regenerative medicine: metal versus synthetic polymers. Polymers 15(12):2601. https://doi.org/10.3390/polym15122601
11. Miri Z, Haugen HJ, Loca D et al (2024) Review on the strategies to improve the mechanical strength of highly porous bone bioceramic scaffolds. J Eur Ceram Soc 44(1):23–42. https://doi.org/10.1016/j.jeurceramsoc.2023.09.003
12. On SW, Cho SW, Byun SH, Yang BE (2020) Bioabsorbable osteofixation materials for maxillofacial bone surgery: a review on polymers and magnesium-based materials. Biomedicines 8(9):300. https://doi.org/10.3390/biomedicines8090300
13. D'Ambrosi R, Meena A, Arora ES, Attri M, Schäfer L, Migliorini F (2023) Reconstruction of the anterior cruciate ligament: a historical view. Ann Transl Med 11(10):364–364. https://doi.org/10.21037/atm-23-87
14. Maintz M, Tourbier C, de Wild M et al (2024) Patient-specific implants made of 3D printed bioresorbable polymers at the point-of-care: material, technology, and scope of surgical application. 3D Print Med 10(1):13. https://doi.org/10.1186/s41205-024-00207-0
15. Roca-Millan E, Jané-Salas E, Marí-Roig A et al (2022) The application of beta-tricalcium phosphate in implant dentistry: a systematic evaluation of clinical studies. Materials 15(2):655. https://doi.org/10.3390/ma15020655
16. Hussain M, Khan SM, Shafiq M, Abbas N (2024) A review on PLA-based biodegradable materials for biomedical applications. Giant. 18:100261. https://doi.org/10.1016/j.giant.2024.100261
17. Rocha CV, Gonçalves V, da Silva MC, Bañobre-López M, Gallo J (2022) PLGA-based composites for various biomedical applications. Int J Mol Sci 23(4):2034. https://doi.org/10.3390/ijms23042034
18. Liu S, Qin S, He M, Zhou D, Qin Q, Wang H (2020) Current applications of poly(lactic acid) composites in tissue engineering and drug delivery. Compos B Eng 199:108238. https://doi.org/10.1016/j.compositesb.2020.108238
19. Khouri NG, Bahú JO, Blanco-Llamero C, Severino P, Concha VOC, Souto EB (2024) Polylactic acid (PLA): properties, synthesis, and biomedical applications—a review of the literature. J Mol Struct 1309:138243. https://doi.org/10.1016/j.molstruc.2024.138243

20. del-Mazo-Barbara L, Gómez-Cuyàs J, Martínez-Orozco L, Santana Pérez O, Bou-Petit E, Ginebra MP (2024) In vitro degradation of 3D-printed polycaprolactone\biomimetic hydroxyapatite scaffolds: impact of the sterilization method. Polym Test 139:108566. https://doi.org/10.1016/j.polymertesting.2024.108566
21. Liang HY, Lee WK, Hsu JT et al (2024) Polycaprolactone in bone tissue engineering: a comprehensive review of innovations in scaffold fabrication and surface modifications. J Funct Biomater 15(9):243. https://doi.org/10.3390/jfb15090243
22. Jin A, del Valle LJ, Puiggalí J (2023) Copolymers and blends based on 3-hydroxybutyrate and 3-hydroxyvalerate units. Int J Mol Sci 24(24):17250. https://doi.org/10.3390/ijms242417250
23. Pulingam T, Appaturi JN, Parumasivam T, Ahmad A, Sudesh K (2022) Biomedical applications of polyhydroxyalkanoate in tissue engineering. Polymers 14(11):2141. https://doi.org/10.3390/polym14112141
24. Fan L, Ren Y, Emmert S et al (2023) The use of collagen-based materials in bone tissue engineering. Int J Mol Sci 24(4):3744. https://doi.org/10.3390/ijms24043744
25. Zheng L, Tseomashko N, Voronova A, Vasil'kov A, Hu X, Wang X (2024) Recent advances of collagen composite biomaterials for biomedical engineering: antibacterial functionalization and 3D-printed architecturalization. Collagen Leather 6(1):22. https://doi.org/10.1186/s42825-024-00164-8
26. Rajinikanth BS, Rajkumar DSR, Keerthika K, Vijayaragavan V (2024) Chitosan-based biomaterial in wound healing: a review. Cureus. Published online February 28, 2024. https://doi.org/10.7759/cureus.55193
27. Shan J, Yu Y, Liu X, Chai Y, Wang X, Wen G (2024) Recent advances of chitosan-based composite hydrogel materials in application of bone tissue engineering. Heliyon 10(19):e37431. https://doi.org/10.1016/j.heliyon.2024.e37431
28. Abka-khajouei R, Tounsi L, Shahabi N, Patel AK, Abdelkafi S, Michaud P (2022) Structures, properties and applications of alginates. Mar Drugs 20(6):364. https://doi.org/10.3390/md20060364
29. Pangjantuk A, Kaokaen P, Kunhorm P, Chaicharoenaudomrung N, Noisa P (2024) 3D culture of alginate-hyaluronic acid hydrogel supports the stemness of human mesenchymal stem cells. Sci Rep 14(1):4436. https://doi.org/10.1038/s41598-024-54912-1
30. Afewerki S, Sheikhi A, Kannan S, Ahadian S, Khademhosseini A (2019) Gelatin-polysaccharide composite scaffolds for 3D cell culture and tissue engineering: towards natural therapeutics. Bioeng Transl Med 4(1):96–115. https://doi.org/10.1002/btm2.10124
31. El-Seedi HR, Said NS, Yosri N et al (2023) Gelatin nanofibers: recent insights in synthesis, bio-medical applications and limitations. Heliyon. 9(5):e16228. https://doi.org/10.1016/j.heliyon.2023.e16228
32. Nogueira DMB, Rosso MP de O, Buchaim DV, Zangrando MSR, Buchaim RL (2024) Update on the use of 45S5 bioactive glass in the treatment of bone defects in regenerative medicine. World J Orthopedics 15(3):204–214. https://doi.org/10.5312/wjo.v15.i3.204
33. Szczodra A, Houaoui A, Agniel R et al (2024) Boron substitution in silicate bioactive glass scaffolds to enhance bone differentiation and regeneration. Acta Biomater 186:489–506. https://doi.org/10.1016/j.actbio.2024.07.053
34. Mondal S, Park S, Choi J et al (2023) Hydroxyapatite: a journey from biomaterials to advanced functional materials. Adv Coll Interface Sci 321:103013. https://doi.org/10.1016/j.cis.2023.103013
35. Damiri F, Fatimi A, Magdalena Musuc A et al (2024) Nano-hydroxyapatite (nHAp) scaffolds for bone regeneration: preparation, characterization and biological applications. J Drug Deliv Sci Technol 95:105601. https://doi.org/10.1016/j.jddst.2024.105601
36. Cinici B, Yaba S, Kurt M, Yalcin HC, Duta L, Gunduz O (2024) Fabrication strategies for bioceramic scaffolds in bone tissue engineering with generative design applications. Biomimetics 9(7):409. https://doi.org/10.3390/biomimetics9070409
37. Cámara-Torres M, Sinha R, Sanchez A et al (2022) Effect of high content nanohydroxyapatite composite scaffolds prepared via melt extrusion additive manufacturing on the osteogenic differentiation of human mesenchymal stromal cells. Biomater Adv 137:212833. https://doi.org/10.1016/j.bioadv.2022.212833

38. Solangi NH, Karri RR, Mubarak NM, Mazari SA (2024) Mechanism of polymer composite-based nanomaterial for biomedical applications. Adv Indus Eng Polym Res 7(1):1–19. https://doi.org/10.1016/j.aiepr.2023.09.002
39. Bikiaris ND, Koumentakou I, Samiotaki C et al (2023) Recent advances in the investigation of poly(lactic acid) (PLA) nanocomposites: incorporation of various nanofillers and their properties and applications. Polymers 15(5):1196. https://doi.org/10.3390/polym15051196
40. Capuana E, Lopresti F, Ceraulo M, la Carrubba V (2022) Poly-l-lactic acid (PLLA)-based biomaterials for regenerative medicine: a review on processing and applications. Polymers 14(6):1153. https://doi.org/10.3390/polym14061153
41. Suamte L, Tirkey A, Babu PJ (2023) Design of 3D smart scaffolds using natural, synthetic and hybrid derived polymers for skin regenerative applications. Smart Mater Med 4:243–256. https://doi.org/10.1016/j.smaim.2022.09.005
42. Gharibshahian M, Salehi M, Beheshtizadeh N et al (2023) Recent advances on 3D-printed PCL-based composite scaffolds for bone tissue engineering. Front Bioeng Biotechnol 11. https://doi.org/10.3389/fbioe.2023.1168504
43. Sreena R, Raman G, Manivasagam G, Nathanael AJ (2024) Bioactive glass–polymer nanocomposites: a comprehensive review on unveiling their biomedical applications. J Mater Chem B 12(44):11278–11301. https://doi.org/10.1039/D4TB01525H
44. Nuthana Kalva S, Ali F, Koç M (2025) Recent advances in the post-processing of magnesium based scaffolds for orthopedic applications. Next Mater 6:100295. https://doi.org/10.1016/j.nxmate.2024.100295
45. Verma A, Ogata S (2023) Magnesium based alloys for reinforcing biopolymer composites and coatings: a critical overview on biomedical materials. Adv Indus Eng Polym Res 6(4):341–355. https://doi.org/10.1016/j.aiepr.2023.01.002
46. He J, Li K, Wu T, Chen J, Li S, Zhang X (2023) Research progress in degradable metal-based multifunctional scaffolds for bone tissue engineering. MedComm—Biomater Appl 2(4). https://doi.org/10.1002/mba2.60
47. Aldhaher A, Shahabipour F, Shaito A et al (2023) 3D hydrogel/bioactive glass scaffolds in bone tissue engineering: status and future opportunities. Heliyon 9(7):e17050. https://doi.org/10.1016/j.heliyon.2023.e17050
48. Khodaverdi K, Naghib SM, Mozafari MR, Rahmanian M (2024) Chitosan/hydroxyapatite hydrogels for localized drug delivery and tissue engineering: a review. Carbohyd Polym Technol Appl:100640. https://doi.org/10.1016/j.carpta.2024.100640
49. Nguyen NH, Lu Z, Elbourne A et al (2024) Engineering antibacterial bioceramics: design principles and mechanisms of action. Materials Today Bio 26:101069. https://doi.org/10.1016/j.mtbio.2024.101069
50. Sánchez ML, Valdez H, Conde M, Viaña-Mendieta P, Boccaccini AR (2023) Polymers and bioactive compounds with a macrophage modulation effect for the rational design of hydrogels for skin regeneration. Pharmaceutics 15(6):1655. https://doi.org/10.3390/pharmaceutics15061655
51. Srivastava A, Bhati P, Singh S et al (2024) A review on polylactic acid-based blends/composites and the role of compatibilizers in biomedical engineering applications. Polym Eng Sci 64(3):1003–1044. https://doi.org/10.1002/pen.26626
52. Farwa U, Lee HY, Lim H et al (2023) Poly(l-lactide)/polycaprolactone based multifunctional coating to deliver paclitaxel/VEGF and control the degradation rate of magnesium alloy stent. Int J Biol Macromol 250:126218. https://doi.org/10.1016/j.ijbiomac.2023.126218
53. Li J, Wang C, Gao G et al (2022) MBG/ PGA-PCL composite scaffolds provide highly tunable degradation and osteogenic features. Bioactive Mater 15:53–67. https://doi.org/10.1016/j.bioactmat.2021.11.034
54. Harish V, Tewari D, Gaur M et al (2022) Review on nanoparticles and nanostructured materials: bioimaging, biosensing, drug delivery, tissue engineering, antimicrobial, and agro-food applications. Nanomaterials 12(3):457. https://doi.org/10.3390/nano12030457
55. Elfaleh I, Abbassi F, Habibi M et al (2023) A comprehensive review of natural fibers and their composites: an eco-friendly alternative to conventional materials. Results Eng 19:101271. https://doi.org/10.1016/j.rineng.2023.101271

56. Simpson FC, Islam MM, Buznyk O et al (2022) Electron-beam irradiated recombinant human collagen-phosphorylcholine corneal implants retain pro-regeneration capacity. Front Bioeng Biotechnol 10. https://doi.org/10.3389/fbioe.2022.883977
57. Altunordu G, Tezcaner A, Evis Z, Keskin D (2023) Improvement of bioactivity with dual bioceramic incorporation to nanofibrous PCL scaffolds. Materialia 27:101699. https://doi.org/10.1016/j.mtla.2023.101699
58. Govindaraju DT, Chen CH, Shalumon KT, Kao HH, Chen JP (2023) Bioactive nanostructured scaffold-based approach for tendon and ligament tissue engineering. Nanomaterials 13(12):1847. https://doi.org/10.3390/nano13121847
59. Omigbodun FT, Oladapo BI, Osa-uwagboe N (2024) Exploring the frontier of Polylactic Acid/Hydroxyapatite composites in bone regeneration and their revolutionary biomedical applications—a review. J Reinforced Plastics Compos. Published online September 6, 2024. https://doi.org/10.1177/07316844241278045
60. Zhao F, Sun J, Xue W et al (2021) Development of a polycaprolactone/poly(p-dioxanone) bioresorbable stent with mechanically self-reinforced structure for congenital heart disease treatment. Bioactive Mater 6(9):2969–2982. https://doi.org/10.1016/j.bioactmat.2021.02.017
61. Zhang M, An H, Zhang F et al (2023) Prospects of using chitosan-based biopolymers in the treatment of peripheral nerve injuries. Int J Mol Sci 24(16):12956. https://doi.org/10.3390/ijms241612956
62. Yang C, Zhang Z, Gan L, Zhang L, Yang L, Wu P (2023) Application of biomedical microspheres in wound healing. Int J Mol Sci 24(8):7319. https://doi.org/10.3390/ijms24087319
63. la Monica F, Campora S, Ghersi G (2024) Collagen-based scaffolds for chronic skin wound treatment. Gels 10(2):137. https://doi.org/10.3390/gels10020137
64. Piatti E, Miola M, Verné E (2024) Tailoring of bioactive glass and glass-ceramics properties for in vitro and in vivo response optimization: a review. Biomater Sci 12(18):4546–4589. https://doi.org/10.1039/D3BM01574B
65. Mukherjee C, Varghese D, Krishna JS et al (2023) Recent advances in biodegradable polymers—properties, applications and future prospects. Eur Polymer J 192:112068. https://doi.org/10.1016/j.eurpolymj.2023.112068
66. Pires PC, Mascarenhas-Melo F, Pedrosa K et al (2023) Polymer-based biomaterials for pharmaceutical and biomedical applications: a focus on topical drug administration. Eur Polymer J 187:111868. https://doi.org/10.1016/j.eurpolymj.2023.111868
67. Rouf S, Malik A, Singh N et al (2022) Additive manufacturing technologies: Industrial and medical applications. Sustain Oper Comp 3:258–274. https://doi.org/10.1016/j.susoc.2022.05.001
68. Makepa DC, Chihobo CH (2024) Barriers to commercial deployment of biorefineries: a multi-faceted review of obstacles across the innovation chain. Heliyon 10(12):e32649. https://doi.org/10.1016/j.heliyon.2024.e32649
69. Chen Y, Li X (2022) The utilization of carbon-based nanomaterials in bone tissue regeneration and engineering: respective featured applications and future prospects. Med Novel Technol Dev 16:100168. https://doi.org/10.1016/j.medntd.2022.100168
70. Xu X, Liu Y, Fu W et al (2020) Poly(N-isopropylacrylamide)-based thermoresponsive composite hydrogels for biomedical applications. Polymers 12(3):580. https://doi.org/10.3390/polym12030580
71. Wang Z, Sun Y, Li C (2024) Advances in 3D printing technology for preparing bone tissue engineering scaffolds from biodegradable materials. Front Bioeng Biotechnol 12. https://doi.org/10.3389/fbioe.2024.1483547
72. Wistner SC, Rashad L, Slaughter G (2023) Advances in tissue engineering and biofabrication for in vitro skin modeling. Bioprinting. 35:e00306. https://doi.org/10.1016/j.bprint.2023.e00306
73. Guo X, Song P, Li F et al (2023) Research progress of design drugs and composite biomaterials in bone tissue engineering. Int J Nanomed 18:3595–3622. https://doi.org/10.2147/IJN.S415666

74. Skwira A, Szewczyk A, Sądej R, Prokopowicz M (2023) Bioglass obtained via one-pot synthesis as osseointegrative drug delivery system. Int J Pharm 633:122610. https://doi.org/10.1016/j.ijpharm.2023.122610
75. Low DYS, Supramaniam J, Leong WD et al (2023) Self-healing synthetic rubber composites: review of recent progress and future directions towards sustainability. Mater Today Sustain 24:100545. https://doi.org/10.1016/j.mtsust.2023.100545
76. Kibrete F, Trzepieciński T, Gebremedhen HS, Woldemichael DE (2023) Artificial intelligence in predicting mechanical properties of composite materials. J Compos Sci 7(9):364. https://doi.org/10.3390/jcs7090364
77. Alaneme KK, Kareem SA, Ozah BN, Alshahrani HA, Ajibuwa OA (2022) Application of finite element analysis for optimizing selection and design of Ti-based biometallic alloys for fractures and tissues rehabilitation: a review. J Market Res 19:121–139. https://doi.org/10.1016/j.jmrt.2022.05.001

Chapter 2
Processing and Characterization of Resorbable Composites

Maneesh Dubey, Ruchin Kacker, Sanjay Kumar Singh, and Amit Arora

1 Introduction to Resorbable Composites

1.1 Definition and Importance

Overview of Resorbable Materials: Resorbable materials are a class of biomaterials designed to degrade and be absorbed by the body over a predetermined period. These materials offer several advantages over permanent implants, including reduced risk of infection, foreign body reactions, and the need for secondary surgical procedures to remove them [15]. The resorption process occurs through mechanisms such as hydrolysis, enzymatic degradation, or cellular uptake, allowing for seamless integration and elimination from the body.

Key Applications Areas: Resorbable composites have found widespread applications in various medical and technological fields due to their unique properties and benefits. Some key applications include:

Orthopedic Implants: Resorbable screws, plates, and pins are commonly used to fix fractures and stabilize bone during the healing process. These implants gradually

M. Dubey
Department of Mechanical Engineering, IIT Madras, Chennai, India

R. Kacker (✉) · S. K. Singh
SISTec GN, Bhopal, Madhya Pradesh, India
e-mail: ruchinkacker@sistec.ac.in

S. K. Singh
e-mail: snajaysingh@sistec.ac.in

A. Arora
Galgotias University, Greater Noida, Uttar Pradesh, India
e-mail: amit06arora@gmail.com

A. Prasad and V. Katiyar (eds.), *Resorbable Polymers for Bioimplants and Fixation Devices*, Materials Horizons: From Nature to Nanomaterials,
https://doi.org/10.1007/978-981-96-9306-1_2

degrade, allowing the bone to regenerate and heal naturally without the need for surgical removal [15].

Tissue Engineering: Resorbable scaffolds provide a temporary structure for cell growth and tissue regeneration. These scaffolds are designed to degrade as new tissue forms, ultimately being replaced by the regenerated tissue [77].

Drug Delivery: Resorbable composites can be used to deliver drugs in a controlled and sustained manner. The drug is incorporated into the composite, and as the material degrades, the drug is gradually released into the body [75].

Dental Implants: In dentistry, resorbable materials are utilized for temporary fillings and as anchors for dental implants. These materials provide support until the permanent implant integrates with the bone, after which they are resorbed [61].

Electronic Devices: Resorbable composites are being explored for use in temporary electronic devices, such as biosensors and drug delivery systems. These devices are designed to degrade harmlessly after fulfilling their intended purpose.

The versatility and biocompatibility of resorbable composites make them promising materials for a wide range of applications. By understanding their properties and characteristics, researchers and engineers can continue to develop innovative and effective solutions for various medical and technological challenges.

1.2 Types of Resorbable Composites

Biodegradable vs. Bioresorbable Composites: While the terms "biodegradable" and "bioresorbable" are often used interchangeably, there is a subtle distinction between them. Biodegradable composites are capable of breaking down into smaller components, but the degradation products may not necessarily be absorbed by the body. Some biodegradable materials may leave behind residual byproducts that require clearance by the immune system. On the other hand, bioresorbable composites are designed to degrade completely and be absorbed by the body without leaving any harmful residues. The degradation products are typically metabolized and eliminated through natural pathways [40].

In medical applications, bioresorbable composites are generally preferred because they offer more complete integration with the body, reducing the risk of adverse reactions. This characteristic makes bioresorbable materials ideal for applications such as temporary implants and drug delivery systems, where complete degradation and resorption are necessary.

Commonly Used Polymer Matrices: Resorbable composites are typically composed of a polymer matrix reinforced with inorganic or organic fillers. The choice of polymer matrix significantly influences the mechanical properties, degradation rate, and biocompatibility of the composite. Common polymer matrices include:

Natural Polymers [48]

Collagen: A protein found in various tissues, collagen offers good biocompatibility and mechanical properties.

Gelatin: A denatured form of collagen, gelatin is valued for its gel-forming properties and biodegradability.

Chitosan: Derived from chitin, chitosan is known for its antimicrobial properties and biocompatibility.

Alginate: Extracted from seaweed, alginate is used for its gel-forming ability and biodegradability.

Synthetic Polymers [10]

Polylactic acid (PLA): A biodegradable polyester derived from lactic acid, PLA offers good mechanical properties and biocompatibility.

Polyglycolic acid (PGA): PGA is a biodegradable polyester known for its rapid degradation rate.

Polycaprolactone (PCL): PCL has a slower degradation rate compared to PLA and PGA, offering sustained mechanical support.

Polyhydroxyalkanoates (PHAs): These biodegradable polyesters are produced by bacteria and are known for their versatility and biocompatibility.

The selection of a polymer matrix depends on the specific application's requirements, such as the desired mechanical properties, degradation rate, and biocompatibility. In many cases, polymer blends or copolymers are used to achieve a combination of properties.

1.3 Historical Background and Development

Early Research Milestones: The development of resorbable composites can be traced back to the early twentieth century, when researchers explored the use of natural materials like silk and collagen for surgical sutures. These early materials, while biodegradable, lacked the controlled degradation and mechanical properties required for more complex applications.

Significant advancements occurred in the mid-twentieth century with the introduction of synthetic polymers like polylactic acid (PLA) and polyglycolic acid (PGA). Initially developed for medical sutures, these materials demonstrated promising biodegradability and mechanical properties. Research then focused on incorporating these polymers with fillers and reinforcements to develop composites with enhanced performance for specific medical applications.

Evolution of Properties [52]

Over time, research has led to significant improvements in the properties of resorbable composites, with a focus on:

Mechanical properties: Tailoring composites for strength, stiffness, and toughness has enabled their use in applications like orthopedic implants and tissue scaffolds.
Degradation rate: Controlling degradation rates has allowed for precise matching of material properties with biological responses.
Biocompatibility: Efforts have been made to improve biocompatibility, reducing adverse reactions and ensuring integration with the body.
Functionality: Incorporating bioactive molecules such as growth factors and drugs has led to multifunctional composites with enhanced therapeutic properties.
Manufacturing techniques: Advances in techniques like 3D printing and electrospinning have enabled the production of complex, customized resorbable composite structures.

These advancements have broadened the applications of resorbable composites in fields like medicine, biotechnology, and environmental science, contributing to their increasing prominence.

2 Materials for Resorbable Composites

2.1 Polymer Matrices

Polylactic Acid (PLA): Polylactic acid (PLA) is a biodegradable polyester derived from lactic acid, which is found naturally in various organisms. It is well-known for its excellent biocompatibility, mechanical properties, and versatility, being processed through techniques like injection molding and 3D printing [80]. PLA homopolymers and copolymers can be tailored to achieve specific degradation rates by adjusting molecular weight and stereochemistry. It has excellent mechanical strength and stiffness. It can be processed using various methods such as injection molding and 3D printing. PLA's degradation rate can be adjusted to meet specific application requirements. PLA's versatility has led to its widespread use in various fields. In the medical sector, it is employed for implants like screws, plates, and stents. It also serves as a scaffold for tissue engineering, facilitating the growth of new tissues. Additionally, PLA is utilized in drug delivery systems to control the release of medications. Beyond medical applications, PLA finds use in packaging and consumer goods, offering a sustainable alternative to traditional plastics.

Polyglycolic Acid (PGA): As a biodegradable polyester with a faster degradation rate compared to PLA, polyglycolic acid (PGA) is well-suited for applications requiring temporary support [33]. PGA's mechanical properties, including high tensile strength and modulus, combined with its rapid degradation, make it an ideal choice for absorbable sutures. PGA exhibits high tensile strength, rapid biodegradability, and biocompatibility. Its unique characteristics have led to its widespread use in surgical sutures, absorbable meshes, and temporary tissue scaffolds.

Polycaprolactone (PCL): Known for its slower degradation rate compared to PLA and PGA, polycaprolactone (PCL) is ideal for applications demanding long-term mechanical support. PCL's flexibility, toughness, and processability have made it a popular choice in various biomedical applications [8]. PCL offers excellent mechanical properties, a slower degradation rate, biocompatibility, and can be processed through injection molding and 3D printing. Its unique characteristics have led to its use in tissue engineering scaffolds, drug delivery systems, and medical implants such as stents and sutures.

2.2 Reinforcements and Fillers

Ceramic-Based Reinforcements: Ceramic reinforcements, such as hydroxyapatite (HA) and tricalcium phosphate (TCP), significantly enhance the mechanical properties and biocompatibility of resorbable composites [27]. Bioactive glass and carbon-based materials further contribute to improved tissue regeneration and osteoconductivity. For example, hydroxyapatite (HA) promotes bone tissue regeneration, while tricalcium phosphate (TCP) is biocompatible and resorbable. Bioactive glass encourages tissue bonding, and carbon fibers/nanotubes offer high strength and conductivity. The incorporation of ceramic reinforcements leads to improved mechanical strength and biocompatibility, tailored degradation rates, and the potential for functionalization with bioactive molecules.

Natural Fiber Composites: Natural fibers, such as cellulose, collagen, and chitosan, are increasingly employed as reinforcements due to their sustainability, biodegradability, and biocompatibility [31]. These fibers offer improved mechanical properties and are often functionalized with bioactive molecules to enhance their performance. For instance, cellulose exhibits good mechanical properties and biocompatibility, while collagen offers excellent biocompatibility and osteoinductivity. Silk is known for its high tensile strength, and chitosan possesses antimicrobial and biocompatible properties. The use of natural fiber reinforcements promotes sustainability and biodegradability, enhances mechanical properties, and offers potential for bioactive functionalization.

Hybrid Fillers: Hybrid fillers, which combine different types of reinforcements, offer a means to optimize the mechanical and functional properties of resorbable composites [6]. These fillers allow for the tailoring of degradation rates and mechanical performance to meet specific application requirements. Like, hybrid fillers include ceramic-fiber composites, ceramic-polymer composites, and natural fiber-polymer composites. The use of hybrid fillers enables the tailoring of mechanical properties, enhances biocompatibility, and opens up the potential for multifunctional applications.

2.3 Additives and Surface Modifications

Biodegradation Accelerants: To tailor the degradation rate of resorbable composites, additives that accelerate biodegradation can be employed. These additives promote hydrolysis or enzymatic degradation, ensuring that the degradation rate aligns with the healing process or the release of therapeutic agents [69]. Citric acid, a natural acid, accelerates the degradation of PLA and PGA. Enzymes like lipase and esterase catalyze the hydrolysis of ester bonds in polymers. Inorganic salts such as magnesium chloride and calcium chloride enhance the degradation of certain polymers. The use of biodegradation accelerants enables controlled degradation rates, tailored release of bioactive molecules, and a reduction in foreign body reactions.

Antibacterial Coatings: Antibacterial coatings are essential in preventing infections associated with resorbable implants, particularly in surgical applications [51]. Various approaches to antibacterial coatings have been developed, each offering distinct levels of efficacy against pathogens. Antibiotic-releasing coatings incorporate antibiotics for slow release. Silver-based coatings utilize silver ions for broad-spectrum antimicrobial properties. Antimicrobial peptides, such as LL-37, are natural peptides effective against various bacteria. Self-disinfecting coatings release antimicrobial agents upon bacterial contact. Antibacterial coatings reduce the risk of infection, improve patient outcomes, and enhance biocompatibility.

Surface Treatments for Enhanced Biocompatibility: Surface treatments are crucial in improving tissue integration and reducing foreign body reactions in resorbable composites [9]. By modifying surface chemistry and topography, these treatments can significantly enhance cellular interactions. Examples include plasma treatment, chemical modification with functional groups or bioactive molecules like RGD peptides, bio-functionalization with growth factors or proteins, and texturing with micro- or nanostructures. These treatments offer numerous benefits, including enhanced biocompatibility, improved tissue integration, reduced foreign body reactions, and the ability to tailor the biological response.

Incorporating these additives and surface modifications allows for fine-tuning the properties of resorbable composites, making them more suited for specific applications. These strategies play a key role in advancing material performance, ensuring biocompatibility, and enhancing clinical efficacy. Figure 1 illustrates the evolving characteristics of resorbable composites.

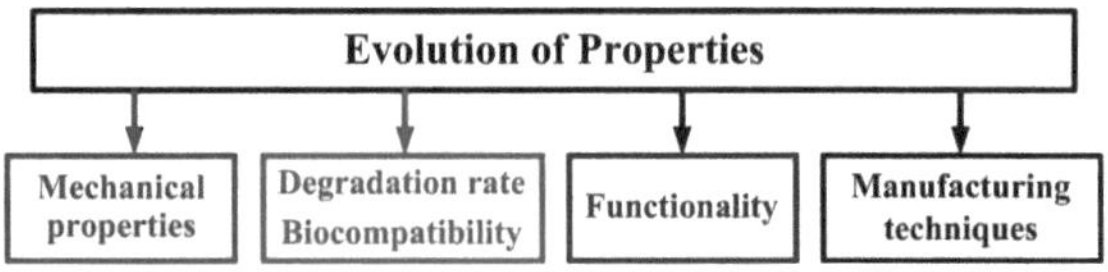

Fig. 1 Evolution properties of resorbable composites

3 Processing Techniques for Resorbable Composites

3.1 Manufacturing Methods

Extrusion and Injection Molding: Extrusion and injection molding are widely used traditional manufacturing techniques for producing resorbable composites in various shapes and sizes [56]. Extrusion involves forcing a polymer melt through a die to create continuous profiles like rods, tubes, and sheets. Reinforcements such as ceramic particles or natural fibers can be added to the polymer melt to form composite materials. Extrusion is well-suited for large-scale production of resorbable composites, especially those with simple geometries. Injection molding involves injecting a polymer melt into a mold cavity to create complex parts. Reinforcements can be integrated by mixing them into the polymer melt or placing them within the mold cavity before injection. Injection molding is commonly used for small and medium-sized resorbable composites with intricate designs, such as medical implants.

3D Printing and Additive Manufacturing: 3D printing, also known as additive manufacturing, represents a significant advancement in the fabrication of resorbable composites [34]. It enables the creation of complex, highly customized structures. Fused Deposition Modeling (FDM) is a common 3D printing technique that extrudes heated polymer filament layer by layer to form a 3D object. Reinforcements can be embedded in the filament or added during printing. Stereo-lithography (SLA) uses a laser to cure liquid resin layer by layer, forming solid structures. Reinforcements may be added to the resin to improve mechanical properties. Selective Laser Sintering (SLS) involves using a laser to fuse powdered materials together to create a solid object. Reinforcements can be incorporated into the powder mix, enhancing strength and functionality. 3D printing offers numerous benefits, including the ability to fabricate complex and highly customized designs, a reduction in material waste during production, applicability in personalized medicine, and enhanced design flexibility for functional implants and scaffolds. Figure 2 illustrates the schematic representation of FDM 3D printing.

Electrospinning: Electrospinning is a versatile processing technique that produces nanofibrous materials from polymer solutions, offering excellent surface area and porosity [64]. In electrospinning, a polymer solution is subjected to a high voltage electric field, causing the solution to form a fine jet that elongates into nanofibers as it dries. Reinforcements can be introduced either directly into the polymer solution or incorporated within the nanofibers during spinning. Electrospinning offers the ability to produce nanofibrous materials with high surface area and porosity, tunable mechanical properties and enhanced biocompatibility, potential for bioactive molecule integration and functionalization, and applications in tissue engineering scaffolds, drug delivery systems, and filtration materials.

The choice of manufacturing method for resorbable composites depends on the intended application, desired shape, size, mechanical properties, and production scale. In many cases, combining multiple processing techniques allows for greater

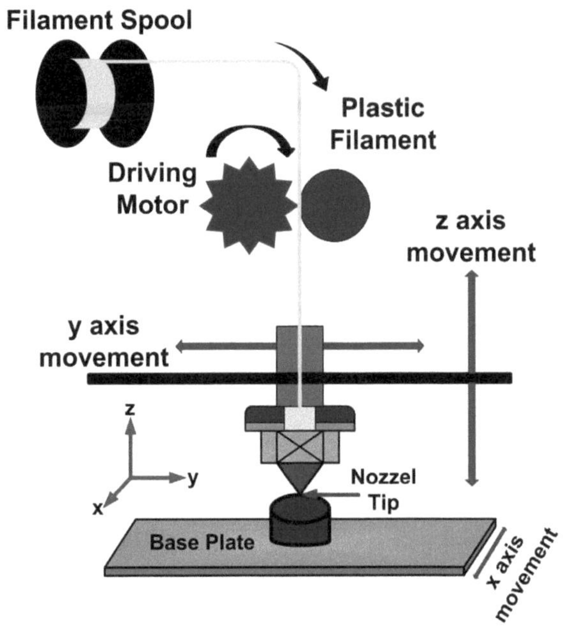

Fig. 2 Schematic diagram of FDM 3D printer

optimization and the production of advanced composite materials tailored to specific needs.

3.2 Parameters Influencing Processing

Temperature and Pressure Controls: Temperature and pressure are critical parameters in the processing of resorbable composites, as they directly influence material properties, processing efficiency, and product quality [11, 42]. The temperature of the polymer melt or solution determines its viscosity and flow behavior. Higher temperatures can improve processability but may accelerate degradation, affecting the composite's mechanical properties. Precise temperature control is essential for consistent processing and product quality. Pressure is used to force polymer material through the die or mold cavity during extrusion or injection molding. Higher pressure enhances the density and uniformity of the composite, but excessive pressure can cause premature degradation or material damage.

Processing Time and Degradation: Processing time significantly influences the degradation rate and mechanical properties of resorbable composites [42]. Longer processing times can increase the material's temperature, leading to accelerated degradation. However, insufficient processing time may result in incomplete curing or improper bonding of composite components. Optimized processing times are crucial to balance the material's properties and processing efficiency. The degradation of

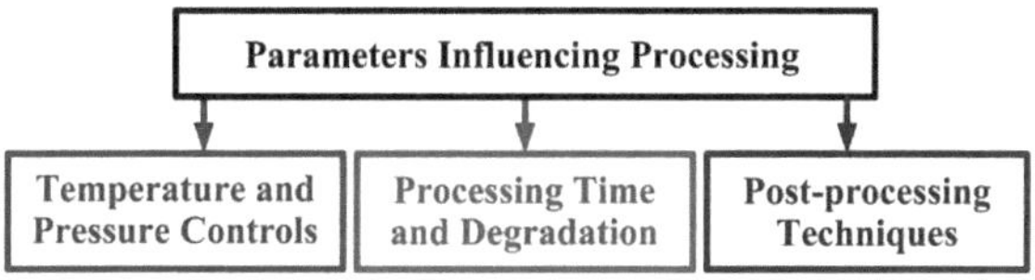

Fig. 3 Parameters influencing processing of resorbable composites

resorbable composites is influenced by processing conditions such as temperature and pressure. High temperatures and pressures tend to accelerate degradation, while lower temperatures and pressures retard degradation. Understanding the relationship between processing conditions and degradation is essential for designing composites with the desired properties and lifespan.

Post-processing Techniques: Post-processing techniques can modify the properties and performance of resorbable composites after manufacturing [47]. Resorbable composites need sterilization before implantation to prevent infection. Common methods include autoclaving, gamma irradiation, and ethylene oxide gas sterilization. Surface treatments can improve biocompatibility, functionality, or mechanical properties. Examples include plasma treatment, chemical modification, and biofunctionalization. Proper packaging is essential to protect resorbable composites during storage and transportation. Packaging materials must be compatible with the composite and prevent contamination.

Careful control of temperature, pressure, processing time, and post-processing techniques optimizes resorbable composite processing, ensuring high-quality products with desired properties and performance. Figure 3 illustrates the factors that affect the processing of resorbable composites.

3.3 Challenges in Processing

Thermal Sensitivity: Many resorbable polymers are thermally sensitive, making them prone to degradation or structural changes when exposed to elevated temperatures [59]. This poses challenges in processing techniques like extrusion and injection molding. High temperatures can accelerate degradation, reducing mechanical properties and causing premature failure. The thermal sensitivity of these polymers limits available processing techniques, as some require temperatures that exceed the polymer's stability. Strategies to address thermal sensitivity include using lower processing temperatures when feasible and considering alternative methods, such as 3D printing or electrospinning, which typically operate at lower temperatures. Additionally, incorporating additives or fillers that enhance thermal stability can be an effective approach.

Maintaining Structural Integrity: Maintaining structural integrity during processing can be challenging, especially for complex shapes or large-scale production of resorbable composites [12]. Defects like porosity, voids, or delamination compromise mechanical properties and biocompatibility. Scaling production introduces new complexities in maintaining consistency and quality control. Optimize processing parameters to minimize defects. Implement quality control to monitor properties throughout production. Develop scalable processes that maintain product quality.

Processing of Hybrid Composites: Processing hybrid composites, which combine different types of reinforcements or fillers, presents challenges due to the heterogeneous nature of these materials [28]. Ensuring compatibility between various components is difficult, as different materials may have divergent processing requirements. Achieving uniform dispersion of reinforcements or fillers in the polymer matrix can be challenging, particularly when dealing with materials of different densities or surface properties. The mechanical properties of hybrid composites are sensitive to the composition and distribution of components, requiring optimized processing conditions. Select compatible components and optimize processing conditions. Use advanced mixing techniques to ensure uniform dispersion. Characterize the mechanical properties to assess the effectiveness of the processing method.

Addressing these challenges requires a deep understanding of material properties, careful design of processing techniques, and strict quality control. By overcoming these obstacles, it is possible to produce high-quality resorbable composites with reliable performance and consistent properties.

4 Characterization of Resorbable Composites

Characterization of resorbable composites is crucial for understanding their mechanical properties, structure, and suitability for various applications. This chapter focuses on techniques to assess durability, functionality, and optimize resorbable materials in biomedical and other critical fields.

4.1 Mechanical Characterization

Mechanical characterization assesses strength, elasticity, toughness, and resistance to mechanical stresses, crucial for load-bearing and dynamic environments. Common tests include tensile, compression, fatigue, creep, and impact tests.

Tensile and Compression Testing: Tensile and compression testing evaluate strength and deformation under load. These tests measure tensile strength, compressive strength, and modulus of elasticity. Tensile testing stretches a sample to assess resistance to pulling forces, while compression testing subjects it to compressive forces. Understanding both tensile and compressive responses is crucial for resorbable composites in load-bearing applications like bone fixation devices or tissue scaffolds [3]. These tests ensure structural integrity to support the healing process while degrading at a controlled rate as tissue regenerates.

Fatigue and Creep Behavior: Fatigue and creep behavior are critical for resorbable composites under repetitive or long-term loading [66]. Fatigue testing evaluates performance under cyclic loading, simulating conditions like orthopedic implants or cardiovascular stents. Fatigue failure occurs due to repeated loading cycles. Creep testing measures deformation under constant stress over time. This is important for resorbable composites subjected to continuous pressure, such as implants or scaffolds. For example, creep resistance is crucial for maintaining the dimensional stability of bone plates or pins. Understanding fatigue and creep ensures resorbable composites can withstand long-term use without premature failure, providing reliability throughout their functional life before resorption.

Impact Resistance: Impact resistance is important for resorbable composites, especially in applications where the material may be subjected to sudden forces or shocks [75]. Impact testing evaluates the ability of a material to absorb energy and resist fracture under high-impact forces. This is crucial for resorbable composites used in dynamic environments, such as sports injuries or trauma-related implants. For instance, resorbable plates or screws used in bone fixation must withstand accidental impacts without breaking, ensuring structural integrity during the healing process. Impact resistance is also important for resorbable packaging materials.

4.2 Thermal and Rheological Characterization

Thermal and rheological characterization techniques are essential for understanding the thermal stability, degradation, and flow properties of resorbable composites, predicting their performance in biomedical and environmental applications [29].

Differential Scanning Calorimetry (DSC): Differential scanning calorimetry (DSC) is used to assess thermal properties, including melting points, crystallization behavior, and glass transition temperatures [55]. DSC measures heat flow into or out of a material during a controlled temperature program, detecting phase transitions and thermal events.

For resorbable polymers like polylactic acid (PLA) and polycaprolactone (PCL), DSC reveals crystallinity, affecting mechanical properties and degradation rates [7]. Higher crystallinity typically results in slower degradation, requiring optimization of thermal properties for specific applications like tissue scaffolding or biodegradable packaging.

Thermogravimetric Analysis (TGA): Thermogravimetric analysis (TGA) measures weight loss as a material is heated at a controlled rate, providing insight into thermal degradation profiles and decomposition temperatures [23]. Understanding thermal degradation temperature is crucial for ensuring stability during processing and use in biomedical applications, where sterilization often involves elevated temperatures [72]. TGA also provides data on composition, as different polymers and additives decompose at different temperatures, allowing for identification of components within the composite.

Rheometry: Rheometry measures flow properties and viscosity of resorbable composites, essential for understanding how materials behave under different shear stresses [36]. Rheological properties are important for processing techniques like extrusion or 3D printing, where the material must flow smoothly for shaping into complex geometries. For resorbable composites used in biomedical applications, such as tissue engineering scaffolds, controlling viscosity is crucial for ensuring the material can be printed or molded without defects. Rheometry helps optimize the formulation of composites to ensure stability during the manufacturing process and achieve desired mechanical properties once fabricated.

4.3 Morphological and Structural Characterization

Morphological and structural characterization is essential for understanding surface topography, internal structure, and chemical composition in resorbable composites [49]. These properties predict and optimize performance in various applications. Characterization methods provide insights into component distribution, crystallinity, and chemical bonding. This section focuses on SEM, XRD, and FTIR.

Scanning Electron Microscopy (SEM): Scanning electron microscopy (SEM) analyzes surface morphology and microstructural details at high magnifications [24]. SEM produces detailed, high-resolution images by scanning the surface with a focused electron beam. The interaction between the beam and the sample generates signals that form an image, revealing surface features, grain structure, and particle dispersion. Figure 4 illustrates the methods used for the structural characterization of resorbable composites.

Applications in Resorbable Composites: SEM examines fiber-matrix interfaces, surface roughness, porosity, and particle distribution within resorbable composites [50]. This is crucial for tissue engineering, where surface morphology influences cell attachment and proliferation. SEM detects defects or irregularities like voids, cracks, or phase separation, affecting mechanical properties and degradation behavior.

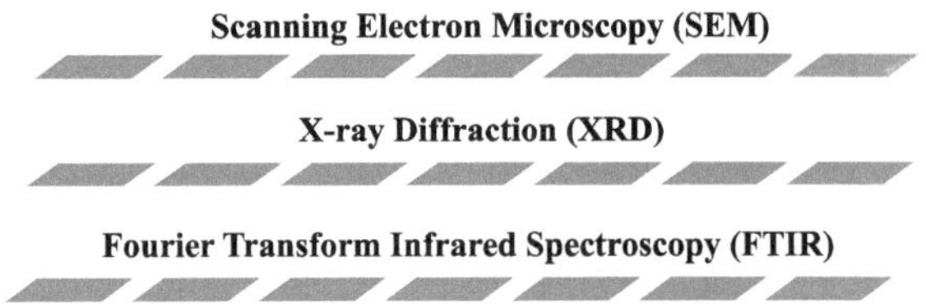

Fig. 4 Means for structural characterization of resorbable composites

Advantages and Limitations: This technique provides high resolution and depth of field, allowing for detailed surface imaging, and it can analyze coated, uncoated, and fractured surfaces. It is particularly useful for studying particle dispersion and fiber orientation in composite materials. However, it requires sample preparation, such as coating with a conductive material like gold or carbon to prevent charging, and primarily offers surface-level information, with limited insights into the internal structure of the composite.

X-ray Diffraction (XRD): X-ray diffraction (XRD) is a non-destructive technique determining crystallographic structure, phase composition, and crystallinity in materials [22]. When an X-ray beam interacts with a crystalline structure, it is diffracted in specific directions based on atomic plane spacing. Analyzing the diffraction pattern identifies the material's crystal structure and quantifies crystalline and amorphous regions.

Applications in Resorbable Composites: XRD characterizes crystalline phases in resorbable composites, particularly in polymer matrices like PLA and PCL [4]. Crystallinity impacts mechanical properties and degradation behavior, with crystalline regions degrading more slowly than amorphous ones. XRD also detects filler materials like calcium phosphates or ceramics, providing insights into structural integrity.

Advantages and Limitations: This technique offers advantages such as providing detailed information about crystal structure and phase composition, and it effectively monitors changes in crystallinity over time. It is applicable to a wide range of materials, including polymers, ceramics, and metals. However, its limitations include primarily offering bulk information, which may not be ideal for analyzing surface features, and requiring sufficient crystallinity for accurate results.

Fourier Transform Infrared Spectroscopy (FTIR): Fourier transform infrared spectroscopy (FTIR) analyzes chemical structure and molecular composition by detecting how materials absorb infrared light [21]. FTIR identifies specific functional groups within a polymer matrix based on their characteristic vibrational frequencies. This technique is especially useful for determining chemical bonds present in a material and monitoring chemical changes during degradation.

Applications in Resorbable Composites: FTIR analyzes the chemical composition of polymer matrices in resorbable composites [36]. It identifies specific functional groups, such as ester bonds in PLA and PCL, and tracks changes in these groups during degradation. FTIR examines the interaction between the polymer matrix and reinforcing materials, affecting mechanical properties and biocompatibility. Additionally, FTIR detects additives or surface modifications enhancing material performance, such as coatings or cross-linking agents.

Advantages and Limitations: The technique offers several advantages, such as being non-destructive and providing valuable insights into molecular structure and bonding. It is effective in monitoring degradation processes by detecting changes in chemical bonds and can be applied to a wide range of organic and inorganic materials. However, it has some limitations, including lower spatial resolution compared to techniques like SEM or XRD, and the need for proper sample preparation, as water or impurities can interfere with the infrared spectrum.

5 Biocompatibility and Degradation Behavior

Biocompatibility and degradation behavior are crucial for resorbable composites in biomedical applications [26]. Materials must be safe, compatible with the body, and provide mechanical support. This chapter covers in vitro testing methods for assessing biocompatibility and explores degradation behavior.

5.1 *In Vitro Biocompatibility Testing*

In vitro biocompatibility testing evaluates how resorbable composites interact with living cells and tissues [76]. These tests determine cell growth, cell death, or immune response. Common in vitro tests include cell proliferation and viability assays, cytotoxicity testing, and inflammatory response evaluation.

Cell Proliferation and Viability Assays: Cell proliferation and viability assays assess the ability of resorbable composites to support cell growth and health. These assays typically involve culturing cells on the composite material and measuring their ability to multiply over time. Common methods include the MTT and WST-1 assays, which rely on the reduction of tetrazolium salts by metabolically active cells, indicating cell viability and proliferation [1].

Cytotoxicity Testing: Cytotoxicity testing evaluates whether a material releases harmful substances to cells, leading to cell death or dysfunction [54]. Cytotoxicity is assessed by exposing cells to the material's extracts or degradation products and measuring cell death or membrane integrity. The most common assays for cytotoxicity include the lactate dehydrogenase (LDH) assay and live/dead cell staining. Figure 5 illustrates the cytotoxicity assessment of resorbable composites.

Inflammatory Response Evaluation: The inflammatory response evaluation determines whether the resorbable composite induces an immune response when in contact with cells or tissues [63]. In vitro models typically use immune cells, such

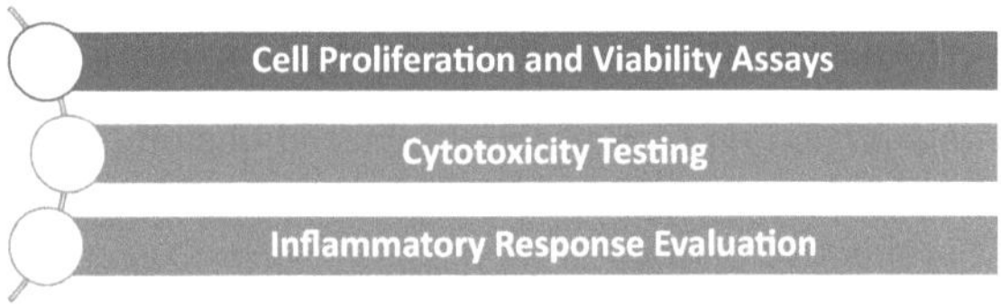

Fig. 5 Cytotoxicity testing of resorbable composites

as macrophages, to assess the release of pro-inflammatory cytokines like IL-6 and TNF-α in response to the material.

5.2 In Vivo Studies

In vivo studies assess resorbable composites' real-world performance in a biological environment [38]. These studies provide insights into how materials interact with living tissues, integrate into the host organism, and degrade over time. Animal models, tissue integration assessment, and long-term resorption analysis evaluate biocompatibility and functional behavior of resorbable composites.

Animal Models for Resorbable Composites: Animal models play a pivotal role in studying resorbable composites in a living organism [53]. These models are selected based on the similarity of their biological and physiological responses to human systems. Common animal models include rodents (e.g., rats and mice), rabbits, and larger mammals such as pigs and sheep, each offering specific advantages depending on the target application.

Tissue Integration and Healing Responses: A primary goal of using resorbable composites is to promote tissue integration and facilitate natural healing processes [71]. In vivo studies assess how well the composite integrates with surrounding tissues and supports tissue regeneration. This integration is often evaluated by examining cellular infiltration, vascularization, and the formation of new tissue around the composite material.

Long-term Resorption Behavior: Long-term resorption behavior is a critical factor in the suitability of resorbable composites for biomedical applications [58]. In vivo studies determine the rate and mechanism of degradation under physiological conditions. Ideally, the composite degrades at a controlled rate, matching tissue regeneration, without harmful byproducts or compromising structural integrity during healing.

Degradation Kinetics: In vivo studies assess how the composite degrades over time, typically through chemical, enzymatic, and mechanical processes. The degradation rate is often monitored by evaluating changes in mass, mechanical properties, and structural integrity over extended periods. For example, resorbable composites used in bone regeneration should maintain mechanical support until newly formed tissue can bear the load independently [73]. Figure 6 illustrates the biological assessment of resorbable composites.

Byproducts and Biological Impact: Long-term studies also evaluate degradation byproducts. It is critical that breakdown products are non-toxic and do not accumulate in surrounding tissues. Histological and biochemical analyses track the presence of degradation products and their potential impact on nearby organs and tissues [78].

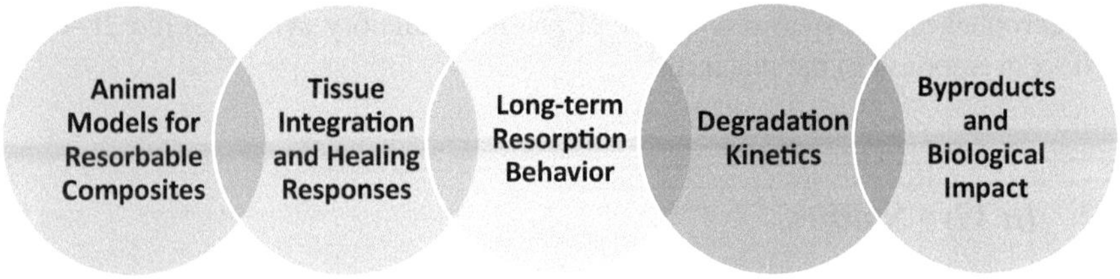

Fig. 6 Biological evaluation of resorbable composites

5.3 Factors Influencing Degradation

The degradation behavior of resorbable composites is influenced by chemical composition, environmental conditions, and design parameters [41]. Understanding these factors is essential for optimizing performance in biomedical applications. This section explores primary degradation mechanisms and how various factors affect the process. Figure 7 illustrates that the degradation behavior of resorbable composites is affected by factors such as chemical composition, environmental conditions, and design parameters.

Hydrolysis and Enzymatic Degradation: Hydrolysis and enzymatic degradation are the two primary mechanisms through which resorbable composites degrade in biological environments [43]. Hydrolysis involves the cleavage of chemical bonds in the polymer matrix due to reaction with water, while enzymatic degradation occurs when specific enzymes in the body break down the composite material.

Hydrolysis: Hydrolysis is the predominant degradation mechanism for most synthetic polymers, such as polylactic acid (PLA), polyglycolic acid (PGA), and polycaprolactone (PCL) [79]. In this process, water molecules penetrate the polymer matrix

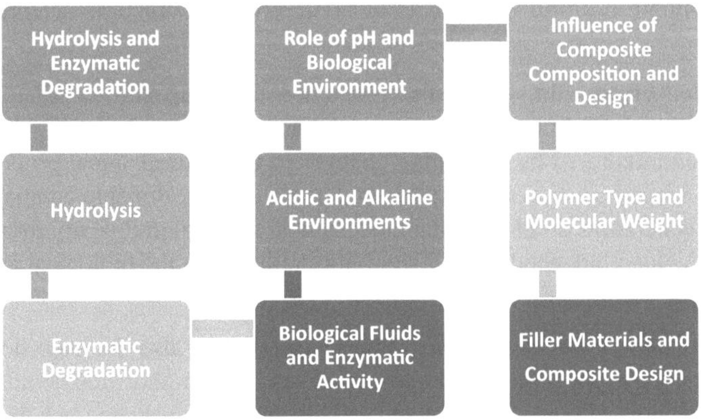

Fig. 7 Degradation behavior of resorbable composites is influenced by chemical composition, environmental conditions and design parameters

and cleave the ester bonds, leading to a reduction in molecular weight and eventual breakdown of the material into smaller, soluble fragments. The rate of hydrolysis depends on factors such as the polymer's crystallinity, molecular weight, and the presence of hydrophilic groups.

Enzymatic Degradation: Enzymatic degradation is more common in natural polymers but can also occur in some synthetic polymers [13]. Enzymes such as lipases and esterases catalyze the hydrolysis of ester bonds in the polymer matrix, accelerating the degradation process. The presence and concentration of specific enzymes in the biological environment significantly influence the rate of enzymatic degradation.

Role of pH and Biological Environment: The pH and the biological environment surrounding the composite material play a crucial role in influencing its degradation rate [16]. Variations in pH levels can accelerate or decelerate hydrolytic and enzymatic degradation processes, affecting the overall performance of the composite.

Acidic and Alkaline Environments: In acidic environments, the degradation of polyesters, such as PLA and PGA, is typically accelerated due to the increased protonation of ester bonds, which makes them more susceptible to hydrolysis [39]. Conversely, alkaline environments can also enhance the hydrolytic degradation of certain polymers, particularly those containing ester linkages. The impact of pH is especially important in biomedical applications, where localized changes in pH occur during inflammation, infection, or tissue regeneration.

Biological Fluids and Enzymatic Activity: The presence of biological fluids and their composition can influence the degradation behavior of composites [49]. For example, in the case of enzyme-sensitive polymers, the concentration and activity of specific enzymes in bodily fluids like blood, interstitial fluid, or synovial fluid can vary, leading to changes in degradation rates. Additionally, the ionic composition of the biological environment, including the presence of salts and proteins, can affect water uptake and subsequent degradation.

Influence of Composite Composition and Design: The composition and design of resorbable composites are critical factors that determine their degradation behavior [75]. Variables such as the type of polymer used, the presence of reinforcing fillers, and the structural design of the composite can all influence how quickly and uniformly the material degrades.

Polymer Type and Molecular Weight: The choice of polymer matrix significantly impacts the degradation rate [5]. Polymers with higher crystallinity, such as PLA, tend to degrade more slowly than amorphous polymers due to the lower accessibility of water molecules to crystalline regions. Similarly, polymers with higher molecular weights degrade more slowly because they require more time for the molecular weight to decrease to a level where the polymer can dissolve.

Filler Materials: The inclusion of fillers, such as hydroxyapatite or calcium phosphate, can either accelerate or slow down degradation depending on the nature of the filler and its interaction with the polymer matrix [18]. Fillers may create micro-galvanic effects or act as barriers to water penetration, thereby influencing the overall degradation process.

Composite Design: The design of the composite, including its porosity, surface area, and thickness, can affect how quickly it degrades [2]. Composites with high porosity and surface area are more likely to undergo rapid degradation due to increased water penetration and greater enzyme accessibility. In contrast, denser composites with fewer pores may degrade more slowly, maintaining mechanical integrity for a longer period.

6 Applications of Resorbable Composites

Resorbable composites are promising biomaterials for biomedical applications, eliminating the need for secondary surgical removal. They have applications in orthopedic implants, cardiovascular stents, and tissue engineering scaffolds due to their biocompatibility, mechanical properties, and controlled degradation rates.

6.1 Biomedical Applications

Resorbable composites are invaluable in biomedical applications where temporary support and gradual degradation are necessary for healing and regeneration processes [20]. These composites offer the advantage of being replaced by native tissue as they degrade, making them highly attractive for specific medical interventions.

Orthopedic Implants: Resorbable composites are extensively used in orthopedic applications, particularly in fracture fixation devices such as plates, screws, and pins [37]. These materials provide temporary mechanical support to damaged bones while promoting natural healing. As the bone regenerates, the composite degrades, reducing the need for surgical removal. Polymers such as PLA and PGA, often reinforced with ceramic fillers like hydroxyapatite, enhance the mechanical properties of these composites.

Bone Fracture Fixation: Resorbable orthopedic implants are designed to maintain structural integrity long enough for bone healing [30]. For example, PLA-based composites degrade over several months, allowing bone to regain strength during the healing process. The slow degradation rate is crucial in load-bearing applications, where premature loss of mechanical strength could lead to complications.

Cartilage Regeneration: Resorbable composites have been used in cartilage repair [25]. Scaffolds made from PCL and PGA support cartilage regeneration in defects, providing temporary support while the cartilage regenerates.

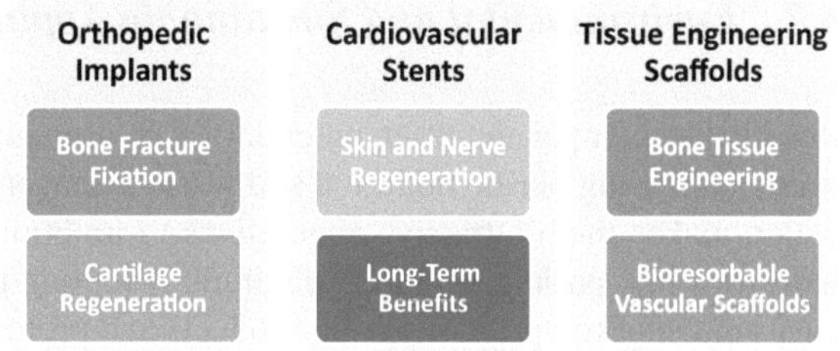

Fig. 8 Resorbable composites in biomedical applications

Cardiovascular Stents: Cardiovascular stents made from resorbable composites provide temporary scaffolding to blood vessels after angioplasty, preventing restenosis while the vessel heals [32]. Unlike traditional metal stents, resorbable stents gradually degrade and are absorbed by the body, reducing long-term risks associated with permanent implants, such as chronic inflammation or thrombosis. Figure 8 shows resorbable composites in biomedical applications.

Bioresorbable Vascular Scaffolds (BVS): Bioresorbable stents are typically made from materials like PLA, often blended with magnesium or other bioresorbable metals to enhance mechanical strength [70]. These stents maintain their structure for six months to a year, after which they begin to degrade, allowing the blood vessel to retain its normal function without the presence of foreign material.

Long-Term Benefits: The use of bioresorbable stents has been associated with improved long-term outcomes, as they reduce the risk of late stent thrombosis, a common complication with permanent stents. The degradation of the composite scaffold allows for vessel vasomotion to return, further enhancing patient recovery.

Tissue Engineering Scaffolds: Resorbable composites have been widely used in tissue engineering as scaffolds for regenerating tissues such as bone, cartilage, skin, and nerves [26]. These scaffolds provide temporary mechanical support and serve as a substrate for cell attachment and tissue regeneration. As the cells proliferate and the tissue regenerates, the scaffold gradually degrades, leaving behind newly formed tissue.

Bone Tissue Engineering: In bone regeneration, scaffolds made from PLA/PGA blends, reinforced with bioactive ceramics like calcium phosphate, demonstrate excellent biocompatibility and mechanical properties [74]. These composites support osteogenesis while slowly degrading as the bone regenerates, allowing for a seamless transition from scaffold to native bone.

Skin and Nerve Regeneration: Resorbable scaffolds are also used in skin and nerve tissue engineering [26]. For skin regeneration, composite scaffolds made from collagen or other natural polymers provide a favorable environment for cell growth and healing. In nerve regeneration, tubular scaffolds made from PCL or chitosan are used to bridge gaps in damaged nerves, supporting axon growth as the scaffold degrades.

6.2 Environmental and Sustainable Applications

Resorbable composites offer potential in environmental and sustainable applications, addressing plastic pollution and waste management [57]. Their natural degradation makes them attractive alternatives to traditional non-degradable plastics in industries like packaging and agriculture. Adopting these materials can reduce the environmental footprint of plastic products and promote a circular economy.

Biodegradable Packaging: The packaging industry is a major contributor to global plastic waste, with millions of tons of non-biodegradable plastics ending up in landfills and oceans each year [46]. Resorbable composites, particularly those made from biodegradable polymers such as polylactic acid (PLA) and polyhydroxyalkanoates (PHA), offer a promising solution to this issue. These materials can break down into harmless byproducts under natural environmental conditions, thus reducing the environmental impact of plastic waste.

PLA-Based Packaging: PLA is one of the most widely used biodegradable polymers in packaging applications [44]. Its properties, such as transparency, rigidity, and ease of processing, make it suitable for a variety of packaging formats, including food containers, films, and bottles. PLA-based composites can be engineered to degrade in industrial composting facilities or under specific environmental conditions, making them a sustainable alternative to traditional petroleum-based plastics. However, the degradation rate of PLA can vary depending on the temperature, humidity, and microbial activity in the environment.

Compostable Packaging Solutions: In addition to PLA, other biodegradable polymers like PHA and starch-based composites are gaining traction in the packaging industry [62]. These materials are fully compostable, breaking down into water, carbon dioxide, and biomass in composting environments. Their use in packaging applications can help reduce the accumulation of plastic waste in landfills and marine ecosystems, promoting a more sustainable approach to plastic use.

Agricultural Films: Resorbable composites are also finding applications in agriculture, particularly in the production of biodegradable mulch films [67]. Traditional polyethylene-based mulch films, widely used to enhance crop yields, pose significant environmental challenges due to their persistence in soil and difficulty in recycling. Biodegradable films made from resorbable composites, such as PLA, PCL, and starch-based polymers, offer an eco-friendly alternative that can degrade in soil after fulfilling their function.

Biodegradable Mulch Films: Mulch films made from biodegradable polymers provide the same benefits as conventional plastics, such as weed suppression, moisture retention, and soil temperature regulation, but without the negative environmental impact [45]. These films degrade into non-toxic byproducts through microbial activity in the soil, eliminating the need for film removal and disposal at the end of the growing season. The degradation rate of these films can be tailored by adjusting the polymer composition, ensuring that the film lasts for the desired duration of the crop cycle.

Environmental Benefits: The use of biodegradable films in agriculture not only reduces plastic waste but also helps prevent soil contamination [68]. Conventional plastic films can break down into microplastics, which persist in the soil and can have detrimental effects on soil health and crop productivity. In contrast, biodegradable films decompose fully, leaving no harmful residues behind. This makes them an important tool for promoting sustainable agricultural practices.

6.3 Other Emerging Applications

Resorbable composites have potential in emerging applications beyond traditional biomedical and environmental sectors. Their biodegradability, biocompatibility, and tunable degradation rates make them ideal for cutting-edge fields like drug delivery and wound healing. These materials address complex challenges in healthcare and related fields. Figure 9 shows the emerging applications of resorbable composites.

Drug Delivery Systems: Resorbable composites revolutionize drug delivery by providing controlled release mechanisms and eliminating the need for device retrieval. These systems are designed to gradually release therapeutic agents as the composite material degrades within the body, offering a targeted and sustained drug delivery approach.

Polymer-Based Drug Delivery: Polymers such as PLA, PGA, and their copolymers (PLGA) are commonly used in drug delivery systems. These polymers can be fabricated into microspheres, nanoparticles, or hydrogels that encapsulate drugs, proteins, or growth factors. As the polymer degrades, the drug is released in a controlled manner, allowing for localized and sustained therapeutic effects [35]. The rate of drug release can be tailored by adjusting the composition and molecular weight of the polymer, providing flexibility in meeting specific medical needs.

Applications in Cancer Therapy: Resorbable composites show great promise in cancer therapy [14], where localized drug delivery is crucial for maximizing treatment efficacy while minimizing systemic side effects. For example, PLA-based

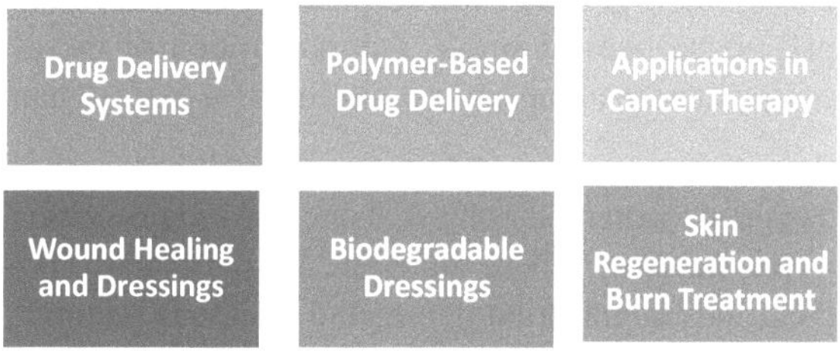

Fig. 9 Emerging applications of resorbable composites

nanoparticles loaded with chemotherapeutic agents can be injected directly into tumor sites, where they release the drug over an extended period. This targeted approach improves the therapeutic index and reduces the toxic effects commonly associated with traditional cancer treatments.

Wound Healing and Dressings: The field of wound healing has benefited greatly from the development of resorbable composite dressings, which provide both physical protection and therapeutic benefits [19]. These dressings are designed to promote tissue regeneration while degrading naturally, eliminating the need for dressing removal and reducing the risk of infection or additional trauma to the healing tissue.

Biodegradable Dressings: Resorbable wound dressings made from polymers like chitosan, collagen, or PCL are increasingly used to accelerate the healing process. These materials provide a moist environment, essential for optimal wound healing, and gradually degrade as the tissue regenerates. Some composite dressings are also loaded with antimicrobial agents or growth factors to further enhance the healing process. The use of biodegradable dressings simplifies wound care, as there is no need to replace or remove the dressing, reducing patient discomfort and the risk of infection.

Skin Regeneration and Burn Treatment: Resorbable composite scaffolds have also been employed in the treatment of severe skin injuries, such as burns. These scaffolds support the regeneration of skin tissue by providing a temporary matrix that encourages cell attachment and proliferation. As the scaffold degrades, it is gradually replaced by new skin tissue, offering a seamless transition from artificial support to natural healing.

7 Future Trends and Innovations in Resorbable Composites

As resorbable composites evolve, there is growing interest in developing advanced materials with enhanced functionality. Innovations in smart and functional composites pave the way for next-generation applications in various industries. These materials respond to external stimuli or repair themselves, making them ideal for complex biomedical and environmental uses.

Smart and Functional Composites: Smart composites refer to materials that can change their properties in response to external stimuli, such as temperature, light, pH, or mechanical stress. The integration of these functionalities into resorbable composites opens up new possibilities for more dynamic and interactive applications.

Stimuli-Responsive Materials: Stimuli-responsive resorbable composites are designed to alter their physical or chemical properties when exposed to specific environmental cues. These materials are particularly promising for applications where controlled degradation or targeted release is required, such as drug delivery systems or advanced wound dressings.

Temperature-Responsive Composites: One key area of research involves the development of temperature-sensitive materials. For example, some composites are engineered to undergo phase transitions at body temperature, making them useful in drug delivery systems where controlled release is essential. Polymers like poly (N-isopropylacrylamide) (PNIPAAm) are known for their temperature-responsive behavior, where they exhibit a sharp transition in solubility at a specific temperature, allowing for precise control over material properties.

pH-Responsive Materials: Another promising area is the development of pH-responsive composites, which can change their structure or degradation rate depending on the pH of the surrounding environment. These materials are particularly useful in drug delivery, where the acidic environment of a tumor or an infection site can trigger the release of therapeutic agents. The ability to tailor degradation and drug release based on local conditions offers a more targeted approach to treatment.

Self-Healing Composites: Self-healing composites are materials that can autonomously repair damage, extending their functional lifespan and reducing the need for replacement or maintenance. In resorbable composites, the incorporation of self-healing capabilities is particularly relevant for biomedical implants and tissue scaffolds, where long-term durability and functionality are critical.

Mechanism of Self-Healing: Self-healing resorbable composites often rely on microencapsulated healing agents or reversible chemical bonds that can reform after damage. For instance, some self-healing polymers use dynamic covalent bonds, which can break and reform in response to mechanical stress, allowing the material to repair itself over time. These materials are particularly useful in load-bearing applications, such as orthopedic implants, where maintaining mechanical integrity is essential for healing.

Applications in Biomedical Devices: In biomedical devices, self-healing resorbable composites can enhance the longevity and reliability of implants or scaffolds, reducing the need for surgical interventions due to material failure. For example, self-healing bone scaffolds can repair microcracks that form during the healing process, ensuring the scaffold maintains its structural integrity until the bone has fully regenerated. This capability is especially beneficial in applications where the material must remain functional for extended periods before degradation.

7.1 Advances in Bioprinting and Fabrication

Recent advancements in fabrication techniques, especially bioprinting, have enabled the development of resorbable composites with precise control over structure and functionality, leading to complex and customized applications. Two key innovations are 4D printing and nano- and micro-structured composites, offering exciting possibilities for improving resorbable materials in biomedical and other fields.

4D Printing of Resorbable Materials: 4D printing is an advanced form of 3D printing that incorporates the dimension of time. This technology involves the fabrication of materials that can change shape, properties, or functionality in response to environmental stimuli over time. When applied to resorbable composites, 4D printing enables the creation of smart materials that can dynamically adjust their properties in response to changes in their environment, such as temperature, moisture, or pH levels.

Shape-Memory and Stimuli-Responsive Materials: A key innovation in 4D printing is the use of shape-memory polymers (SMPs) in resorbable composites. These polymers can be programmed to change their shape or configuration when exposed to specific stimuli, such as heat or light, making them ideal for applications like self-fitting implants or tissue scaffolds. For example, a resorbable stent printed with SMPs could be designed to expand once inside the body, eliminating the need for complex surgical procedures. Over time, as the material resorbs, the stent would disappear, leaving no long-term foreign body behind.

Biomedical Applications: 4D-printed resorbable composites are particularly promising for tissue engineering and regenerative medicine. These materials can be printed in complex shapes that adapt to the body's natural environment, supporting tissue growth and regeneration. For example, a 4D-printed scaffold for bone regeneration could be designed to change its shape in response to the body's healing process, providing tailored support as the bone regenerates. The combination of 4D printing with resorbable materials offers exciting possibilities for developing dynamic implants that adjust to the body's needs over time.

Nano- and Micro-Structured Composites: Advances in nanotechnology have enabled the creation of nano- and micro-structured resorbable composites, which exhibit enhanced mechanical, biological, and chemical properties compared to their bulk counterparts. These materials, with structures designed at the nanoscale or microscale, offer greater control over degradation rates, mechanical strength, and bioactivity, making them ideal for a variety of advanced applications.

Nanocomposites for Controlled Degradation: Nano-structured resorbable composites, often incorporating nanoparticles or nanofibers, allow for precise control over degradation behavior. By adjusting the size, shape, and composition of the nanomaterials, the degradation rate of the composite can be finely tuned to meet the requirements of specific biomedical applications. For example, a nanocomposite scaffold for tissue engineering can be designed to degrade at a rate that matches the growth of new tissue, providing continuous support without leaving harmful residues.

Enhanced Mechanical Properties: Micro-structured resorbable composites, which incorporate microscale features such as pores or channels, offer improved mechanical properties, such as increased toughness and elasticity. These enhancements are particularly important for load-bearing applications, such as bone implants, where the composite material must provide adequate support while gradually resorbing. The microstructuring process can also be used to create hierarchical structures that mimic

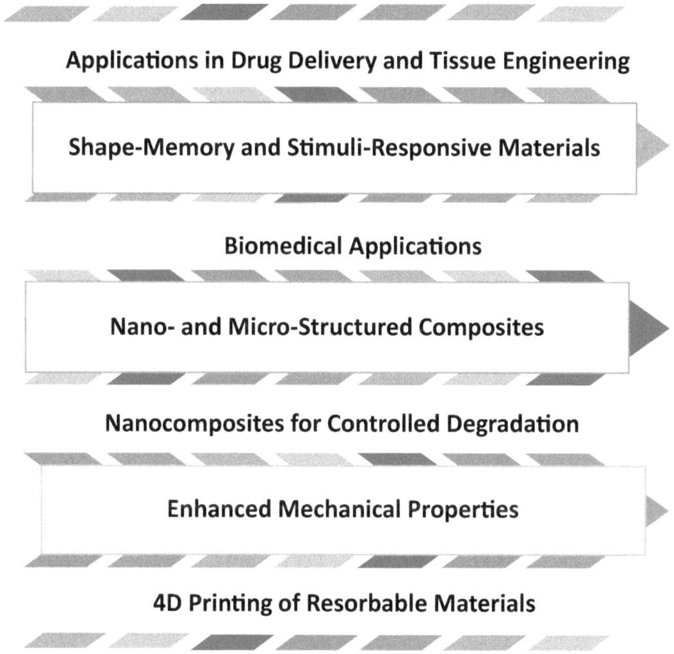

Fig. 10 Advanced functional materials for next-generation biomedical applications

natural tissues, improving the integration of the composite with the surrounding biological environment.

Applications in Drug Delivery and Tissue Engineering: Nano- and micro-structured composites have significant potential in drug delivery systems and tissue engineering. For drug delivery, nano-sized particles embedded within the composite can provide controlled release of therapeutic agents over time. In tissue engineering, the nanoscale or microscale features of the composite can enhance cell adhesion, proliferation, and differentiation, promoting faster and more effective tissue regeneration. These materials offer exciting possibilities for creating more effective and functional biomedical devices. Figure 10 illustrates advanced functional materials designed for next-generation biomedical applications.

7.2 *Regulatory and Ethical Considerations*

As resorbable composites gain prominence in biomedical and environmental applications, regulatory and ethical considerations must be addressed. These include legal frameworks for product safety and broader societal impacts on sustainability and environmental stewardship.

Standards and Approvals in Biomedical Applications: The development and use of resorbable composites in biomedical applications are subject to stringent regulatory standards to ensure patient safety and product efficacy. Regulatory bodies such as the U.S. Food and Drug Administration (FDA) and the European Medicines Agency (EMA) play a crucial role in setting guidelines for testing, manufacturing, and approving biomedical devices [17].

FDA Approval Process: In the U.S., the FDA classifies medical devices based on the risk they pose to patients, with resorbable composites typically falling under Class II or Class III, depending on their intended use and invasiveness [60]. The approval process requires manufacturers to provide comprehensive data on the material's biocompatibility, degradation behavior, and mechanical performance, often through preclinical studies and clinical trials. For instance, resorbable orthopedic implants must demonstrate long-term safety and effectiveness before receiving FDA clearance.

ISO Standards: On a global scale, the International Organization for Standardization (ISO) provides a framework for quality assurance in the development of biomedical devices, including resorbable composites. ISO 10993, which outlines the biological evaluation of medical devices, is particularly relevant for ensuring that materials are biocompatible and safe for clinical use [65]. Meeting ISO standards is often a prerequisite for market approval in many regions and helps ensure the material's safety across different regulatory environments.

Sustainability and Environmental Impact: The increasing use of resorbable composites also raises important questions about sustainability and environmental impact. While resorbable materials are designed to degrade after fulfilling their function, the broader life cycle of these materials—ranging from production to post-use degradation—must be considered to ensure they align with principles of environmental sustainability.

Sustainable Manufacturing: The production of resorbable composites, particularly those used in biomedical applications, requires careful consideration of resource use and environmental impact. Sustainable practices, such as using renewable or bio-based polymers like PLA and PCL, can reduce the carbon footprint associated with material production. Additionally, adopting energy-efficient fabrication techniques, such as 3D printing, can minimize waste and energy consumption, further contributing to sustainable development goals.

Environmental Degradation: Another key factor is the impact of resorbable composites on the environment after use. While these materials are designed to degrade naturally in biological settings, their behavior in non-biological environments is less well understood. For instance, resorbable composites used in agricultural films or biodegradable packaging may release microplastics or other degradation byproducts into the environment, raising concerns about long-term ecological effects. Therefore, understanding the full degradation pathways and environmental fate of these materials is crucial for ensuring they do not contribute to pollution or harm ecosystems.

8 Conclusion

Resorbable composites represent a revolutionary class of biomaterials with significant potential in various biomedical applications. Their composition, structure, and processing techniques play crucial roles in determining their mechanical and biological performance.

Key Points

Composition and Structure: The selection of polymer matrices and reinforcements influences mechanical and biological properties.

Processing Techniques: Methods like extrusion, injection molding, 3D printing, and electrospinning offer tailored geometries and properties.

Challenges: Thermal sensitivity and maintaining structural integrity are important considerations in processing.

Biodegradation: Understanding biodegradation mechanisms is crucial for predicting behavior in biological environments.

Mechanical Properties: Balancing strength, flexibility, and biocompatibility is essential for supporting tissue regeneration without causing adverse reactions.

Biocompatibility: Surface modifications enhance cell adhesion and proliferation, promoting tissue integration.

Clinical Applications: Resorbable composites have diverse applications, from orthopedic implants to drug delivery systems.

Future Directions: Continued research into hybrid materials, biodegradation kinetics, and functionalization techniques will improve performance.

Innovations in bioprinting, nanotechnology, and bioactive agents will lead to next-generation resorbable composites.

Overall, resorbable composites offer a promising platform for advancing medical science and transforming healthcare by providing personalized, efficient, and effective solutions to a wide range of medical challenges.

References

1. Adan A, Kiraz Y, Baran Y (2016) Cell proliferation and cytotoxicity assays. Curr Pharm Biotechnol 17(14):1213–1221

2. Al-Maharma AY, Patil SP, Markert B (2020) Effects of porosity on the mechanical properties of additively manufactured components: a critical review. Mater Res Exp 7(12):122001
3. Ansari MAA, Golebiowska AA, Dash M, Kumar P, Jain PK, Nukavarapu SP, Nanda HS et al (2022) Engineering biomaterials to 3D-print scaffolds for bone regeneration: practical and theoretical consideration. Biomater Sci 10(11):2789–2816
4. Archer E, Torretti M, Madbouly S (2023) Biodegradable polycaprolactone (PCL) based polymer and composites. Phys Sci Rev 8(11):4391–4414
5. Armentano I, Dottori M, Fortunati E, Mattioli S, Kenny JM (2010) Biodegradable polymer matrix nanocomposites for tissue engineering: a review. Polym Degrad Stab 95(11):2126–2146
6. Bahrami M, Abenojar J, Martínez MÁ (2020) Recent progress in hybrid biocomposites: mechanical properties, water absorption, and flame retardancy. Materials 13(22):5145
7. Barral V, Dropsit S, Cayla A, Campagne C, Devaux É (2021) Study of the influence of pcl on the in vitro degradation of extruded pla monofilaments and melt-spun filaments. Polymers 13(2):171
8. Bartnikowski M, Dargaville TR, Ivanovski S, Hutmacher DW (2019) Degradation mechanisms of polycaprolactone in the context of chemistry, geometry and environment. Prog Polym Sci 96:1–20
9. Batool F, Özçelik H, Stutz C, Gegout PY, Benkirane-Jessel N, Petit C, Huck O (2021) Modulation of immune-inflammatory responses through surface modifications of biomaterials to promote bone healing and regeneration. J Tissue Eng 26(12):20417314211041428. https://doi.org/10.1177/20417314211041428. PMID: 34721831; PMCID: PMC8554547
10. Benatti ACB, Pattaro AF, Rodrigues AA, Xavier MV, Kaasi A, Barbosa MIR, Kharmandayan P et al (2019) Bioreabsorbable polymers for tissue engineering: PLA, PGA, and their copolymers. In: Materials for biomedical engineering. Elsevier, pp 83–116
11. Betanzos FB, Gimeno-Fabra M, Segal J, Grant D, Ahmed I (2016) Cyclic pressure on compression-moulded bioresorbable phosphate glass fibre reinforced composites. Mater Des 100:141–150
12. Brandt-Wunderlich C, Ruppelt P, Zumstein P, Schmidt W, Arbeiter D, Schmitz KP, Grabow N (2019) Mechanical behavior of in vivo degraded second generation resorbable magnesium scaffolds (RMS). J Mech Behav Biomed Mater 91:174–181
13. Chen CC, Dai L, Ma L, Guo RT (2020) Enzymatic degradation of plant biomass and synthetic polymers. Nat Rev Chem 4(3):114–126
14. Chew SA, Danti S (2017) Biomaterial-based implantable devices for cancer therapy. Adv Healthcare Mater 6(2):1600766
15. Chocron Y, Azzi AJ, Cugno S (2019) Resorbable implants for mandibular fracture fixation: a systematic review and meta-analysis. Plastic Reconstr Surgery-Global Open 7(8):e2384
16. Cifuentes SC, Gavilán R, Lieblich M, Benavente R, González-Carrasco JL (2016) In vitro degradation of biodegradable polylactic acid/magnesium composites: relevance of Mg particle shape. Acta Biomater 32:348–357
17. Darrow JJ, Avorn J, Kesselheim AS (2021) FDA regulation and approval of medical devices: 1976–2020. JAMA 326(5):420–432
18. Demina VA, Krasheninnikov SV, Buzin AI, Kamyshinsky RA, Sadovskaya NV, Goncharov EN, Chvalun SN et al (2020) Biodegradable poly (l-lactide)/calcium phosphate composites with improved properties for orthopedics: effect of filler and polymer crystallinity. Mater Sci Eng, C 112:110813
19. Deng X, Gould M, Ali MA (2022) A review of current advancements for wound healing: biomaterial applications and medical devices. J Biomed Mater Res B Appl Biomater 110(11):2542–2573
20. Dobrzyńska-Mizera M, Dodda JM, Liu X, Knitter M, Oosterbeek RN, Salinas P, Sadiku ER et al (2024) Engineering of bioresorbable polymers for tissue engineering and drug delivery applications. Adv Healthc Mater:2401674
21. Dutta A (2017) Fourier transform infrared spectroscopy. In: Spectroscopic methods for nanomaterials characterization, pp 73–93

22. Epp J (2016) X-ray diffraction (XRD) techniques for materials characterization. In: Materials characterization using nondestructive evaluation (NDE) methods. Woodhead Publishing, pp 81–124
23. Fateh T, Richard F, Rogaume T, Joseph P (2016) Experimental and modelling studies on the kinetics and mechanisms of thermal degradation of polymethyl methacrylate in nitrogen and air. J Anal Appl Pyrol 120:423–433
24. Goldstein JI, Newbury DE, Michael JR, Ritchie NW, Scott JHJ, Joy DC (2017) Scanning electron microscopy and X-ray microanalysis. Springer
25. Gulati K, Meher MK, Poluri KM (2017) Glycosaminoglycan-based resorbable polymer composites in tissue refurbishment. Regen Med 12(4):431–457
26. Hajebi S, Mohammadi Nasr S, Rabiee N, Bagherzadeh M, Ahmadi S, Rabiee M, Hamblin MR et al (2021) Bioresorbable composite polymeric materials for tissue engineering applications. Int J Polym Mater Polym Biomater 70(13):926–940
27. Hajiali F, Tajbakhsh S, Shojaei A (2018) Fabrication and properties of polycaprolactone composites containing calcium phosphate-based ceramics and bioactive glasses in bone tissue engineering: a review. Polym Rev 58(1):164–207
28. Heimbach B, Tonyali B, Zhang D, Wei M (2018) High performance resorbable composites for load-bearing bone fixation devices. J Mech Behav Biomed Mater 81:1–9
29. Huang B, Bártolo PJ (2018) Rheological characterization of polymer/ceramic blends for 3D printing of bone scaffolds. Polym Testing 68:365–378
30. Ibrahim H, Esfahani SN, Poorganji B, Dean D, Elahinia M (2017) Resorbable bone fixation alloys, forming, and post-fabrication treatments. Mater Sci Eng, C 70:870–888
31. Ilyas RA, Aisyah HA, Nordin AH, Ngadi N, Zuhri MYM, Asyraf MRM, Ibrahim R et al (2022) Natural-fiber-reinforced chitosan, chitosan blends and their nanocomposites for various advanced applications. Polymers 14(5):874
32. Indolfi C, De Rosa S, Colombo A (2016) Bioresorbable vascular scaffolds—basic concepts and clinical outcome. Nat Rev Cardiol 13(12):719–729
33. Jem KJ, Tan B (2020) The development and challenges of poly (lactic acid) and poly (glycolic acid). Adv Industr Eng Polym Res 3(2):60–70
34. Jindal S, Manzoor F, Haslam N, Mancuso E (2021) 3D printed composite materials for craniofacial implants: current concepts, challenges and future directions. Int J Adv Manuf Technol 112(3):635–653
35. Kamaly N, Yameen B, Wu J, Farokhzad OC (2016) Degradable controlled-release polymers and polymeric nanoparticles: mechanisms of controlling drug release. Chem Rev 116(4):2602–2663
36. Kenny EK, Gately NM, Killion JA, Devine DM, Higginbotham CL, Geever LM (2016) Melt extruded bioresorbable polymer composites for potential regenerative medicine applications. Polym-Plast Technol Eng 55(4):432–446
37. Kim T, See CW, Li X, Zhu D (2020) Orthopedic implants and devices for bone fractures and defects: Past, present and perspective. Eng Regener 1:6–18
38. La Mattina AA, Mariani S, Barillaro G (2020) Bioresorbable materials on the rise: from electronic components and physical sensors to in vivo monitoring systems. Adv Sci 7(4):1902872
39. Larrañaga A, Lizundia E (2019) A review on the thermomechanical properties and biodegradation behaviour of polyesters. Eur Polymer J 121:109296
40. Lau AKT, Hung APY (eds) (2017) Natural fiber-reinforced biodegradable and bioresorbable polymer composites
41. Lehtonen TJ, Tuominen JU, Hiekkanen E (2013) Resorbable composites with bioresorbable glass fibers for load-bearing applications. In vitro degradation and degradation mechanism. Acta Biomaterialia 9(1):4868–4877
42. Little H, Clarke SA, Cunningham E, Buchanan F (2018) Process-induced degradation of bioresorbable PDLGA in bone tissue scaffold production. J Mater Sci—Mater Med 29:1–10
43. Macdougall L, Culver H, Lin CC, Bowman C, Anseth K (2020) Degradable and resorbable polymers. In: Biomaterials science. Academic Press, pp 167–190

44. Mangaraj S, Yadav A, Bal LM, Dash SK, Mahanti NK (2019) Application of biodegradable polymers in food packaging industry: a comprehensive review. J Pack Technol Res 3:77–96
45. Mansoor Z, Tchuenbou-Magaia F, Kowalczuk M, Adamus G, Manning G, Parati M, Khan H et al (2022) Polymers use as mulch films in agriculture—a review of history, problems and current trends. Polymers 14(23):5062
46. Mazhandu ZS, Muzenda E, Mamvura TA, Belaid M, Nhubu T (2020) Integrated and consolidated review of plastic waste management and bio-based biodegradable plastics: challenges and opportunities. Sustainability 12(20):8360
47. Morra EM (2022) Sterilization and post-processing of bioresorbable polymers for cardiovascular stent applications. Mechanical Engineering
48. Naghshineh N, Tahvildari K, Nozari M (2019) Preparation of chitosan, sodium alginate, gelatin and collagen biodegradable sponge composites and their application in wound healing and curcumin delivery. J Polym Environ 27:2819–2830
49. Navarro M, Aparicio C, Charles-Harris M, Ginebra MP, Engel E, Planell JA (2006) Development of a biodegradable composite scaffold for bone tissue engineering: physicochemical, topographical, mechanical, degradation, and biological properties. In: Ordered polymeric nanostructures at surfaces, pp 209–231
50. Osorio-Arciniega R, García-Hipólito M, Alvarez-Fregoso O, Alvarez-Perez MA (2021) Composite fiber spun mat synthesis and in vitro biocompatibility for guide tissue engineering. Molecules 26(24):7597
51. Paiva JC, Oliveira L, Vaz MF, Costa-de-Oliveira S (2022) Biodegradable bone implants as a new hope to reduce device-associated infections—a systematic review. Bioengineering 9(8):409
52. Pappalardo D, Mathisen T, Finne-Wistrand A (2019) Biocompatibility of resorbable polymers: a historical perspective and framework for the future. Biomacromol 20(4):1465–1477
53. Pinto PO, Pinto PO, Atayde LM, Atayde LM, Campos JM, Campos JM, Maurxhtício AC et al (2016) Therapeutic strategies for bone regeneration: the importance of biomaterials testing in adequate animal models. Adv Compos Mater:275–319
54. Pintor AVB, Queiroz LD, Barcelos R, Primo LSG, Maia LC, Alves GG (2020) MTT versus other cell viability assays to evaluate the biocompatibility of root canal filling materials: a systematic review. Int Endod J 53(10):1348–1373
55. Polak-Kraśna K, Abaei AR, Shirazi RN, Parle E, Carroll O, Ronan W, Vaughan TJ (2021) Physical and mechanical degradation behaviour of semi-crystalline PLLA for bioresorbable stent applications. J Mech Behav Biomed Mater 118:104409
56. Prabhu B, Karau A, Wood A, Dadsetan M, Liedtke H, DeWitt T (2018) Bioresorbable materials for orthopedic applications (Lactide and glycolide based). In: Orthopedic biomaterials: progress in biology, manufacturing, and industry perspectives, pp 287–344
57. Rai P, Mehrotra S, Priya S, Gnansounou E, Sharma SK (2021) Recent advances in the sustainable design and applications of biodegradable polymers. Biores Technol 325:124739
58. Ramakrishna S, Mayer J, Wintermantel E, Leong KW (2001) Biomedical applications of polymer-composite materials: a review. Compos Sci Technol 61(9):1189–1224
59. Rizzarelli P, La Carta S, Rapisarda M, Valenti G (2019) Analytical methods in resorbable polymer development and degradation tracking. In: Materials for biomedical engineering. Elsevier, pp 351–408
60. Ronquillo JG, Zuckerman DM (2017) Software-related recalls of health information technology and other medical devices: implications for FDA regulation of digital health. Milbank Q 95(3):535–553
61. Ryoo KS, Kim PJ, Kim S, Cho YD, Ku Y (2023) A retrospective study of the long-term survival of RESTORE® dental implants with resorbable blast media surface. J Periodontal Implant Sci 53(6):444
62. Sabet M (2024) Exploring biodegradable polymer composites for sustainable packaging: a review on properties, manufacturing techniques, and environmental impacts. Iranian Polym J:1–20
63. Sadowska JM, Ginebra MP (2020) Inflammation and biomaterials: role of the immune response in bone regeneration by inorganic scaffolds. J Mater Chem B 8(41):9404–9427

64. Senthamizhan A, Balusamy B, Uyar T (2017) Electrospinning: a versatile processing technology for producing nanofibrous materials for biomedical and tissue-engineering applications. In: Electrospun materials for tissue engineering and biomedical applications. Woodhead Publishing, pp 3–41
65. Sharma A, Luthra G (2023) Significance of ISO 10993 standards in ensuring biocompatibility of medical devices: a review. J Pharm Res Int 35(8):23–34
66. Sindelar J (2022) Mechanical testing and in vitro degradation of composite materials for bone tissue engineering (Master's thesis)
67. Sintim HY, Bary AI, Hayes DG, Wadsworth LC, Anunciado MB, English ME, Flury M et al (2020) In situ degradation of biodegradable plastic mulch films in compost and agricultural soils. Sci Total Environ 727:138668
68. Somanathan H, Sathasivam R, Sivaram S, Kumaresan SM, Muthuraman MS, Park SU (2022) An update on polyethylene and biodegradable plastic mulch films and their impact on the environment. Chemosphere 307:135839
69. Song R, Murphy M, Li C, Ting K, Soo C, Zheng Z (2018) Current development of biodegradable polymeric materials for biomedical applications. Drug Des Dev Therapy:3117–3145
70. Sousa AM, Amaro AM, Piedade AP (2022) 3D printing of polymeric bioresorbable stents: a strategy to improve both cellular compatibility and mechanical properties. Polymers 14(6):1099
71. Tahmasebi E, Alam M, Yazdanian M, Tebyanian H, Yazdanian A, Seifalian A, Mosaddad SA (2020) Current biocompatible materials in oral regeneration: a comprehensive overview of composite materials. J Market Res 9(5):11731–11755
72. Torgbo S, Sukyai P (2020) Biodegradation and thermal stability of bacterial cellulose as biomaterial: the relevance in biomedical applications. Polym Degrad Stab 179:109232
73. Tozzi G, De Mori A, Oliveira A, Roldo M (2016) Composite hydrogels for bone regeneration. Materials 9(4):267
74. Turnbull G, Clarke J, Picard F, Riches P, Jia L, Han F, Shu W et al (2018) 3D bioactive composite scaffolds for bone tissue engineering. Bioactive Mater 3(3):278–314
75. Vacaras S, Baciut M, Lucaciu O, Dinu C, Baciut G, Crisan L, Bran S et al (2019) Understanding the basis of medical use of poly-lactide-based resorbable polymers and composites–a review of the clinical and metabolic impact. Drug Metab Rev 51(4):570–588
76. Vergnol G, Ginsac N, Rivory P, Meille S, Chenal JM, Balvay S, Hartmann DJ et al (2016) In vitro and in vivo evaluation of a polylactic acid-bioactive glass composite for bone fixation devices. J Biomed Mater Res B Appl Biomater 104(1):180–191
77. Wu T, Ding M, Shi C, Qiao Y, Wang P, Qiao R, Zhong J et al (2020) Resorbable polymer electrospun nanofibers: history, shapes and application for tissue engineering. Chin Chem Lett 31(3):617–625
78. Yancheva V, Velcheva I, Stoyanova S, Georgieva E (2016) Histological biomarkers in fish as a tool in ecological risk assessment and monitoring programs: a review. Appl Ecol Environ Res 14(1):47–75
79. Zaaba NF, Jaafar M (2020) A review on degradation mechanisms of polylactic acid: hydrolytic, photodegradative, microbial, and enzymatic degradation. Polym Eng Sci 60(9):2061–2075
80. Zhao X, Liu J, Li J, Liang X, Zhou W, Peng S (2022) Strategies and techniques for improving heat resistance and mechanical performances of poly (lactic acid)(PLA) biodegradable materials. Int J Biol Macromol 218:115–134

Chapter 3
Mechanical and Biological Performance of Resorbable Composite Materials in Hard Tissue Engineering

S. Ram Prasanth, Syed Naveed Ul Meiraj, Hemnath Anandan Kumar, and Varatharajan Prasannavenkadesan

1 Introduction

Bioactive and bioresorbable composite materials are widely used in tissue engineering. The utilization of bioactive biodegradable composite materials in tissue engineering is extensive, especially regarding new material development combined with biodegradable plastics and bioactive ceramics like tricalcium phosphate, hydroxyapatite and special glass. These composite materials are promising for application as scaffolds to regenerate bone and cartilage tissues [1]. Adding bioactive components to biodegradable plastics enhances their bioactivity, provides control of scaffold degradation rates, and improves the mechanical properties. This trend aims to make much stronger, more stable and fully tissue-compatible scaffolds for tissue engineering. This may lead to enhanced repair and regeneration of bones and cartilage with this new approach. In medicine, biodegradable materials are widely used as temporary media that support tissue regeneration and thus avoid performing additional operations.

S. Ram Prasanth
Indian Institute of Science, Bangalore, Karnataka, India

S. N. U. Meiraj
Indian Institute of Information Technology, Design and Manufacturing, Kancheepuram, Chennai, Tamilnadu, India

H. A. Kumar
Department of Mechanical Engineering, Amrita School of Engineering, Amrita Vishwa Vidyapeetham, Chennai, India

V. Prasannavenkadesan (✉)
Trinity College Dublin, Dublin, Ireland
e-mail: p.varatharajan@tcd.ie

A. Prasad and V. Katiyar (eds.), *Resorbable Polymers for Bioimplants and Fixation Devices*, Materials Horizons: From Nature to Nanomaterials,
https://doi.org/10.1007/978-981-96-9306-1_3

The design of biomedical biodegradable polymers has come a long way over the past few years. Researchers have explored the advantages of 3D and 4D printing for fabricating bioresorbable composites and polymers [2]. Biodegradable materials are inherently degraded through natural forces and thus should be good options for biomedical applications such as tissue engineering, drug delivery and temporary implants [3]. The aim is to design an artificial material that will safely take the place of or repair damaged tissue to enhance the rate of survival and the quality of life. Different applications require specific properties of strength, compatibility and durability. Biodegradable polymers are both naturally derived and synthetically produced. With synthetic polymers, there is more control over desired properties; with natural polymers, there are possibilities for the induction of immune responses. Using biodegradable materials in medicine temporarily supports tissues that need regeneration without further surgeries. The degradation rate may also be tuned to assist the healing processes further.

The demand for resorbable composites is increasing because of their advantages and non-toxicity, which contribute to fabricating new devices in various biomedical technology fields even after 50 years of the success of implants. Due to continuous research, resorbable composite materials have been found, which are believed to be a suitable alternative to the traditional non-resorbable biomaterials [4]. Also, these bioresorbable polymers will degrade naturally over time depending on the resorption rate when implanted into the human body. Bone screws, plates, clips, etc., can be made out of these resorbable biocomposites for regenerative medicine/tissue engineering applications [5–9]. Therefore, the resorbable composites help the patient and the surgeon reduce the cost of a second surgery and physical pain. Due to the progress in the research, bioresorbable passive implants [10, 11], wearable and implantable electronic medical devices are being utilized [12, 13].

2 Overview of Resorbable Composites

2.1 Definition and Properties of Resorbable Composites

The term *bioresorbable* implies that material of any type can be degraded in bulk and resorbed completely in vivo. Therefore, the biomaterial can be removed from our body through excretion or, in other words, the removal of the biomaterial and its by-products entirely out of the body without leaving any residue by some means [14]. From the aspect of chemistry, the word resorption means the removal of a material from its original position through a chemical or physical method [15]. In the field of biology, resorption can be defined as the absorption of substance into the cells or tissues like tooth resorption, bone resorption [16]. The ISO defines "bioresorption" as the ability of a biomaterial to be degraded in the physiological environment, where the by-products are removed or absorbed by the body.

The conditions for a material/composite to be bioresorbable are as follows: (i) the material/composite must undergo degradation, (ii) the degradation rate should be controllable and (iii) the degraded products must be resorbed [14]. Further, the rate of resorption of the biomaterial utilized must match the speed of the tissue that is newly generated. Also, the biomaterial should possess better mechanical strength than the polymer or ceramics alone to provide durability and flexibility simultaneously [17, 18]. Another significant function of bioresorbable materials is to carry out the function of the tissue inside the body, which is the property that is lacking in the case of non-resorbable materials [19]. These bioresorbable polymers were employed initially as bioresorbable sutures in surgery and find their applications in various fields like screws for repairing bone [20], stents for arteries [21] and tissue healing [22].

2.2 *Types of Resorbable Polymers and Composites*

Chitosan, collagen, poly(amides), aliphatic polyesters and poly(anhydrides) are a few of the bioresorbable polymeric materials employed in bio-based applications [23]. The idea of using a bioresorbable polymer is that it can be used as an implant until a particular tissue heals. Then the implant can be absorbed in the body without causing any problems to human physiology. Therefore, the absorption takes place by cleavage of chains (degradation), attributing to a decrease in the chain length of the polymer. The period in which the specific chain length of the polymer is attained below which the implant loses the structural integrity is known as the strength retention period [24]. The implant starts to collapse after this strength retention period which is evident from the decrease in the chain length and mass of the material [23]. Bioresorbable polymeric materials are grouped into two categories depending upon the origin of the polymer: natural and synthetic, as shown in Fig. 1.

2.2.1 Natural Polymers

As the title indicates, the natural polymers are obtained from natural sources which are utilized in various tissue engineering (TE) applications. These natural polymers possess certain merits like biodegradability and biocompatibility and some polymers may be discovered in extracellular matrix (ECM). Consequently, the demerits of these polymers are low mechanical properties and difficulty to process. Because of these limitations, the commercial use of these materials is hampered [26]. The natural polymers were categorized into three types depending on their genesis: (i) polysaccharide-based, (ii) polypeptide and (iii) bacterial polyesters.

Polysaccharides

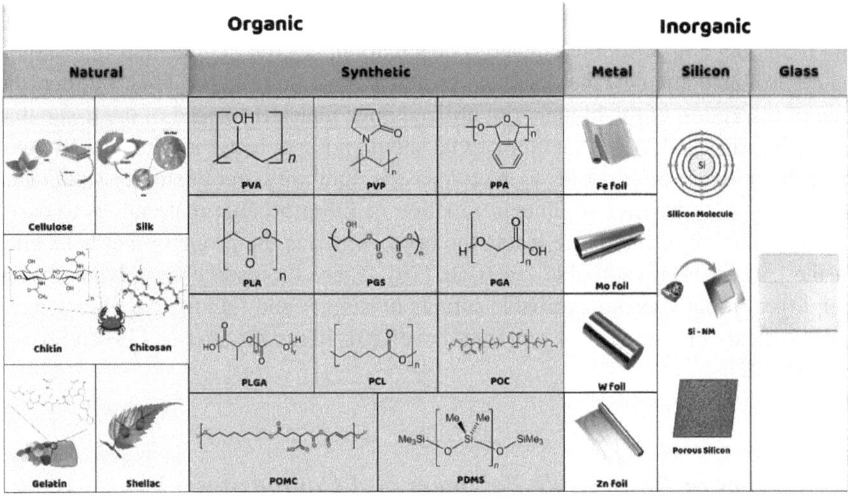

Fig. 1 Various types of bioresorbable materials. Organic materials: natural and synthetic polymeric materials depending upon their origin, and inorganic materials: bioresorbable metallic materials and various forms of glass and silicon [25] License number: 5892460728730

Polysaccharides are considered to be a class of naturally occurring polymers which can be employed as bioresorbable because of their bioresorbability, biocompatibility, cheap production, ease of derivation and offering nearly ECM characteristics preferred for tissue engineering [27]. Chitin is obtained from the exoskeleton of crustaceans, crabs, shrimps and other microorganisms [28]. Chitin is deacetylated to form chitosan, which seems to be an ideal biomaterial for cartilage regeneration. However, chitosan possesses poor solubility with solvents, which inhibits its commercial application in tissue engineering [29]. The properties of the chitosan could be altered chemically by utilizing the available reactive chemical groups, like hydroxyl and amine groups. Due to its bioresorbability and biocompatibility, chitosan could be utilized in tissue engineering [30]. The chitosan-alginate scaffold seems to be promising over protein matrices and other synthetic ones. This chitosan-alginate scaffold shows enhanced mechanical strength because of the synergism between the carboxyl group of the alginate and the hydroxyl group present in the chitosan [31].

Starch is one of the natural polymers which is widely used in tissue engineering applications. Because starch encompasses several favourable properties like bioresorbability, biocompatibility, abundance, recyclability and easy to process. Starch contains several glucose units bridged by glycosidic bonds [32]. It is hydrophilic in nature and can be exploited to control the rate of degradation of a hydrophobic polymer when blended. Also, starch can't be used alone for scaffold application because it possesses low mechanical strength [33]. But it can be used along with another polymer matrix like hydroxyapatite (HA) in bone tissue repair which enhances the mechanical strength and biological properties of the scaffold [34]. When added with HA, the hydroxyl groups in the starch make the biocomposite hydrophilic, and the hydroxyl groups also attract the phosphate, and calcium ions, enhancing the apatite nucleation and bone mineralization efficiently [35].

Polypeptides

The most readily abundant protein in the human body is collagen. It is the main constituent in several body parts, like bone, ligament, blood vessels, etc. Collagen is a biodegradable polymer with excellent biocompatibility, mechanical strength, and enzymatic degradability, which is utilized in tissue engineering applications [36]. Catgut is a collagen-based biocomposite used as a suture material in medical applications. When collagen is blended with HA, which is nothing but the composition of human bone, it can be exploited as a bone implant/transplant accordingly. Collagraft is one example that got FDA approval to be used clinically [23]. Fu et al. made a novel collagen-chitosan-based biocomposite employed as a tissue-engineered heart valve. Also, the mesenchymal stem cells were seeded and found to be differentiated from endothelial cells, which is confirmed through in vitro studies [37]. Gurumurthy et al. produced biocomposites by using elastin polypeptide as a reinforcing filler in the collagen matrix. These biocomposites seem to possess improved mechanical strength, and also, during the course of their culture time, the human adipose-derived stromal cells in the produced composites were shown to be viable for over 22 days [38].

Bacterial Polyesters

Polyhydroxyalkanoates are considered to be a class of aliphatic polyesters which are mostly produced by microbes [39]. PHAs are hydrophobic, bioresorbable, and biocompatibility. These PHAs were employed in various fields of biomedical applications like sutures, scaffolds in TE, tendon repair, adhesion patches, and so on [40]. Polyhydroxybutyrate (PHB), a low molecular-weight linear aliphatic polyester, can be made through various microorganisms owing to its bacterial origin [41]. Albeit, this polymer is employed in biomedical applications because of its biocompatibility, bioresorbability, and mechanical properties due to its semicrystalline nature (isotactic behaviour) [42]. The performance properties of the scaffolds are greatly affected by the PHB nanoparticle. Nevertheless, the crystallinity not only influences the mechanical strength but also affects the bioactivity of the seeded cells/cell proliferation [43]. PHB shows high stiffness and brittleness and low elongation made into textiles which is exclusively utilized in soft tissue engineering applications [44]. Also, bone plates and fillers were made using PHB because of its good mechanical properties [41]. However, PHB has several disadvantages, such as poor bioresorbability rate, high stiffness and brittleness, less hydrophilic [45].

2.2.2 Synthetic Polymers

Synthetic bioresorbable polymers are preferred over natural ones in making scaffolds for TE due to some advantages such as no risk of immunogenicity, reproducible process of fabrication and cheap, easy, and simple to process, degradability and mechanical properties that could be modified based on the end-use application. In comparison with natural polymers, synthetic polymers are less biocompatible and also the products of degradation do not fall under the category of natural metabolites, which may pose complications if accumulated [46, 47]. Some of the widely employed synthetic bioresorbable polymers are aliphatic polyesters like Poly(glycolic acid), Poly(l-lactic acid), Poly(d-lactic acid), Poly(lactic-co-glycolic acid) [48].

Polylactic acid

PLA is a bioresorbable thermoplastic obtained from biotic resources like sugarcane, starch, corn, etc. [49]. It is synthetically produced from the lactide monomer through ring-opening polymerization, polycondensation and azeotropic dehydration reactions [50]. PLA has been utilized in various fields, such as drug delivery [51], orthopaedic [52], tissue engineering [53] and so on because of its excellent biocompatibility, bioresorbability, suitable mechanical property [27] and not to mention that the FDA approves it for different applications [54]. PLA can be semicrystalline or amorphous depending upon the composition of the *l* or *d* isomers present in the polymer, respectively [55]. With respect to tissue engineering, PLA can be employed without any hesitation because of its bioresorbability because the ester bonds in the main chain of the polymer help in cleaving it hydrolytically, and the resulting product of degradation is non-toxic to the human body [56]. It is reported that sometimes PLA alone can lead to some adverse effects [57]. Therefore, HA or tricalcium phosphate is added to the PLA polymer to reduce or remove the adverse effects due to the response of the system [58–60]. Xu et al. prepared a biocomposite-based scaffold using PLA/HA to enhance its mechanical strength. Due to the incorporation of HA into the PLA matrix, the mechanical property and cell adhesion seem to be improved [61]. Dai et al. added pearl powder, a naturally occurring biocomposite of calcium carbonate, to the PLA matrix and electrospunned, where HA is accumulated to fabricate a nanofibrous scaffold. Cell proliferation and adhesion seem to be enhanced due to the inclusion of pearl powder, which is evident from the in vitro studies [62].

Poly(ε-caprolactone)

PCL is a bioresorbable aliphatic polyester which is utilized in a wide range of biomedical applications [63]. Poly(ε-caprolactone) is polymerized from the ε-caprolactone using ring-opening polymerization. Polycaprolactone has ester linkages, making the polymer degradable through hydrolysis [64]. PCL shows good mechanical properties, thermal behaviour, biocompatibility, chemical inertness and toughness. It also exhibits some limitations, such as poor cell adhesion and proliferation due to its hydrophobic nature, which can be improved by some chemical modification [63]. On hydrolytic cleavage of ester bonds, PCL forms caproic acid, which does not threaten the human body condition at lower concentrations. The implant/scaffold made out of PCL can withstand its structural integrity for up to 2 years due to its comparatively sluggish degradation rate, which could be modified by the co-polymerization technique to control its degradation rate [49, 64]. Composites based on PLLA and PCL were developed by Qi et al. by adding n-HA to emulate the ECM of the bone, where the HA ceramic played a crucial part in bone tissue formation. Also, the osteoblast cells were sown in the scaffolds which attributed towards enhanced differentiation, proliferation and bone mineralization [65]. Park et al. worked on the incorporation of β-Tri calcium phosphate (TCP) to modify the characteristics of the PCL-based composite, which is 3D printed as a scaffold. They have improved the mechanical property and osteogenic differentiation by optimizing TCP content at 30%, making it a satisfactory material for bone tissue engineering [66]. Generally, PCL alone isn't recommended to be used in hard tissue applications because of its relatively lower mechanical strength. To solve this issue, Scaffaro et al., for the first time, employed Graphene oxide grafted Polyethylene glycol into the PCL matrix via an electrospinning process, which significantly enhanced the mechanical properties. Also, the osteogenic cells were sown onto the electrospun scaffolds leading towards enhancement in cell proliferation and adhesion [67].

Poly(lactic-co-glycolic acid)

PLGA is a co-polymer of polylactic acid and polyglycolic acid, which has been employed in cell scaffolds, implants and other TE applications stimulating cell growth and organ healing/repair [68]. The PLGA co-polymer offers certain advantages like biocompatibility and good mechanical strength. Also, the degradation in this co-polymer takes place by non-enzymatic hydrolysis, where the products of degradation will get out from the body by forming water and carbon dioxide [69]. Hence, PLGA is an appropriate polymeric composite material utilized in bone tissue engineering applications because of its exceptional balance between toughness and mechanical strength [70]. Polyglycolic acid has a fast rate of degradation and can be completely resorbed in the body. Therefore, Becker et al. utilized a blend of PLLA and PGA where the outer layer is made of PGA, which can degrade relatively fast. Because of these different layers, PGA degrades first, which will prevent the harm of stress shielding, which is a crucial issue in bone implants [71]. Freire et al. employed a PLGA implant to replace the tibia of a rabbit, which unexpectedly led to new bone formation and transmitted healing for the bone tissues [72]. Jose et al. developed nanocomposite scaffolds by incorporating nano-HA into PLGA through an electrospinning process, where the mechanical strength hiked up at low concentrations of nano-HA [73].

2.3 *Advantages of Resorbable Composites for Hard Tissue Engineering*

The increase in bone ailments and the limitations of the traditional grafting techniques resulted in the development of novel implants capable of regeneration/healing of bone. Hence, bioresorbable polymers have been found which could be utilized as a replacement for non-resorbable materials like metals, ceramics and some polymers [23, 49]. The primary benefit of the resorbable polymers lies in where the degradation products will be imbibed by the body or removed through some means (either through excretion or through lungs) after serving its purpose as a scaffold/implant to regenerate a tissue that has been damaged/repaired [15]. Also, once the surgery is done with resorbable biocomposites, another surgery to detach the implant from the patient after a specific time period can be avoided, which takes more energy, time, pain and so on [49]. Another crucial significance of these resorbable biocomposites is their capability to carry out the same task of the tissue that has been damaged/repaired temporarily and appropriately until healing [19]. By controlling the resorption rate, we can control the degradation rate of the resorbable implant, which will prohibit stress shielding by transferring load to the newly forming bone and highly favour the bone regeneration [74].

3 Mechanical Properties of Resorbable Composites

Resorbable polymers and composite materials have shown great interest in applications related to tissue engineering, by means of possible replacement with the traditional ones. Their main role is the mechanical support of the bone-healing process while maintaining the integrity of the structure of the damaged bone. The mechanical behaviour of the biocompatible composite material is of crucial importance, as it defines the suitability of materials for various applications.

3.1 Mechanical Strength and Elasticity Modulus

Properties of interest for mechanical behaviour in biodegradable materials include tensile strength, Young's modulus, compressive strength, flexural strength, toughness, and others. Tensile strength describes the resistance of the material to tensile forces, while Young's modulus provides information on stiffness. Compressive and flexural strength refers to material resistance to compressive and bending forces, respectively. Toughness expresses the energy that a material is able to absorb up to failure. These properties have a determining influence on the material's capability to maintain structural integrity, support tissue ingrowth, and provide a route for degradation.

Mechanical behaviour for biodegradable materials depends on many factors. Microstructure, especially fibre orientation and pore size, impacts mechanical properties. Degradation mechanisms: Hydrolysis and enzymatic degradation significantly alter the mechanical performance of the material as a function of time [75]. Processing alters microstructure and, hence, affects mechanical properties. Other sterilization methods may affect mechanical behaviour, including gamma radiation or ethylene oxide.

Biodegradable polymers, such as polylactic acid (PLA), polyglycolic acid (PGA), and polycaprolactone (PCL), exhibit unique mechanical properties used in tissue engineering [76]. PLLA is a high-strength and high-stiffness material. PGA has high strength but degrades fast, while PCL has a low modulus, is highly tough, and degrades slowly. The stress and strain plots of some biodegradable composites are shown in Fig. 2. This knowledge of the mechanical properties of the polymers allows the researchers to choose appropriate materials for different applications in tissue engineering. These materials would combine the mechanical properties of polymers with ceramics or nanoparticles to enhance their features. Polymer-ceramic composites are prepared for enhanced mechanical properties, bioactivity, and osteoconductivity. In this context, polymer-nanoparticle composites have shown improved mechanical properties, biocompatibility, and cell interaction. Such composites can be tailored according to specific needs in tissue engineering applications. Table 1 gives the mechanical properties of various biodegradable composites used in tissue engineering.

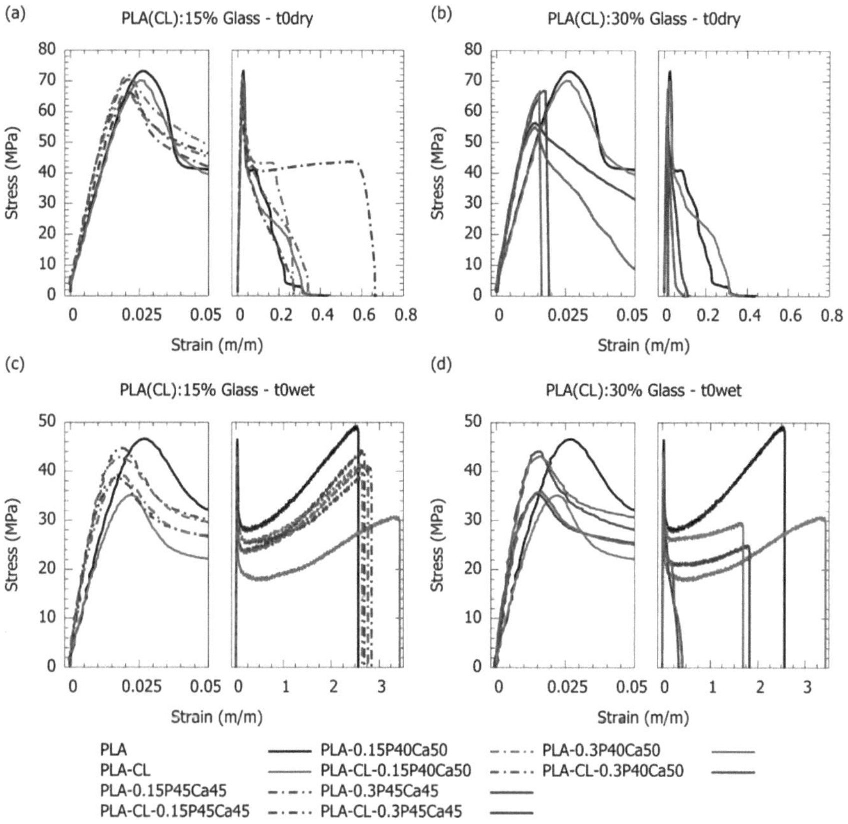

Fig. 2 Mechanical properties of various bioresorbable composites with fillers. **a** composites with 15 wt% of glass in dry conditions, **b** composites with 30% glass in dry conditions, **c** composites with 15% of glass in wet conditions, and **d** composites with 30% of glass in wet conditions. Reprinted with permission taken from [77] Copyright 2024, Elsevier License number: 5878620909562

Table 1 Mechanical properties of various biodegradable composites [75]

Composite	Compressive strength (MPa)	Tensile strength (MPa)	Young's modulus (GPa)
PLLA	80–500	45–70	2.7
PGA	340–920	55–80	5–7
L(PLA)	20–30	100–150	2.7
D,L(PLA)	15–25	90–103	1.9
PCL	20–40	10–35	0.4–0.6
Hydroxyapatite	500–1000	40–200	80–110
Monetite	15–25	10–15	22–35
Brushite	35–60	15–25	40–55
Magnesium	65–1000	135–285	41–45

3.2 *Effect of Bioactive Fillers on Mechanical Performance*

Incorporating bioactive fillers has been one of the effective methods to enhance the mechanical properties of polylactide-based composites for biomedical applications, especially for tissue engineering and biodegradable stents. Judicious selection of bioactive fillers was recently reported to result in enhanced mechanical properties of the composites.

Several mechanisms can explain the mechanical reinforcement of polylactide-based composites through bioactive fillers. The fillers act as a stress transfer agent, where the applied load is transferred to the whole composite, causing the burden or stress on the polymer matrix to decrease. The high aspect ratio and surface area of the fillers also facilitate this by providing them with an excellent interfacial adhesion with the polymer matrix, enhancing the effect of load transfer and improving mechanical properties.

For instance, bioactive fillers can induce crystallization in the polymer matrix, thereby increasing its stiffness and strength. Several studies showed the improved properties of current composites with incorporated fibreglass materials, as shown in Fig. 2, such as polylactide-based polymers with phosphate glasses, to overcome the main drawbacks of current bioabsorbable stents [77].

Hydroxyapatite HA, $Ca_{10}(PO_4)_6(OH)_2$, is a bio-ceramic filler that finds wide acceptance in applications related to medical composites due to its chemical resemblance to the natural mineral of bone, hence guaranteeing its excellent biocompatibility and bioactivity by substitution of species such as Na^+, Mg^{2+}, CO_3^{2-}, and Cl^-. Apart from that, other crystalline calcium compounds useful in developing resorbable composites are calcium carbonate and several calcium orthophosphates: tetracalcium phosphate, α-tricalcium phosphate, dicalcium phosphate dihydrate, dicalcium phosphate anhydrous, octocalcium phosphate, and β-tricalcium phosphate-all of which improve biocompatibility, bioactivity, and resorbability in medical composites [76].

The realization of improved mechanical properties is determined by the properties and dispersion of bioactive fillers. Filler contents ranging between 10 and 30 wt% have been proven to offer a good compromise between mechanical reinforcement and processability that could be maintained [77]. For instance, homogeneous filler dispersion via melt blending or solvent casting allows good stress transfer and avoids filler agglomeration. Long-term stability and degradation behaviour concerning polylactide-based composites with bioactive fillers need to be further investigated. Investigation of the effects of filler surface modification and polymer-filler interfacial interaction will be very useful in the optimization of mechanical properties. Materials of such type could find applications in biomedical field development of high performances and longer lifetime devices.

3.3 *Influence of Degradation on Mechanical Integrity*

The degradation rate of polymers used in implantable biomaterials is a critical parameter that significantly influences the biological response and mechanical property requirements. Controlled degradation allows mechanical properties to be tuned to the dynamic needs of healing tissue, optimizing biocompatibility by reducing inflammatory responses. In addition, uncontrolled degradation results in premature failure, injury to the tissue, and other adverse immune reactions. While chemical structure, material composition, and environmental conditions strongly influence polymer degradation kinetics, understanding such influences in the design of an optimally performing implantable biomaterial becomes essential. Degradation rate control includes polymer synthesis, material blending, surface modification, and enzyme-mediated degradation [78].

Kharzi et al. [79] conducted a study to investigate the degradation rate of a resorbable composite bone plate for mechanical properties and bioactivity. The study concluded that the composite showed high strength and mechanical properties to bear a significant load for orthopaedic applications. Additionally, the composite maintained 65% of the strength for 8 weeks when soaked in simulated body fluid (Fig. 3). Their study also concluded that the composite mimics the bone by releasing biocompatible ions like silicon, calcium, magnesium, etc. Different degradation rates may be preferred depending on the application, tissue type, and desired outcome. For example, the degradation rate for orthopaedic implants should match bone remodelling rates, while it should be a rate that allows for tissue regeneration in tissue engineering scaffolds. By optimizing the rates of degradation, researchers can implant biomaterials that act in concert with the host's healing process for more predictable results and minimize the risk of complications. Control of polymer degradation rates is a significant factor in the successful regeneration of tissues, implant integration, and the long-term functionality of any device.

The biomaterials to be used to fabricate a scaffold must be non-chronic inflammatory upon implantation and biocompatible with a complete absence of any immune response. The cell ingrowth inside the scaffold requires the porous nature of the

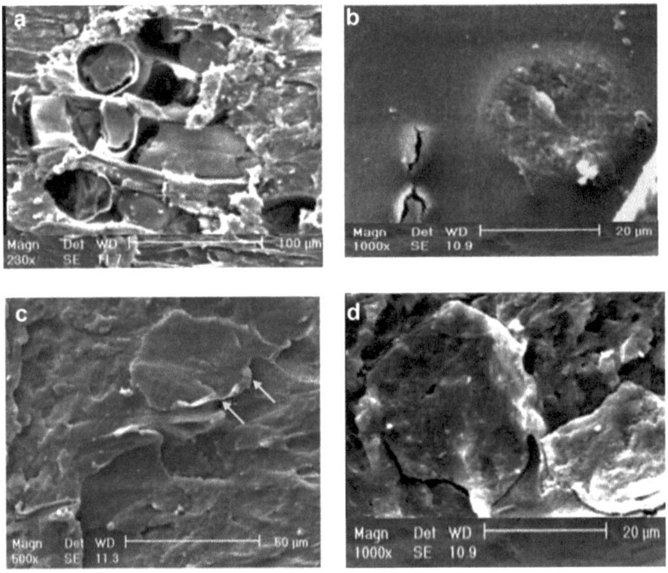

Fig. 3 SEM analysis of composite sample during soaking. **a** Analysis after 3 days of non-soaking, **b** analysis after 3 days of soaking treated, **c** analysis after 10 days of soaking treated, and **d** analysis after 14 days of soaking treated. Reprinted with permission taken from [79] Copyright 2024, Elsevier License number: 5881730800850

construct to allow the cell growth or at least pores for nutrient diffusion. Another essential feature is that the construct has to integrate with the host tissue; hence, vascularization becomes a key factor. Material physical, mechanical, and chemical properties are important, depending on the specific biomechanical requirements of the tissue under consideration. Further, malleability, ability to be sterilized, and cost have to be considered while fabricating a scaffold. The biodegradation rate of the scaffold is an important issue, so degradable scaffolds with different degradation rates are required. However, since most of the polymers in their alone state lack the required mechanical strength or properties to fabricate scaffolds, researchers are now investigating degradable composites as alternatives to polymer scaffolds (Fig. 4).

4 Biological Performance and Osteointegration

The biological performance of the materials lies in the cytotoxicity/biocompatibility, Cell adhesion and proliferation. These are the critical parameters that define the biological performance of the biomaterials that have been used in biomedical applications. By utilizing bioresorbable materials, firstly, we have to study the host response and the material response to qualify for the biocompatibility test. [80] The medical devices or implants made out of bioresorbable composites must undergo a sequence

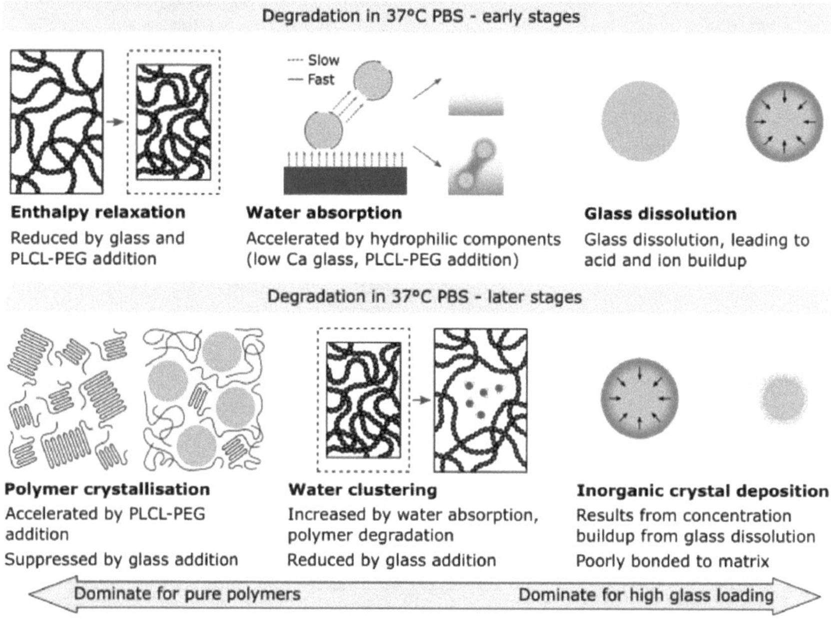

Fig. 4 Structural changes during the degradation process in polymer-glass composites with time. Reprinted with permission taken from [77] Copyright 2024, Elsevier License number: 5878620909562

of tests for utilization in clinic to confirm their safety as regulated by the International Organisation for Standardisation (ISO) [81, 82].

4.1 *In Vitro Biocompatibility*

In vitro studies were done initially before in vivo studies predicting the material's behaviour, where the material is tested at a lab, after which the material is subjected to animal trials. Therefore, in vitro trials help us to analyze the short-term effects, whereas in vivo help in analyzing the long-term impact of the biomaterial inside the body [9]. The first step in the testing of in vitro biocompatibility involves the cytotoxicity analysis. This test utilizes tissue cells in vitro to monitor the propagation, proliferation and morphology of the implants [83]. ISO 10993-5 suggests that countries create their regulations for the cytotoxicity in vitro test according to their local situation [84]. This cytotoxicity test through in vitro is simple, fast, susceptible and protects the animals from adverse effects/toxicity [85, 86]. According to the ISO 10993-5 standard, we have three tests to evaluate cytotoxicity: Extract, direct contact and indirect contact tests (which also include the agar overlay assay and filter diffusion).

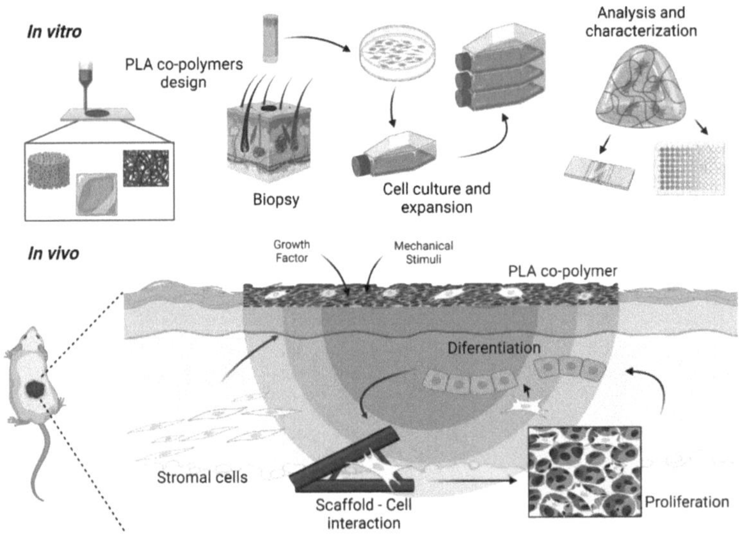

Fig. 5 PLA and PLA-based resorbable biocomposites developed for in vitro and in vivo testing [91]

The cytotoxicity of the implants/composites, which are soluble, can be determined by using the extractable tests, which should be in accordance with the animal cytotoxicity tests. The most sensitive technique for assessing the toxicity of the implant/biocomposite is the direct contact assay, through which we can even measure deficient toxicity levels [87]. An agar overlay assay is employed for the devices exhibiting considerable toxicity and bulk filtration [88]. For small molecular-weight devices/implants, the toxicity is evaluated using a molecular filtration technique [89]. The cause of cellular death cannot be determined via in vitro cytotoxicity analysis. Therefore, analyzing the in vitro cell response to the implant/biocomposite helps us to identify any detrimental reaction. The interaction of biomaterial with various types of cells within the foreign body reaction can be qualified using protein-based arrays [90]. Enzyme-linked immunosorbent assays and antibody arrays are used to profile the cytokines, made by cells interacting with the biomaterial/implant [9]. Figure 5 shows in vitro and in vivo analyses of PLA-based composites.

4.2 *In Vivo Bone Regeneration and Osteointegration*

Resorbable composite biomaterials combine biodegradable polymers with bioactive fillers for renovation in bone tissue engineering. Novel materials degrade at a rate that allows the progression of tissue replacement to coincide with that of bone growth, while the release of bioactive ions and growth factors stimulates osteogenesis. This synchrony in degradation and growth allows functional bone tissue to form,

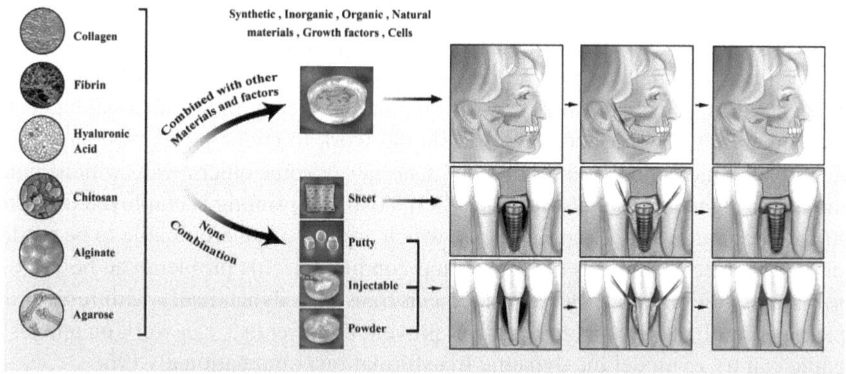

Fig. 6 Various biocompatible composites used in bone healing. Reprinted with permission taken from [93] Copyright 2024, Elsevier

thereby avoiding major problems associated with the use of non-resorbable materials. The properties that influence the effectiveness of resorbable composite materials in bone regeneration include degradation rate, mechanical strength, porosity, surface topography, and bioactivity [92].

Controlled degradation will ensure synchronized tissue replacement and bone growth while sufficient mechanical integrity supports initial tissue loading. Optimized porosity allows for cell infiltration and vascularization with nutrient exchange. Micro- and nano-scale surface features enhance cell adhesion, proliferation, and differentiation. Incorporation of bioactive fillers, such as hydroxyapatite or bioglass, stimulates osteogenesis. Figure 6 represents the various biodegradable composites used in bone regeneration processes.

A number of preclinical studies have been performed to establish the feasibility and efficacy of resorbable composite materials for bone regeneration. Increased bone growth and density enhanced osteoid formation and mineralization, and improved bone-implant interface and mechanical anchorage were reported by various studies. For instance, bone ingrowth increased by 30% for rabbit femoral defects treated with composite materials based on polylactide hydroxyapatite, while composite materials based on polyglycolide-bioglass enhanced the bone regeneration for rat calvarial defect by 25%. Similarly, composites based on polycaprolactone-calcium phosphate exhibited 40% higher osteointegration in sheep tibial defects [94].

5 Summary and Future Perspectives

The potential of resorbable biocomposites has revolutionized the biomedical sector (especially tissue engineering applications), which is expected to progress more in the near future. When placed inside the body, these resorbable biocomposites will be exposed to different stimuli such as temperature, pressure, flow, variable stresses & strains and electric fields. The material should have resistance towards these variables, which are very interesting and tough where researchers can work. In recent years, scientists have been working on producing multifunctional biocomposite materials, but further, scientists should work to create novel biocomposites either by chemical modification of these materials or some other means which could emulate or regenerate bone/soft tissues [95]. Also, 3D printing is employed to make intricate structures for TE applications, which are static and aren't able to be modified with respect to various environmental conditions. This problem can be solved by utilizing 4D printed material, which can transform dynamically with respect to external stimuli. This concept is not yet proven, but scientists can work on and also people can try to model the dynamic transformation computationally [96].

References

1. Boccaccini AR, Roelher JA, Hench LL et al (2002) A composites approach to tissue engineering. In: Lin H-T, Singh M (eds) Ceramic engineering and science proceedings. John Wiley & Sons Inc., Hoboken, NJ, USA, pp 805–816
2. Saska S, Pilatti L, Blay A, Shibli JA (2021) Bioresorbable polymers: advanced materials and 4d printing for tissue engineering. Polymers 13:563. https://doi.org/10.3390/polym13040563
3. Deb S (2009) Degradable polymers and polymer composites for tissue engineering. In: Cellular response to biomaterials. Elsevier, pp 28–60
4. Barth J, Akritopoulos P, Graveleau N et al (2016) Efficacy of osteoconductive ceramics in bioresorbable screws for anterior cruciate ligament reconstruction: a prospective intrapatient comparative study. Orthop J Sports Med 4:2325967116647724. https://doi.org/10.1177/2325967116647724
5. Bose D, Prasannavenkadesan V, Katiyar V (2024) Tailored material for bioresorbable internal fixation devices: a novel approach using nanohydroxyapatite and chitosan in polylactic acid. J Mater Sci 59:215–227. https://doi.org/10.1007/s10853-023-09110-1
6. Prasad A, Bhasney SM, Prasannavenkadesan V et al (2023) Nano-hydroxyapatite reinforced polylactic acid bioabsorbable cancellous screws for bone fracture fixations. J Appl Polymer Sci 140:e54577. https://doi.org/10.1002/app.54577
7. Prasad A, Bhasney SM, Prasannavenkadesan V et al (2023) Polylactic acid reinforced with nano-hydroxyapatite bioabsorbable cortical screws for bone fracture treatment. J Polym Res 30:177. https://doi.org/10.1007/s10965-023-03542-8
8. Prasannavenkadesan V, Naveed Ul Meiraj S, Narayanan KBB et al (2024) 4 Chitosan in orthopedics: current advancements and future prospects. In: Prasad A, Chakraborty G, Davim JP (eds) Sustainable bio-based composites. De Gruyter, pp 59–78
9. Geddes L, Themistou E, Burrows JF et al (2021) Evaluation of the in vitro cytotoxicity and modulation of the inflammatory response by the bioresorbable polymers poly(D, L-lactide-co-glycolide) and poly(L-lactide-co-glycolide). Acta Biomater 134:261–275. https://doi.org/10.1016/j.actbio.2021.07.049
10. Maurus PB, Kaeding CC (2004) Bioabsorbable implant material review. Oper Tech Sports Med 12:158–160. https://doi.org/10.1053/j.otsm.2004.07.015
11. Tenekecioglu E, Farooq V, Bourantas CV et al (2016) Bioresorbable scaffolds: a new paradigm in percutaneous coronary intervention. BMC Cardiovasc Disord 16:38. https://doi.org/10.1186/s12872-016-0207-5
12. Someya T, Bao Z, Malliaras GG (2016) The rise of plastic bioelectronics. Nature 540:379–385. https://doi.org/10.1038/nature21004

13. Hwang S, Song J, Huang X et al (2014) High-performance biodegradable/transient electronics on biodegradable polymers. Adv Mater 26:3905–3911. https://doi.org/10.1002/adma.201306050
14. Vert M, Li SM, Spenlehauer G, Guerin P (1992) Bioresorbability and biocompatibility of aliphatic polyesters. J Mater Sci: Mater Med 3:432–446. https://doi.org/10.1007/BF00701240
15. Vert M, Doi Y, Hellwich K-H et al (2012) Terminology for biorelated polymers and applications (IUPAC Recommendations 2012). Pure Appl Chem 84:377–410. https://doi.org/10.1351/PAC-REC-10-12-04
16. Liu Y, Zheng Y, Hayes B (2017) Degradable, absorbable or resorbable—what is the best grammatical modifier for an implant that is eventually absorbed by the body? Sci China Mater 60:377–391. https://doi.org/10.1007/s40843-017-9023-9
17. Ramakrishna S, Mayer J, Wintermantel E, Leong KW (2001) Biomedical applications of polymer-composite materials: a review. Compos Sci Technol 61:1189–1224. https://doi.org/10.1016/S0266-3538(00)00241-4
18. Shikinami Y, Okuno M (2001) Bioresorbable devices made of forged composites of hydroxyapatite (HA) particles and poly l-lactide (PLLA). Part II: practical properties of miniscrews and miniplates. Biomaterials 22:3197–3211. https://doi.org/10.1016/S0142-9612(01)00072-2
19. Del Campo R, Savoini B, Muñoz A et al (2017) Processing and mechanical characteristics of magnesium-hydroxyapatite metal matrix biocomposites. J Mech Behav Biomed Mater 69:135–143. https://doi.org/10.1016/j.jmbbm.2016.12.023
20. Chazono M, Tanaka T, Komaki H, Fujii K (2004) Bone formation and bioresorption after implantation of injectable β-tricalcium phosphate granules–hyaluronate complex in rabbit bone defects. J Biomed Mater Res 70A:542–549. https://doi.org/10.1002/jbm.a.30094
21. Eberhart RC, Su S-H, Nguyen KT et al (2003) Review: bioresorbable polymeric stents: current status and future promise. J Biomater Sci Polym Ed 14:299–312. https://doi.org/10.1163/156856203321478838
22. Ignjatović N, Wu V, Ajduković Z et al (2016) Chitosan-PLGA polymer blends as coatings for hydroxyapatite nanoparticles and their effect on antimicrobial properties, osteoconductivity and regeneration of osseous tissues. Mater Sci Eng, C 60:357–364. https://doi.org/10.1016/j.msec.2015.11.061
23. Maman P, Nagpal M, Aggarwal G (2019) Resorbable polymer fiber reinforced composites in biomedical application. In: Materials for biomedical engineering. Elsevier, pp 121–166
24. Ayala P, Caves J, Dai E et al (2015) Engineered composite fascia for stem cell therapy in tissue repair applications. Acta Biomater 26:1–12. https://doi.org/10.1016/j.actbio.2015.08.012
25. Morsada Z, Hossain MM, Islam MT et al (2021) Recent progress in biodegradable and bioresorbable materials: from passive implants to active electronics. Appl Mater Today 25:101257. https://doi.org/10.1016/j.apmt.2021.101257
26. Wang J, Wang L, Zhou Z et al (2016) Biodegradable polymer membranes applied in guided bone/tissue regeneration: a review. Polymers 8:115. https://doi.org/10.3390/polym8040115
27. Vach Agocsova S, Culenova M, Birova I et al (2023) Resorbable biomaterials used for 3d scaffolds in tissue engineering: a review. Materials 16:4267. https://doi.org/10.3390/ma16124267
28. Kaur S, Dhillon GS (2014) The versatile biopolymer chitosan: potential sources, evaluation of extraction methods and applications. Crit Rev Microbiol 40:155–175. https://doi.org/10.3109/1040841X.2013.770385
29. Wu Q-X, Lin D-Q, Yao S-J (2014) Design of chitosan and its water soluble derivatives-based drug carriers with polyelectrolyte complexes. Mar Drugs 12:6236–6253. https://doi.org/10.3390/md12126236
30. Hein S, Wang K, Stevens WF, Kjems J (2008) Chitosan composites for biomedical applications: status, challenges and perspectives. Mater Sci Technol 24:1053–1061. https://doi.org/10.1179/174328408X341744
31. Li Z, Leung M, Hopper R et al (2010) Feeder-free self-renewal of human embryonic stem cells in 3D porous natural polymer scaffolds. Biomaterials 31:404–412. https://doi.org/10.1016/j.biomaterials.2009.09.070

32. Mohamad AH, Abdullah OG, Saeed SR (2020) Effect of very fine nanoparticle and temperature on the electric and dielectric properties of MC-PbS polymer nanocomposite films. Results Phys 16:102898. https://doi.org/10.1016/j.rinp.2019.102898
33. Ragunathan S, Govindasamy G, Raghul DR et al (2020) Hydroxyapatite reinforced natural polymer scaffold for bone tissue regeneration. Mater Today: Proc 23:111–118. https://doi.org/10.1016/j.matpr.2019.07.712
34. Mohd Roslan MR, Mohd Kamal NL, Abdul Khalid MF et al (2021) The state of starch/hydroxyapatite composite scaffold in bone tissue engineering with consideration for dielectric measurement as an alternative characterization technique. Materials 14:1960. https://doi.org/10.3390/ma14081960
35. Mahmoud EM, Sayed M, El-Kady AM et al (2020) In vitro and in vivo study of naturally derived alginate/hydroxyapatite bio composite scaffolds. Int J Biol Macromol 165:1346–1360. https://doi.org/10.1016/j.ijbiomac.2020.10.014
36. Kim OV, Litvinov RI, Weisel JW, Alber MS (2016) Compressive mechanics of collagen-fibrin composites and their structural alterations. Biophys J 110:338a. https://doi.org/10.1016/j.bpj.2015.11.1820
37. Fu J-H, Zhao M, Lin Y-R et al (2017) Degradable chitosan-collagen composites seeded with cells as tissue engineered heart valves. Heart Lung Circ 26:94–100. https://doi.org/10.1016/j.hlc.2016.05.116
38. Gurumurthy B, Bierdeman PC, Janorkar AV (2016) Composition of elastin like polypeptide–collagen composite scaffold influences in vitro osteogenic activity of human adipose derived stem cells. Dent Mater 32:1270–1280. https://doi.org/10.1016/j.dental.2016.07.009
39. Werner C (2006) Polymers for regenerative medicine. Springer, Berlin Heidelberg, Berlin, Heidelberg
40. Furrer P, Panke S, Zinn M (2007) Efficient recovery of low endotoxin medium-chain-length poly([R]-3-hydroxyalkanoate) from bacterial biomass. J Microbiol Methods 69:206–213. https://doi.org/10.1016/j.mimet.2007.01.002
41. Chen Q, Liang S, Thouas GA (2013) Elastomeric biomaterials for tissue engineering. Prog Polym Sci 38:584–671. https://doi.org/10.1016/j.progpolymsci.2012.05.003
42. Soleymani Eil Bakhtiari S, Karbasi S, Toloue EB (2021) Modified poly(3-hydroxybutyrate)-based scaffolds in tissue engineering applications: a review. Int J Biol Macromol 166:986–998. https://doi.org/10.1016/j.ijbiomac.2020.10.255
43. Dhania S, Bernela M, Rani R et al (2023) Polyhydroxybutyrate (PHB) in nanoparticulate form improves physical and biological performance of scaffolds. Int J Biol Macromol 236:123875. https://doi.org/10.1016/j.ijbiomac.2023.123875
44. Martin DP, Williams SF (2003) Medical applications of poly-4-hydroxybutyrate: a strong flexible absorbable biomaterial. Biochem Eng J 16:97–105. https://doi.org/10.1016/S1369-703X(03)00040-8
45. Mohan A, Girdhar M, Kumar R et al (2021) Polyhydroxybutyrate-based nanocomposites for bone tissue engineering. Pharmaceuticals 14:1163. https://doi.org/10.3390/ph14111163
46. Cameron RE, Kamvari-Moghaddam A (2008) Synthetic bioresorbable polymers. In: Degradation rate of bioresorbable materials. Elsevier, pp 43–66
47. Guan J, Sacks MS, Beckman EJ, Wagner WR (2004) Biodegradable poly(ether ester urethane)urea elastomers based on poly(ether ester) triblock copolymers and putrescine: synthesis, characterization and cytocompatibility. Biomaterials 25:85–96. https://doi.org/10.1016/S0142-9612(03)00476-9
48. Albertsson A, Varma IK (2002) Aliphatic polyesters. In: Steinbüchel A (ed) Biopolymers online, 1st ed. Wiley
49. Mulchandani N, Prasad A, Katiyar V (2019) Resorbable polymers in bone repair and regeneration. In: Materials for biomedical engineering. Elsevier, pp 87–125
50. Salerno A, Guarino V, Oliviero O et al (2016) Bio-safe processing of polylactic-co-caprolactone and polylactic acid blends to fabricate fibrous porous scaffolds for in vitro mesenchymal stem cells adhesion and proliferation. Mater Sci Eng, C 63:512–521. https://doi.org/10.1016/j.msec.2016.03.018

51. Chen J, Sheu AY, Li W et al (2014) Poly(lactide-co-glycolide) microspheres for MRI-monitored transcatheter delivery of sorafenib to liver tumors. J Control Release 184:10–17. https://doi.org/10.1016/j.jconrel.2014.04.008
52. Persson M, Lorite GS, Kokkonen HE et al (2014) Effect of bioactive extruded PLA/HA composite films on focal adhesion formation of preosteoblastic cells. Colloids Surf, B 121:409–416. https://doi.org/10.1016/j.colsurfb.2014.06.029
53. Lih E, Park KW, Chun SY et al (2016) Biomimetic porous PLGA scaffolds incorporating decellularized extracellular matrix for kidney tissue regeneration. ACS Appl Mater Interfaces 8:21145–21154. https://doi.org/10.1021/acsami.6b03771
54. Tyler B, Gullotti D, Mangraviti A et al (2016) Polylactic acid (PLA) controlled delivery carriers for biomedical applications. Adv Drug Deliv Rev 107:163–175. https://doi.org/10.1016/j.addr.2016.06.018
55. Maadani AM, Salahinejad E (2022) Performance comparison of PLA- and PLGA-coated porous bioceramic scaffolds: mechanical, biodegradability, bioactivity, delivery and biocompatibility assessments. J Control Release 351:1–7. https://doi.org/10.1016/j.jconrel.2022.09.022
56. Singhvi MS, Zinjarde SS, Gokhale DV (2019) Polylactic acid: synthesis and biomedical applications. J Appl Microbiol 127:1612–1626. https://doi.org/10.1111/jam.14290
57. Hong Z, Zhang P, He C et al (2005) Nano-composite of poly(-lactide) and surface grafted hydroxyapatite: mechanical properties and biocompatibility. Biomaterials 26:6296–6304. https://doi.org/10.1016/j.biomaterials.2005.04.018
58. Mathieu LM, Montjovent M-O, Bourban P et al (2005) Bioresorbable composites prepared by supercritical fluid foaming. J Biomed Mater Res 75A:89–97. https://doi.org/10.1002/jbm.a.30385
59. Lee MR, Lee GW, Kim JE et al (2019) Biocompatibility of a PLA-based composite containing hydroxyapatite derived from waste bones of dolphin Neophocaena asiaeorientalis. J Aust Ceram Soc 55:269–279. https://doi.org/10.1007/s41779-018-0232-1
60. Rakmae S, Ruksakulpiwat Y, Sutapun W, Suppakarn N (2011) Physical properties and cytotoxicity of surface-modified bovine bone-based hydroxyapatite/poly(lactic acid) composites. J Compos Mater 45:1259–1269. https://doi.org/10.1177/0021998310377934
61. Xu G, Mou S, Yin Z et al (2007) Biomimetic strengthening polylactide scaffold materials for bone tissue engineering. Front Chem China 2:27–30. https://doi.org/10.1007/s11458-007-0006-8
62. Dai J, Yang S, Jin J, Li G (2016) Electrospinning of PLA/pearl powder nanofibrous scaffold for bone tissue engineering. RSC Adv 6:106798–106805. https://doi.org/10.1039/C6RA21796F
63. Kannan R, Wei G, Ma PX (2022) Synthetic polymeric biomaterials for tissue engineering. In: Tissue engineering using ceramics and polymers. Elsevier, pp 41–74
64. Yoon DM, Fisher JP (2021) Natural and synthetic polymeric scaffolds. In: Narayan R (ed) Biomedical materials. Springer International Publishing, Cham, pp 257–283
65. Qi H, Ye Z, Ren H et al (2016) Bioactivity assessment of PLLA/PCL/HAP electrospun nanofibrous scaffolds for bone tissue engineering. Life Sci 148:139–144. https://doi.org/10.1016/j.lfs.2016.02.040
66. Park SH, Park SA, Kang YG et al (2017) PCL/β-TCP composite scaffolds exhibit positive osteogenic differentiation with mechanical stimulation. Tissue Eng Regen Med 14:349–358. https://doi.org/10.1007/s13770-017-0022-9
67. Scaffaro R, Lopresti F, Maio A et al (2017) Electrospun PCL/GO-g-PEG structures: processing-morphology-properties relationships. Compos A Appl Sci Manuf 92:97–107. https://doi.org/10.1016/j.compositesa.2016.11.005
68. Cheng Y, Deng S, Chen P, Ruan R (2009) Polylactic acid (PLA) synthesis and modifications: a review. Front Chem China 4:259–264. https://doi.org/10.1007/s11458-009-0092-x
69. Makadia HK, Siegel SJ (2011) Poly lactic-co-glycolic acid (PLGA) as biodegradable controlled drug delivery carrier. Polymers 3:1377–1397. https://doi.org/10.3390/polym3031377
70. Zhang R, Ma PX (1999) Poly(?-hydroxyl acids)/hydroxyapatite porous composites for bone-tissue engineering. I. Preparation and morphology. J Biomed Mater Res 44:446–455. https://doi.org/10.1002/(SICI)1097-4636(19990315)44:4%3c446::AID-JBM11%3e3.0.CO;2-F

71. Becker G, Calvis A, Hazlett L et al (2014) Bioabsorbable polymeric fracture fixation devices aim to reduce stress shielding in bone. 2014 40th Annual northeast bioengineering conference (NEBEC). IEEE, Boston, MA, USA, pp 1–2
72. Freire AR, Rossi AC, Queiroz TP et al (2012) Histometric analysis of bone repair in bone-implant interface using a polylactic/polyglycolic acid copolymer associated with implants in Rabbit Tibia. J Oral Implantol 38:449–457. https://doi.org/10.1563/AAID-JOI-D-10-00102
73. Jose M, Thomas V, Johnson K et al (2009) Aligned PLGA/HA nanofibrous nanocomposite scaffolds for bone tissue engineering. Acta Biomater 5:305–315. https://doi.org/10.1016/j.actbio.2008.07.019
74. Oliveira JM, Reis RL (2017) Regenerative strategies for the treatment of knee joint disabilities. Springer International Publishing, Cham
75. Sheikh Z, Najeeb S, Khurshid Z et al (2015) Biodegradable materials for bone repair and tissue engineering applications. Materials 8:5744–5794. https://doi.org/10.3390/ma8095273
76. Dobrzyńska-Mizera M, Dodda JM, Liu X, et al (2024) Engineering of bioresorbable polymers for tissue engineering and drug delivery applications. Adv Healthc Mater:2401674. https://doi.org/10.1002/adhm.202401674
77. Oosterbeek RN, Zhang XC, Best SM, Cameron RE (2022) The evolution of the structure and mechanical properties of fully bioresorbable polymer-glass composites during degradation. Compos Sci Technol 218:109194. https://doi.org/10.1016/j.compscitech.2021.109194
78. Hutmacher DW, Goh JCH, Teoh SH. An introduction to biodegradable materials for tissue engineering applications. Ann Acad Med Singapore 30
79. Zargar Kharazi A, Fathi MH, Bahmani F, Fanian H (2011) Partially resorbable composite bone plate with controlled degradation rate, desired mechanical properties and bioactivity. Polym Degrad Stab 96:2055–2063. https://doi.org/10.1016/j.polymdegradstab.2011.10.004
80. Barbucci R, Lamponi S, Magnani A (2000) Biological performance of materials. In: Fabbrizzi L, Poggi A (eds) Chemistry at the beginning of the third millennium. Springer, Berlin Heidelberg, Berlin, Heidelberg, pp 161–183
81. Li W, Zhou J, Xu Y (2015) Study of the in vitro cytotoxicity testing of medical devices. Biomed Reports 3:617–620. https://doi.org/10.3892/br.2015.481
82. Anderson JM (2016) Future challenges in the *in vitro* and *in vivo* evaluation of biomaterial biocompatibility. Regen Biomater 3:73–77. https://doi.org/10.1093/rb/rbw001
83. Soenen SJ, Manshian B, Montenegro JM et al (2012) Cytotoxic effects of gold nanoparticles: a multiparametric study. ACS Nano 6:5767–5783. https://doi.org/10.1021/nn301714n
84. Osthues RM, Da Silva SN, Zavaglia CAC, Fialho SL (2011) Study of the release potential of the antibiotic gentamicin from microspheres of BCP. KEM 493–494:269–274. https://doi.org/10.4028/www.scientific.net/KEM.493-494.269
85. Kunzmann A, Andersson B, Thurnherr T, et al (2011) Toxicology of engineered nanomaterials: focus on biocompatibility, biodistribution and biodegradation. Biochimica et Biophysica Acta (BBA)—General Subjects 1810:361–373. https://doi.org/10.1016/j.bbagen.2010.04.007
86. He Q, Shi J (2011) Mesoporous silica nanoparticle based nano drug delivery systems: synthesis, controlled drug release and delivery, pharmacokinetics and biocompatibility. J Mater Chem 21:5845. https://doi.org/10.1039/c0jm03851b
87. De Melo WM, Maximiano WMA, Antunes AA et al (2013) Cytotoxicity testing of methyl and ethyl 2-cyanoacrylate using direct contact assay on osteoblast cell cultures. J Oral Maxillofac Surg 71:35–41. https://doi.org/10.1016/j.joms.2012.09.002
88. Sjögren G, Sletten G, Dahl JE (2000) Cytotoxicity of dental alloys, metals, and ceramics assessed by Millipore filter, agar overlay, and MTT tests. J Prosthet Dent 84:229–236. https://doi.org/10.1067/mpr.2000.107227
89. Jin C-Y, Zhu B-S, Wang X-F, Lu Q-H (2008) Cytotoxicity of titanium dioxide nanoparticles in mouse fibroblast cells. Chem Res Toxicol 21:1871–1877. https://doi.org/10.1021/tx800179f
90. Lock A, Cornish J, Musson DS (2019) The role of in vitro immune response assessment for biomaterials. JFB 10:31. https://doi.org/10.3390/jfb10030031
91. Castañeda-Rodríguez S, González-Torres M, Ribas-Aparicio RM et al (2023) Recent advances in modified poly (lactic acid) as tissue engineering materials. J Biol Eng 17:21. https://doi.org/10.1186/s13036-023-00338-8

92. Lewandrowski K-U, D. Gresser J, Wise DL, Trantolo DJ (2000) Bioresorbable bone graft substitutes of different osteoconductivities: a histologic evaluation of osteointegration of poly(propylene glycol-co-fumaric acid)-based cement implants in rats. Biomaterials 21:757–764. https://doi.org/10.1016/S0142-9612(99)00179-9
93. Tahmasebi E, Alam M, Yazdanian M et al (2020) Current biocompatible materials in oral regeneration: a comprehensive overview of composite materials. J Market Res 9:11731–11755. https://doi.org/10.1016/j.jmrt.2020.08.042
94. Shi J, Dai W, Gupta A et al (2022) Frontiers of hydroxyapatite composites in bionic bone tissue engineering. Materials 15:8475. https://doi.org/10.3390/ma15238475
95. Bajpai AK, Bajpai J, Saini RK et al (2016) Smart biomaterial devices. CRC Press
96. Miao S, Zhu W, Castro NJ et al (2016) 4D printing smart biomedical scaffolds with novel soybean oil epoxidized acrylate. Sci Rep 6:27226. https://doi.org/10.1038/srep27226

Chapter 4
Advanced 3D-Printed Drug-Eluting Resorbable Implants and Fixation Devices

Sudipto Datta, Ranjit Barua, Bidyanand Mahto, and Arbind Prasad

1 Introduction

New opportunities can be utilized to develop novel technologies for differential medical devices, especially those in drug-releasing absorbable implants and fixation devices [1]. These devices, which serve both as scaffold and local drug delivery systems, have great potential to augment clinical results in many types of surgeries. Due to the incorporation of both the mechanical actions of conventional implants and the medicinal properties of controlled drug delivery, 3D-printed drug-eluting implants are a significant improvement in the field of targeted medicine [2]. Resorbable implants are not an invention, but the extent to which such devices can be tailored through 3D printing is unparalleled. This technology can produce complex geometries and characteristics, which are unique to a patient or a specific application [3]. Furthermore, drugs may be encapsulated and incorporated in the implant matrix to allow slow-release local delivery, this, in turn, eliminates the use of high concentrations of systemically administered drugs and their possible side effects [4]. This chapter discusses the recent developments in 3D-printed drug-eluting resorbable implants and fixation devices incorporating the material choice and used fabrication

S. Datta
Department of Materials Engineering, Indian Institute of Science, Bangalore, Karnataka, India

R. Barua
Omdayal Group of Institutions, Uluberia, Harinarayan Chak, West Bengal, India

B. Mahto
Government Engineering College, Vaishali (Under Department of Science, Technology and Technical Education, Government of Bihar), Vaishali, Bihar, India

A. Prasad (✉)
Mechanical Engineering Department, Katihar Engineering College (Under Department of Science, Technology and Technical Education, Government of Bihar), Katihar, Bihar, India
e-mail: arbind.iitg@gmail.com

A. Prasad and V. Katiyar (eds.), *Resorbable Polymers for Bioimplants and Fixation Devices*, Materials Horizons: From Nature to Nanomaterials,
https://doi.org/10.1007/978-981-96-9306-1_4

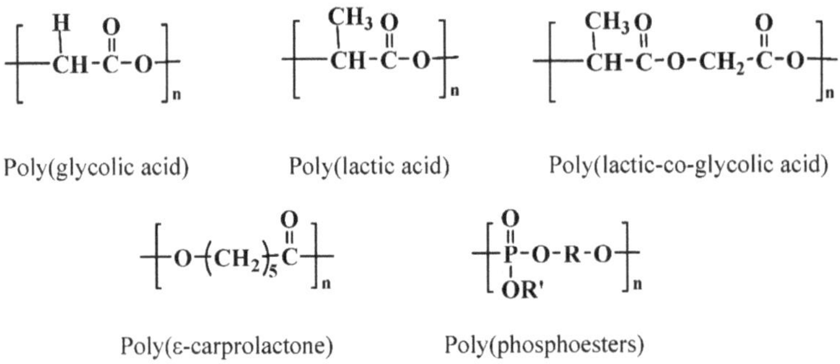

Fig. 1 Chemical composition of several biodegradable polymers (*Picture courtesy* Park et al. [7])

methods and various approaches for drug loading and release [5]. It also looks at the use of these devices in clinical situations and it establishes that these devices can be used in orthopedic surgery, spinal fusion, and craniofacial reconstruction among others [6]. Also, it focuses on the opportunities for and problems with these new implants and how they are regulated, giving an idea as to where the field of growing implants is headed (Fig. 1).

2 Materials for 3D-Printed Drug-Eluting Resorbable Implants

The selection of the material that can undergo a 3D printing process for the fabrication of drug-eluting resorbable implants is of paramount importance to their performance, biocompatibility, and therapeutic effectiveness [8]. There should be biocompatibility, mechanical properties, degradation, and drug loading and delivery capacities. This section deals with the basic components that are employed in the fabrication of these implants; the biodegradable polymers, the bioactive ceramics, and the composites.

2.1 *Biodegradable Polymers*

Non-resorbable implants are of non-biodegradable polymers but they are not as referred to by the term resorbable implants because they do not have to be surgically removed [9]. The selection and synthesis of these polymers can be manipulated to ensure the mechanical needs and the controlled release pattern of the drug.

2.1.1 Polylactic Acid (PLA)

There is also another BPA, biodegradable thermoplastic that is quite popular added to 3D printing is poly (lactic acid) otherwise known as PLA. There is a bio-based origin of polyurethane which consists of cornstarch, it is characterized by several properties for example biocompatibility and mechanical [10]. PLA simply dissolves in water, breaks the polymer into the lactic acid and the body uses the lactic acid [11].

Advantages

- High mechanical strength and rigidity.
- Biocompatible and biodegradable.
- Appropriate for various printing techniques.

Drug Delivery Applications: PLA scaffolds may be preloaded with drugs like antibiotics or anti-inflammatory agents that deliver localized therapy whereby the scaffold degrades while releasing the drug [12].

2.1.2 Polyglycolic Acid (PGA)

Another biodegradable polymer that exists in the 3D printing technology is polyglycolic acid better referred to as PGA. It is more highly degraded than PLA and, thus is appropriate in areas that require a quicker degradation rate [13]. It means that PGA is usually copolymerized with other polymers to control the degradation rate and mechanics of the material.

Advantages

- Quick degradation rate.
- Decent tensile strength.

Drug Delivery Applications: PGA-based implants also have data on the ability to release growth factors or anti-inflammatory medications in a short period and such cases may be useful in certain conditions where wanting of immediate release of drugs is necessary [14].

2.1.3 Polycaprolactone (PCL)

The concern has been made and understood that PCL is also a biodegradable polyester, however, this is acknowledged to degrade at a slower rate than both PLA and PGA. With such characteristics, it is highly mobile and can support a steady status for a definite period, which proves highly beneficial, especially in regions that require a little more time to mend the tissues [15].

Advantages

- Extended degradation time.
- Elastic and hard.

Drug Delivery Applications: PCL is used in the formulation of implants that show diverse release rates of the drug in the body hence enhanced therapeutic duration. More often it is combined with other polymers or bioactive substances to alter the profile of drug release and to enhance the mechanical characteristics [16].

2.2 *Bioactive Ceramics*

Hydroxyapatite (HA) and tricalcium phosphate (β–TCP) is used in resorbable 3D-printed implants because of their osteoconductivity and bone-regenerating capacity. They can be blended with biodegradable polymers to fabricate composite scaffolds which can provide mechanical inspiration as well as drug release.

2.2.1 Hydroxyapatite (HA)

HA, which stands for hydroxyapatite is a member of the calcium apatite group that is associated with bone and teeth. Moreover, it is one of the bio-friendliest materials using which one can promote bone regrowth by stimulating osteoblasts.

Advantages

- Outstanding biocompatibility.
- Promotes bone development and integration.

Drug Delivery Applications: The HA-based scaffold may be further co-localized with osteogenic drugs, and growth factors, that help in bone formation and the healing process. HA integrated with 3D-printed implants enhances bioactivity and localized drug release applications.

2.2.2 Tricalcium Phosphate (TCP)

The other bioactive ceramic commonly in bone tissue engineering is Tricalcium Phosphate (TCP). TCP shares most characteristics of HA but has differences in solubility and resorption rate and is hence used where required.

Advantages

- Decent osteoconductivity.
- Adjustable resorption rates.

Drug Delivery Applications: TCP-based implants allow for the design of controlled release drug-forming systems that promote osteogenesis or decrease inflammation levels for better clinical outcomes in repairing and regenerating bone tissue [17].

2.3 *Composite Materials*

These can have biodegradable polymer properties with the use of bioactive ceramics to produce desirable characteristics. These composites can enhance the characteristics of the mechanical reinforcement, controllable degradation rates, and the functions of drug release.

2.3.1 Polymer-Ceramic Composites

A few polymer-ceramic composites such as PLA-HA or PCL-TCP are a blend of polymers and ceramics. The polymer network is used to fabricate the base of the scaffold and to control the degradation behaviors of the scaffold and the ceramic phase is used to improve the biocompatibility and promote the formation of bone tissue.

Advantages

- Stable biological and mechanical properties.
- Tunable degradation and drug release profiles.

Drug Delivery Applications: It makes them ideal for aggressive clinical applications, as these composites can release a mix of, for example, antibiotics, anti-inflammatory drugs, and growth factors in a regulated manner [18].

2.3.2 Gradient Composites

In the scaffold of gradient composites, some areas consist of varying concentrations of polymers and ceramics; it has a gradient that can be altered to fit the required clinical application. This strategy helps in the variation and control of mechanical support as well as the drug release characteristics all around the implant.

Advantages

- Modified biological and mechanical properties.
- Aptitude to solve complex clinical needs.

Drug Delivery Applications: Another advantage of gradient composites is that one can design a multilevel drug release system where individual layers release drugs at different concentrations to treat different stages of tissue repair [19].

3 3D Printing Techniques for Drug-Eluting Implants

3D printing or additive manufacturing has become a preferred technique in the fabrication of drug-eluting implants due to the ability to produce structures with high density and well-defined characteristics. In this section, 3D printing methods employed for developing drug-releasing implant designs incorporated with the bulk material are explained with focus on Fused Deposition Modeling (FDM), Selective Laser Sintering (SLS), Stereolithography (SLA), and Digital Light Processing (DLP). Thus, each method is the best and most appropriate with its advantages and disadvantages better for aesthetics of implants and their performance (Fig. 2).

3.1 *Fused Deposition Modeling (FDM)*

Fused Deposition Modeling (FDM) is perhaps one of the most well-known 3D printing techniques, largely because it is extremely cost-efficient and has a very high applicability range [20]. One of these methods of fabrication is where a thermoplastic material in the form of a filament which when heated oozes out from a nozzle and in a layer at a time is built up into a structure.

3.1.1 Principle of Operation

FDM employs a basic material which is PLA or PCL and is melted to its softening point, later extruding through a thin nozzle when it is in a semi-liquid state [21]. The material undergoes four stages of extrusion and is placed on the build platform to form a layer-by-layer buildup of the implant [22]. The shape of the structure can

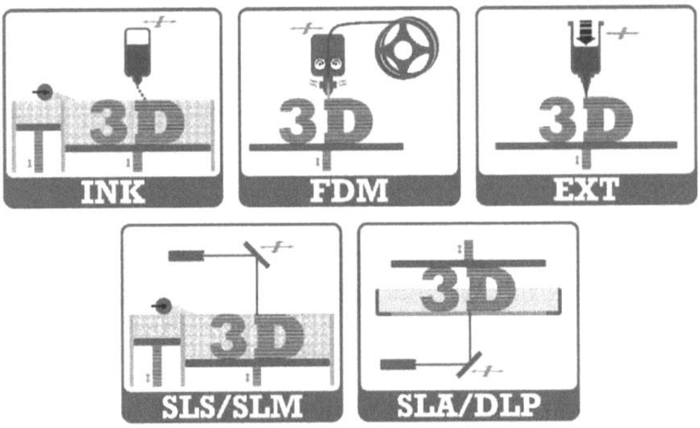

Fig. 2 Diagrammatic representation of the many printing methods. Fused deposition modeling (FDM), extrusion 3D printing (EXT), stereolithographic 3D printing/digital light projection (SLA/DLP), selective laser sintering/selective laser melting (SLS/SLM), and 3D inkjet printing (INK) (*Picture courtesy* Domsta et al. [8])

be manually regulated to intricate geometries and variability in porosity is another significant factor when using the process for drug-eluting purposes.

3.1.2 Advantages and Limitations

Advantages

- Extensive variety of materials accessible.
- Comparatively low cost and ease of use.
- Accomplished creating intricate geometries and structures.

Limitations

- Limited resolve associated with other methods.
- Likely problems with material adhesion and surface finish.
- Encounters in mixing drugs consistently through the structure.

Applications: One real-life use of FDM is in the creation of implants with a stepwise buildup of drugs that degrade at variable rates in the body. For instance, PLA-based implants can be programmed to release antibiotics or even anti-inflammatory drugs, giving localized therapy at the implantation site [23].

3.2 Selective Laser Sintering (SLS)

Selective Laser Melting is also under the family of powder bed technologies where a laser light is used to fuse layers of powder material to create a single part [24]. This method is preferable when internal structure, as well as mechanical characteristics of the implant material, are important.

3.2.1 Principle of Operation

Selectively, a laser traces and melts powdered material mainly polyamide or bioactive ceramics by the digital model in SLS [25]. Next layers of the powder are bonded together to form a solid structure layer upon layer until the complete implant is built.

3.2.2 Advantages and Limitations

Advantages

- High resolution and accuracy.
- Ability to create complex internal structures and porosities.
- Wide range of material options, including bioactive ceramics.

Limitations

- High cost of equipment and materials.
- Post-processing may be required to remove excess powder.
- Limited ability to incorporate certain drug types into the powder matrix.

Applications: Electively, a laser draws and sinters the powdered material predominantly polyamide or bioactive ceramics following the digital model in SLS. Subsequent layers of the powder are then fused and deposited in an orderly fashion to manufacture an entire implant.

3.3 Stereolithography (SLA)

Stereolithography (SLA) is a Lithographic process that polymerizes photo-sensitive materials to generate geometrically accurate and detailed three-dimensional structures [26]. In this technique, the liquid photopolymers are hardened using UV light successively layer by layer until the final implant is formed.

3.3.1 Principle of Operation

SLA is a lithographic technique that uses photo-sensitive materials to solidify three-dimensional structures with accurate geometric patterns [27]. In this case, the liquid photopolymers are cured through UV light and are brought to bear onto one another in each layer to form the final shape of the implant.

3.3.2 Advantages and Limitations

Advantages

- High surface finish and resolution.
- Skill to generate complex geometries and fine details.
- Appropriate for both composite and polymeric materials.

Limitations

- Less material selections compared to SLS and FDM.
- Advanced cost of equipment and materials.
- Challenges in achieving uniform drug dispersion within the resin.

Applications: Due to the fine planning for both the implant and the drug, SLA is useful in the construction of delicate drug-eluting implants with controlled release patterns. For example, by incorporating anti-inflammatory drugs, SLA-printed PCL scaffolds can improve tissue healing where the scaffold delivers the drug in a controlled manner [28].

3.4 *Digital Light Processing (DLP)*

It is like the second technique called the SLA uses a laser to cure the photopolymer resin; however, a digital light projector is used instead. It favors the curing of entire layers at once thus increasing the distinct advancement in printing speeds.

3.4.1 Principle of Operation

DLP deals with the use of a digital mask of each layer to be used to pour a vat of liquid photopolymer resin. Choi et al. [29] described that through the projected light, the layers of resin are solidified to build the structure. The build platform is advanced in controlled steps and the whole process is continued up to the fabrication of the final implant [30].

3.4.2 Advantages and Limitations

Advantages

- Quicker printing speeds related to SLA.
- High accuracy and resolution.
- Aptitude to generate intricate detailed structures and geometries.

Limitations

- Less material choices compared to SLS and FDM.
- Higher material costs and equipment.
- Possible matters with uniform drug integration.

4 Drug Loading and Release Mechanisms

The factors for the success of drug-eluting resorbable implants lie in the proper implantation and subsequent release of the drug dosage. In the case of these implants' drugs containing polymeric structures must be developed in such a way that the drug-loading techniques are appropriate and the drug release mechanisms planned to produce the requisite therapeutic effects [31]. This section looks at the different approaches to drug loading and drug delivery which include surface coating, encapsulation, and matrix incorporation. Both are quite close in terms of usefulness and advantages with concern to strength and weakness and are behind the functionality and usage of the Implants.

4.1 Surface Coating

The factors for the success of drug-eluting resorbable implants lie in the proper implantation and subsequent release of the drug dosage. In the case of these implants' drugs containing polymeric structures must be developed in such a way that the drug-loading techniques are appropriate and the drug release mechanisms planned to produce the requisite therapeutic effects. This section looks at the different approaches to drug loading and drug delivery which include surface coating, encapsulation, and matrix incorporation. They work very efficiently, and that is why there is usage and functionality of the implants and the abilities related to them.

4.1.1 Techniques for Surface Coating

- **Electrostatic Spray Deposition**: This technique employs the attachment of an electrostatic charge on polymers or nanoparticles laden with the drug and deposition on the implant surface. It provides a method of applied thickness and gives one control over the rate of drug release [32].
- **Sol–gel coating**: Involves the application of a sol–gel precursor with the active agent, followed by cross-linking to produce a thin layer of the active agent that elutes from the implant surface. This method is ideal when it comes to biocompatible and adherent coatings [33].

4.1.2 Advantages and Limitations

Advantages

- Easy to use and regulate the coating thickness.
- Suitable for implants with complex geometries. Complicated geometries of the implants are suitable.
- Permits for high loading of drugs on the layer of surface.

Limitations

- Limited to surface release; not suitable for implants requiring deep drug penetration.
- Coatings may be prone to wear or degradation, affecting drug release consistency.

Applications: Surface coating is often used in orthopedic implants to provide localized drug delivery, such as coating screws or plates with antibiotics to prevent infection [34].

4.2 Encapsulation

Encapsulation involves enclosing the drug within a polymeric or ceramic matrix, which is then incorporated into the implant. This technique allows for controlled and sustained release of the drug over an extended period.

4.2.1 Techniques for Encapsulation

- **Microencapsulation**: Drugs are encapsulated in micro-sized particles using techniques like spray drying or solvent evaporation. These microcapsules are then incorporated into the implant matrix [35].
- **Nanoparticle Encapsulation**: Drugs are loaded into nanoparticles, which are then mixed with the implant material. This method provides a high surface area for drug release and allows for precise control over the release kinetics [36].

4.2.2 Advantages and Limitations

Advantages

- Enables controlled and sustained drug release over a longer duration.
- Protects sensitive drugs from degradation.

Limitations

- Challenges in production and possibly a challenge in achieving a uniform distribution.
- Kinetics of release may be related to the rate of degradation of the matrix used for encapsulation.

Applications: Encapsulation is especially applied in implants for long-term conditions, like drug-eluting stents where continuous discharge of anticoagulants or anti-inflammatories is desirable [37].

4.3 Matrix Incorporation

Codding of the drug implies the mixing of the drug with the implant material which can be polymeric or ceramic before printing or fabrication. It is incorporated into the implant matrix and dispenses as the material breaks down, thereby releasing the drug.

4.3.1 Techniques for Matrix Incorporation

- **Homogeneous Mixing**: According to this, the drug is incorporated into the implant material and then printed together to minimize the variation of drug distribution within the scaffold [38].
- **Layer-by-Layer Incorporation**: The drug is incorporated into different layers of the implant during the fabrication process using micro-machining, which forms a system that will release the drug gradually [39].

4.3.2 Advantages and Limitations

Advantages

- This is important to make certain the drug is distributed evenly throughout the implant area.
- Allows for the manipulation of the release rate through the formulation of the matrix.

Limitations

- May require additional modification to improve mechanical properties and drug release characteristics.
- May need to be optimized for some reason such as enhancing mechanical characteristics or releasing the drug.

Applications: For example, in bone tissue engineering, matrix incorporation is used in the construction of a tissue-engineered scaffold that can release growth factors or antibiotics with a view of promoting bone formation or eliminating bacteria respectively [40].

4.4 Combination Approaches

However, in some instances merging of the above techniques is used to obtain the best drug release profiles and maximize treatment benefits. For instance, an implant may employ a system that combines the use of surface coating for an early release and then incorporate the drug into a matrix as a long-term release system.

4.4.1 Advantages and Limitations

Advantages

- It has the advantage of offering a desired-release characteristic of the drug to suit clinical requirements.

- It has the advantages of more than one drug delivery system.

Limitations

- Increased difficulty in fabrication and design.
- Some of the difficulties in measuring and achieving equal drug release across different mechanisms.

Applications: Such systems are helpful in cases of implants, especially for cancer therapy whereby the system delivers a higher dosage first and a sustained release to continue the therapy, [41].

5 Future Directions in Advanced 3D-Printed Drug-Eluting Resorbable Implants and Fixation Devices

In the context of the drug-eluting stent and resorbable implants and fixation devices, there are many innovations expected to be seen soon considering advancement in material science printing technology, and biomedical engineering. The future developments in this area will be directed to several aspects that may offer more advantages in the Clark intelligibility of implant such as: Included in this section is a projection and analysis of the trends observed in the growth of this progressive area of study.

5.1 Development of Advanced Materials

5.1.1 Smart Polymers and Composites

It has been noticed that the progress in smart polymers and composites particularly pose a great potential in improving the performance of drug-eluting implants. Environment sensitive smart materials could help support controlled and timed release of drugs. For example, polymers that change their properties when exposed to body temperature and will stimulate subsequent change in the structure may be used to accurately control drug release [42].

5.1.2 Biodegradable and Bioactive Materials

There will be more usage of materials biodegradable and bioactive like poly (lactic-co-glycolic acid) or PLGA composites and bioceramics. They will be able to encourage the regeneration of tissues and will also show moderate biodegradation and elimination in the body. Further research will involve the study of the improvement of the mechanical properties and degradation rates of these materials to suit different medical applications [43].

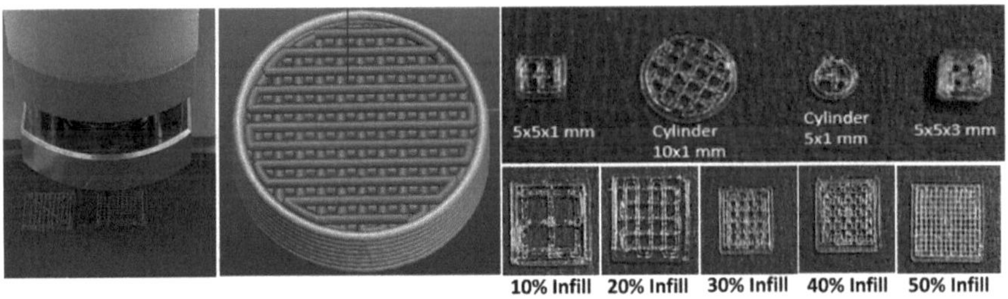

Fig. 3 Images of PLGA scaffolds that were 3D printed with varying infill densities and geometries and that eluted ketoprofen. Green and blue lines, respectively, indicate the X and Y coordinates (*Picture courtesy* Annaji et al. [46])

5.2 Innovations in 3D Printing Technologies

5.2.1 Multi-material and Multi-function Printing

New multi-material and multifunction 3D printing technologies are probably going to be used to change the drug-releasing implant design in the future. These technologies will enable the fabrication of multi-component constructs with combinations of materials having different properties within a single implant and offer structures with desired drug release characteristics and improved mechanical strength [44].

5.2.2 Enhanced Printing Resolution and Speed

Developments in the printing resolution and speeds mean the creation of more detailed and defined implants can be produced. New methods, like two-photon polymerization and micro-stereolithography, can offer better resolution and shorter fabrication time to get a Schrödinger wave in the manufacturing procedures for the underlying geometries of drug-releasing options and devices [45] (Fig. 3).

5.3 Integration of Advanced Drug Delivery Systems

5.3.1 Combination Therapies

Future implants are expected to deliver drugs and therapeutic agents in combination hence making it able to fit several drugs at once. It is most useful in diseases whose management requires a great deal of input and therefore encompasses different management plans. For example, it can easily achieve the goal of designing an implant that will contain the capacity to release an antibiotic and an anti-inflammatory agent in a certain sequence to help overcome an infection and aid healing [47].

5.3.2 Personalized Medicine

The integration of personalized medicine into 3D-printed implants is a promising direction. Advances in patient-specific data, such as genetic information and disease profiles, can guide the design of customized implants with tailored drug release profiles. This personalized approach aims to optimize treatment efficacy and minimize adverse effects [48].

5.4 Improved Understanding of Drug Release Mechanisms

5.4.1 Advanced Computational Modeling

Future works will incorporate the use of more sophisticated modeling methods in the computing process in estimating the drug release rate and the implant's degradation process. These models can be useful in identifying the performance of drug-eluting implants under various physiological states hence enabling design and optimization of drug delivery systems [49].

5.4.2 In Vivo and In Vitro Testing

More sophisticated in vivo and in vitro test procedures will be useful in establishing the real-world performance of drug-eluting implants. Concerning the evaluation of the performance of these implants on tissues and for drug delivery throughout their lifespan, organ-on-a-chip models and advanced imaging techniques will be presented [50].

5.5 Addressing Clinical and Regulatory Challenges

5.5.1 Biocompatibility and Safety

To this end, the safety of administered drugs, biocompatibility of new-generation implantable structures, and absence of adverse outcomes will remain priorities. Future work will direct to enhance the safety factor of such implants to reduce side effects and harmlessness to reside with body tissues and organs. The study will also look at potential chronic effects of implant degradation and drug elution at the site of implantation [51].

5.5.2 Regulatory and Manufacturing Standards

Standardized protocols for the manufacturing and testing of the 3D-printed drug-eluting implants will play a critical role in improving the reliability of such implants. Researchers, manufacturers, and regulatory bodies' collaboration will aid in the formulation of standard procedures as well as the enhancement of the rate at which new innovative implants will be developed and delivered to practice settings [52].

6 Conclusion

High-performance drug-eluting resorbable implants and fixation devices through 3D printing is a breakthrough in the biomedical engineering field, opening new doors for individualized and highly efficient medicine. Therefore, in showcasing the future, and current trends of the technology several factors and issues may be deduced in this field of study. Consequently, the combination of 3D printing with drug-eluting resorbable implants has allowed for the creation of sophisticated tailored medical devices capable of releasing therapeutic products at desired specific locations. These recent implants have incorporated smart polymers and bioactive composites which improved the performance of these implants particularly when releasing drugs and promoting tissue regeneration. There are many improvements in the print technology terms many of them include multi-material printing and high resolution offering better accuracy and utility in the implant design. In the future, the field will have a highly promising platform of the following trends: Responsive to external stimuli, smart materials could improve on existing drug delivery capabilities with time-sensitive and selectively therapeutic applications. The future seems to be even brighter with new advancements in multi-material 3D printing and better molding techniques in the implants, it only seems to add to the list of customizations and benefits in a future context. Applying the concept of individualized medicines and combining the treatments is going to enhance the treatment of many diseases as it should benefit the patients. Thus, it could be stated that the constant improvement of the materials used for implants, the methods for printing, and the methods for delivering the necessary drugs will be quite important in overcoming existing problems and in further development of the discussed implants' clinical applications. For the products of these advances to be easily employed in medical practice, it will be vital to solve questions of biocompatibility, safety, and regulation. All in all, the utilization of 3D-printed drug delivery resorbable implants and fixational devices is a progressive field and is evolving highly with the essential prospect of altering medical therapy. By applying progress in the fields of materials science, printing technologies, and drug delivery systems both researchers and clinicians aim to develop more efficient treatment strategies tailor-made for individual patients. These innovative medical devices will also keep on following promising new directions and solutions to current issues.

Acknowledgements The authors would like to thank Mrs. Nibedita Bardhan for language proofreading. Sudipto Datta acknowledges DST-SERB for supporting him as a National Post-doctoral Fellow.

References

1. Jain N, Waidi YO, Chowdhury S, Barua R, Das S, Prasad A, Datta S (2024) Medical devices tribology. In: Applications of biotribology in biomedical systems. Springer Nature Switzerland, Cham, pp 235–250
2. Chou PY, Lee D, Chen SH, Liao CT, Lo LJ, Liu SJ (2022) 3D-printed/electrospun bioresorbable nanofibrous drug-eluting cuboid frames for repair of alveolar bone defects. Int J Pharm 615:121497
3. Waidi YO, Barua R, Datta S (2023) Metals, polymers, ceramics, composites biomaterials used in additive manufacturing for biomedical applications. In: Modeling, characterization, and processing of smart materials. IGI Global, pp 165–184
4. Liaskoni A, Wildman RD, Roberts CJ (2021) 3D printed polymeric drug-eluting implants. Int J Pharm 597:120330
5. Naseri E, Cartmell C, Saab M, Kerr RG, Ahmadi A (2020) Development of 3d printed drug-eluting scaffolds for preventing piercing infection. Pharmaceutics 12(9):901
6. Barua R, Datta S (2023) Emerging surgical robotic applications for modern Minimally Invasive Surgery (MIS). In: Global perspectives on robotics and autonomous systems: development and applications. IGI Global, pp 314–332
7. Park JH, Ye M, Park K (2005) Biodegradable polymers for microencapsulation of drugs. Molecules 10(1):146–161
8. Domsta V, Seidlitz A (2021) 3D-printing of drug-eluting implants: an overview of the current developments described in the literature. Molecules (Basel, Switzerland) 26(13):4066
9. Kim TH, Lee JH, Ahn CB, Hong JH, Son KH, Lee JW (2019) Development of a 3D-printed drug-eluting stent for treating obstructive salivary gland disease. ACS Biomater Sci Eng 5(7):3572–3581
10. Salamanca E, Choy CS, Aung LM, Tsao TC, Wang PH, Lin WA, Wu YF, Chang WJ (2023) 3D-printed PLA scaffold with fibronectin enhances in vitro osteogenesis. Polymers 15(12):2619
11. Das S, Datta S, Barman A, Barua R (2023) Smart biodegradable and bio-based polymeric biomaterials for biomedical applications. In: Modeling, characterization, and processing of smart materials. IGI Global, pp 56–82
12. Datta S, Barua R, Sarkar R, Barui A, Chowdhury AR, Datta P (2018, September). Design and development of alginate: Poly l lysine scaffolds by 3D bio printing and studying their mechanical, structural and cell viability properties. IOP Conf Ser: Mater Sci Eng 402:012113
13. Aihemaiti P, Jia R, Aiyiti W, Jiang H, Kasimu A (2023) Study on 3D printing process of continuous polyglycolic acid fiber-reinforced polylactic acid degradable composites. Int J Bioprinting 9(4):734
14. Dos Santos J, Balbinot GS, Buchner S, Collares FM, Windbergs M, Deon M, Beck RCR (2022) 3D printed matrix solid forms: can the drug solubility and dose customisation affect their controlled release behaviour? Int J Pharma: X 5:100153
15. Huang H, Qiang L, Fan M, Liu Y, Yang A, Chang D, Li J, Sun T, Wang Y, Guo R, Zhuang H, Li X, Guo T, Wang J, Tan H, Zheng P, Weng J (2023) 3D-printed tri-element-doped hydroxyapatite/polycaprolactone composite scaffolds with antibacterial potential for osteosarcoma therapy and bone regeneration. Bioactive Mater 31:18–37. https://doi.org/10.1016/j.bioactmat.2023.07.004
16. Jiang X, Zhang X, Zeng J (2018) Preparation and characterization of HA/PCL composite scaffolds for bone tissue engineering. J Biomed Mater Res, Part A 106(7):1802–1813. https://doi.org/10.1002/jbm.a.36331

17. Liu J, Chen J, Ma L (2021) Tricalcium phosphate-based scaffolds with controlled drug release for bone regeneration: a review. Mater Sci Eng, C 124:112026. https://doi.org/10.1016/j.msec.2021.112026
18. Zhang X, Zhao W, Liu Y (2021) Advanced 3D printing technologies for drug-eluting implants in regenerative medicine. J Adv Func Mater 31(34):2008034. https://doi.org/10.1002/adfm.202008034
19. Shi J, Kantor B, Lin T (2020) Gradient drug-release systems for biomedical applications: an overview. J Control Release 326:120–138. https://doi.org/10.1016/j.jconrel.2020.03.022
20. Barua R, Datta S, Datta P, Chowdhury AR (2019) Scaffolds and tissue engineering applications by 3D bio-printing process: a new approach. In: Design, development, and optimization of bio-mechatronic engineering products. IGI Global, pp 78–99
21. Datta S, Barua R, Prasad A (2023) Additive manufacturing for the development of artificial organs. In: Advanced materials and manufacturing techniques for biomedical applications, pp 411–427
22. Barua R, Datta S, Roychowdhury A, Datta P (2021) Importance of 3D printing technology in medical fields. In: Research anthology on emerging technologies and ethical implications in human enhancement. IGI Global, pp 704–717
23. Sachs E, Cima M, Williams P (2021) The role of 3D printing in creating complex drug-eluting implants. J Manuf Process 68:155–166. https://doi.org/10.1016/j.jmapro.2021.06.018
24. Barua R, Datta S, Datta P, Roychowdhury A (2022) Study and application of machine learning methods in modern additive manufacturing processes. In: Applications of artificial intelligence in additive manufacturing. IGI Global, pp 75–95
25. Chowdhury S, Jain N, Waidi YO, Barua R, Das S, Prasad A, Datta S (2024) Biological smart biomaterials: materials for biomedical applications. In: Applications of biotribology in biomedical systems. Springer Nature Switzerland, Cham, pp. 313–325
26. Jain N, Waidi YO, Barua R, Vannaladsaysy V, Das S, Prasad A, Datta S et al (2024) 2 Bio-based polymers: processing and applications. Sustain Bio-Based Compos: Biomed Eng Appl 20:25
27. Economidou SN, Pere CPP, Reid A, Uddin MJ, Windmill JFC, Lamprou DA, Douroumis D (2019) 3D printed microneedle patches using stereolithography (SLA) for intradermal insulin delivery. Mater Sci Eng C, Mater Biol Appl 102:743–755
28. Puppi D, Piras AM, Pescarmona PP (2019) 3D printing of drug-eluting implants: a review. Pharm Res 36(12):187
29. Choi KY, Ajiteru O, Hong H, Suh YJ, Sultan MT, Lee H, Lee JS, Lee YJ, Lee OJ, Kim SH, Park CH (2023) A digital light processing 3D-printed artificial skin model and full-thickness wound models using silk fibroin bioink. Acta Biomater 164:159–174
30. Kadry H, Wadnap S, Xu C, Ahsan F (2019) Digital light processing (DLP) 3D-printing technology and photoreactive polymers in fabrication of modified-release tablets. Eur J Pharm Sci: Off J Eur Fed Pharm Sci 135:60–67
31. Bassand C, Siepmann F, Benabed L, Verin J, Freitag J, Charlon S, Soulestin J, Siepmann J (2023) 3D printed PLGA implants: how the filling density affects drug release. J Control Rel: Off J Control Rel Soc 363:1–11
32. Amin A, Wang Q, Zhang W (2022) Electrostatic spray deposition for drug delivery: principles and applications. J Control Release 339:183–199
33. Yang Y, Zheng Y, Zhang C (2021) Sol-gel coating techniques for drug-eluting medical devices. Prog Org Coat 152:106085
34. Hosseini A, Shaterzadeh S, Zhang Z (2021) Drug-eluting coatings for orthopedic implants: current trends and future perspectives. J Orthop Res 39(4):671–681
35. Zhang J, Chen R, Liu X (2020) Microencapsulation techniques for controlled drug release in tissue engineering applications. J Mater Sci—Mater Med 31(3):14
36. Lee J, Kim H, Lee J (2019) Nanoparticle-based drug delivery systems for 3D printed implants. Nanomed: Nanotechnol Biol Med 21:102133
37. Shao L, Ding Z, Zhang Y (2022) Advances in microencapsulation technologies for drug-eluting stents. Adv Mater 34(12):2108095

38. Chen H, Zhang X, Li Z (2021) Homogeneous mixing of drugs with polymeric scaffolds for controlled release applications. Biomater Sci 9(11):3807–3822
39. Sánchez M, Martínez J, Gómez J (2023) Layer-by-layer drug incorporation techniques in additive manufacturing. J Biomed Mater Res, Part A 111(2):245–257
40. Yao C, Xu X, Zhao H (2021) Matrix incorporation strategies for enhanced drug delivery in bone tissue engineering. Biomater Adv 135:112503
41. Li X, Wang J, Huang W (2022) Combination approaches in drug-eluting implants: a comprehensive review. Drug Deliv Transl Res 12(2):260–274
42. Miller K, Park S, Cho Y (2023) Smart polymers for drug delivery applications: advances and perspectives. Macromol Rapid Commun 44(5):2200894
43. Liu Y, Wang X, Zhao Y (2022) Development of biodegradable and bioactive materials for drug-eluting implants. Biomater Sci 10(2):311–326
44. Zhao Y, Liu H, Yang Z (2023) Multi-functional 3D printed implants: innovations and future directions. J Biomed Eng 77(3):115–125
45. Park J, Kim Y, Lee M (2022) Enhancements in 3D printing resolution and speed for medical applications. J Manuf Sci Eng 144(10):100917
46. Annaji M, Mita N, Poudel I, Boddu SH, Fasina O, Babu RJ (2024) Three-dimensional printing of drug-eluting implantable PLGA scaffolds for bone regeneration. Bioengineering 11(3):259
47. Kang Y, Zhang J, Lee S (2024) Combination therapies in drug-eluting implants: Strategies and challenges. Adv Drug Deliv Rev 179:114234
48. Jiang H, Xu Q, Liu T (2023) Personalized medicine in 3D printed drug-eluting implants: current status and future prospects. J Biomed Mater Res B Appl Biomater 111(4):789–801
49. Smith R, Patel V, Robinson N (2022) Computational modeling of drug release mechanisms in 3D printed implants. Biomater Res 26(4):467–480
50. Wang X, Zhang Y, Chen H (2023) Advances in in vivo and in vitro testing methods for drug-eluting implants. Biotechnol Adv 56:107863
51. Nguyen D, Tran T, Chiu H (2023) Biocompatibility and safety of advanced drug-eluting implants: current challenges and future directions. Tissue Eng Part B Rev 29(3):214–230
52. Lee J, Kim H, Smith A (2024) Regulatory and manufacturing standards for 3D printed implants: a comprehensive review. J Med Devices 18(1):015001

Chapter 5
Polymeric Hydrogel and Bioimplants

Sriparna De, Dipankar Das, Subhasish Kamilya, and Agnish Pahari

1 Introduction

1.1 Overview of Hydrogels:

Hydrogels are made of interconnecting chains of natural or synthetic polymers joined by cross-linkers to create a hydrophilic substance with the macromolecular structure of a gel. They have the capacity to grow many times their dry weight and may include as much as 99 percent water or biological fluids [1]. These three-dimensional (3D) highly hydrated porous networks are usually made to store, release, or collect materials and can replicate the natural tissue microenvironment. High intensity radiation, freeze–thaw, and chemical processes can all be used to create hydrogels. Of all the techniques, radiation—such as gamma rays, electron beams, X-rays, or ultraviolet light—is thought to be the most appropriate for creating hydrogels because it is simple to regulate the processing and eliminates the need for potentially hazardous initiators or cross-linkers. Additionally, irradiation has the potential for both sterilization and creation in a single process. Nevertheless, the hydrogels produced with this technique are not very strong mechanically. In order to create hydrogels with high strength and stability without the need for additional cross-linkers and initiators, the freeze–thaw method is now employed. In comparison with the traditional gauze therapy, the versatile applicability of a hydrogel material is quite significant in various biomedical applications. Various polymers including natural polymers such as alginate, chitosan, gelatin, and collagen and synthetic polymers, such as polyurethane, poly(ethylene glycol), polycaprolactone, poly(vinyl pyrrolidone), poly(lactide-co-glycolide), polyacrylonitrile, poly(amino acid), etc., are utilized for hydrogel matrix synthesis [2]. The process of cross-linking creates a polymer network or bulk by

S. De (✉) · D. Das · S. Kamilya · A. Pahari
Department of Allied Health Sciences, Brainware University, Kolkata, West Bengal, India
e-mail: sriparna.de2@gmail.com

A. Prasad and V. Katiyar (eds.), *Resorbable Polymers for Bioimplants and Fixation Devices*, Materials Horizons: From Nature to Nanomaterials,
https://doi.org/10.1007/978-981-96-9306-1_5

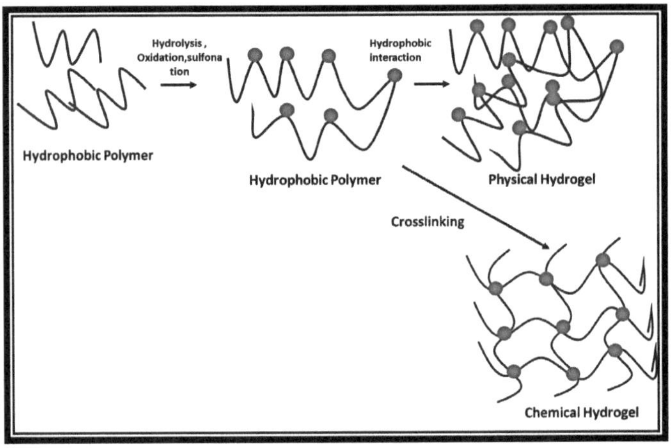

Fig. 1 Hydrogel structure

covalently bonding a linear or branched polymer chain. It is an important step in making hydrogel dressings. To stop the hydrophilic polymer chains from dissolving in an aqueous environment, cross-linking will create a spatial network. Conventional hydrogels have a fracture energy of less than 100 J m^{-2}, and their stretch ability is just a few times greater than their initial length. Figure 1 shows the structure of hydrogels. Chemical hydrogels are often distinguished by improved mechanical qualities and stability. Modern polymer chemical reactions have made it possible to create complex material systems with a wide range of uses [3]

1.2 Significance of Bioimplants

Medical devices are classified differently across the world by several regulatory bodies, each of which has its own definitions and inclusion/exclusion standards. Menstrual hygiene items, for instance, are categorized as consumer goods in India, medical devices in the United States, general products in the EU, and quasi-drugs in South Korea. Drug-eluting medical devices, including stents, syringes, contact lenses, balloon catheters, and vaginal rings laden with antiviral and contraceptive drugs, have been more popular during the last 10 years. The US FDA has classified these devices as drug-device combination products, and work is ongoing to develop appropriate rules and regulations for these combination goods. This evaluation has been organized into a number of subsections using the US Food and Drug Administration's (FDA) thorough categorization system for medical devices. PBHs, which have hydrophilicity and water-swellability, have been explored for fabricating devices, grafts, and implants. They possess properties like hydrophilicity and water-swellability, which mimic soft tissues and ensure the permeation of oxygen and nutrients. PBHs have also been explored as coatings for medical devices for controlled drug release or to inhibit microbial growth. Hydrogels have been used to create a range of flexible and wearable electronic medical devices that track strain and pressure. These gadgets have been increasingly popular in recent years and have a wide range of uses, such as bionic prostheses, body movement tracking, remote health monitoring, human–machine interfaces, and soft robotics [4, 5]. These hydrogels operate on a very straightforward principle: when the hydrogels are subjected to external stress, one or more of its inherent characteristics change. This change may be used to measure the strain or pressure that is being applied [2, 6].

2 Polymeric Hydrogels

Owing to their mechanical strength, versatile ease of shaping, and affordability of manufacture, synthetic polymers are also being employed effectively in biomedical applications. The interconnectedness of the micropores in the upper layers of both varieties of polyurethane wound dressings explains their great drainage capacity. As a result, many channels were created that enable fluids to move from the sublayer's macropores via the top layer's micropores. Additionally, the dressing stayed securely attached to the wound area and could be retained there until the skin had completely healed. A hydrogel composed of PEO for wound dressing was fabricated by amalgamating with PVA and imparting higher toughness for wound remodeling [7].

To create a steady state, the polymeric and nonpolymeric particles in a gel are distributed throughout a liquid medium and cross-linked. The fluid system's 3D cross-linked or interconnected polymeric fiber network creates the sol–gel formless structure. As a result, the materials' stiffness, strength, and stickiness are crucial for gel stability and flow rate. Over the past ten years, bio-polymers, also known as natural polymers, have gained a great deal of attention because of their hydrophilic qualities, biodegradability, safety, and compatibility.18 Three categories are used to categorize the wide range of biomaterials utilized in hydrogel preparation: (1) materials based on proteins, (2) materials based on polysaccharides, and (3) tissue that has been decellularized.

The inimitable capacity of polymer-based hydrogels, which are made of cross-linked hydrophilic polymers, to absorb and capture high volume of water and biological fluids while preserving their three-dimensional structure. The polymer-based hydrogels excel the unique characteristics features that mimic natural tissues, also its swelling capability is particularly significant, for retaining significant amounts of water and biological fluids favoring them as a potent material for biomedical applications. Because of this characteristic feature, the polymer-based hydrogels can offer a special environment that promotes vital biological processes [8, 9], including adhesion, differentiation [6, 10] and cell proliferation [10]. As a result, polymer-based hydrogels have now become prominent for use in regenerative medicine and tissue engineering [6, 9–11].

3 Bioimplants

3.1 Definition and Types of Bioimplants

Bioimplants are of paramount importance in augmenting human longevity and quality of life. They are utilized extensively in various ENT specialties with significant success. The genesis of bioimplants can be traced back four millennia, as evidenced by the use of wooden toe replacement by Egyptians and gold dental applications by Romans [12]. In modern times, they have gained significant importance in contemporary medicine and material science. They are widely used in orthopedics, dentistry, cardiovascular device, drug delivery, and skin tissue engineering. A diverse range of bioimplants has been developed to address the escalating demand for replacing for replacing and augmenting natural tissues [13]. The advancement of bioimplants manufacturing is inextricably linked to the progression of biomaterials. The field of biomaterials has experienced substantial growth since the inaugural world summit in 1969 held at Clemson University. [13] The selection of biomaterials for implantations transcends the traditional confines of hardness, elasticity, and friction. The realm of commercial bioimplants is predominantly populated by permanent metals and their alloys, including stainless steel, titanium, and Co–Cr alloys [14]. Although these bioimplants possess excellent mechanical strength, biocompatibility, and wear resistance, they can induce stress shielding due to a significant disparity in elastic modulus compared to natural bone. As a result, the bone, burdened with diminished load-bearing capacity, becomes susceptible to secondary fractures, implant loosening, and a fading of bio efficacy. Furthermore, a subsequent surgical intervention is necessary for implant removal upon fracture healing, subjecting the patient to considerable distress. Moreover, such a surgical intervention is financially burdensome. Corrosion is a major factor in selecting metals for medical implants. Corrosion can unleash a torrent of harmful substances, including allergenic, toxic/cytotoxic, or carcinogenic species into the human body. Therefore, bioimplants must undergo rigorous corrosion and solubility testing before approval.

Biomaterials are classified into several categories, including metals (e.g., stainless steel, titanium, gold, iron, magnesium), polymers(e.g., PLLA, PGA, PDS, nylon, silicon, polyester), ceramics (e.g., hydroxyapatite, alumina, zirconia), and composites which combine elements from these a forementioned categories. Moreover, biomaterials can be classified into two distinct realms: (i) biodegradable materials, ephemeral entities that dissolve into the surrounding tissue post-implantation, and (ii) nonbiodegradable materials, enduring entities that resist the passage of time [15]. The growth of the bioimplants market has led to challenges in quality and efficiency.

However, the traditional tapestry of fabrication process falters in meeting the stringent demands of bioimplants, which yearn for precision, complex structures, and novel materials [15].

3.2 Cobalt-Based Alloy

In 1936, cobalt-based alloys have recognized prominent recognition in the field of hip arthroplasty. Over the next decade, biomedical Co-based alloys are typically categorized into two groups: one is Co–Cr–Mo alloy, containing 27–30% Cr and 5–7% Mo used for biomedical implants [16]. Co–Cr–Mo alloy conjugated with ultra-high molecular weight polyethylene linings was highly utilized in the field of prosthetic knees and ankles transplantation. In combination with 33–37% Ni, 19–21% Cr, and 9–11% Mo the fabricated Co–Cr–Mo was specifically utilized for high-stress-bearing joints, such as prosthetic stems [17]. In bulk, cobalt alloys exhibit excellent biocompatibility, which is directly linked to their good corrosion resistance. It has been demonstrated that cobalt alloys are extremely resistant to corrosion, even in situations that are high in chloride. These properties are thought to be caused by the passive oxide layers that spontaneously grow on the alloy surface. In settings that are corrosive, these layers act as barriers, preventing corrosion. The increased Cr content is primarily responsible for the production of oxide layers, according to X-ray photoelectron spectroscopy (XPS) studies [18]. Various studies showed that as compared to stainless steel, cobalt cobalt-based alloys exhibit a longer lifespan and a reduced likelihood of experiencing fatigue fracture. The material is around ten times stronger than human cortical bone, with an ultimate tensile strength range from 430 to 1028 MPa and an elastic modulus of 230 GPa. Because of their exceptional mechanical qualities, cobalt alloys can be used in a variety of orthopedic applications. However, the phenomenon known as stress shielding is a significant drawback linked to the high elastic modulus. In particular, the bulk of the load may be supported by the implanted device at the location, preventing the necessary stress transfer to nearby bone tissue.

3.3 Titanium-Based Alloy

Titanium and its alloys are considered favorable biomaterials owing to their combination of low density, high strength, exceptional biocompatibility, and optimal mechanical properties. The demand for Ti-alloys in medical applications has experienced significant growth since the 1970s, and this upward trend in their utilization as bioimplant materials is anticipated to persist. Among all titanium-based products, Ti–6Al–4V is the most widely used material, constituting approximately 45% of total production. Intriguingly, although Ti–6Al–4V alloy was initially developed for aerospace applications, its favorable biocompatibility facilitated its subsequent entry into the field of biomaterials [19]. Although titanium is not a naturally occurring element in the human body, and its specific biological role remains unclear, it is unequivocally nontoxic. It has been extensively reported that titanium exhibits complete inertness within the body environment. In the event of excessive titanium within a living organism, the excess would be excreted without undergoing absorption or digestion. The inert nature of Ti and Ti-alloys renders them highly suitable for implant applications. Notwithstanding its inert nature, Ti-alloy implants have been associated with adverse effects such as osteomalacia, allergic reactions, and peripheral neuropathy, which are believed to be attributable to the release of aluminum and vanadium ions from the alloy [19, 20]. The development of a new generation of beta-titanium alloys is currently underway. Efforts have been undertaken to replace Al and V with relatively safer elements such as Mo, Ta, Nb, and Zr. Nevertheless, it should be noted that long-term follow-up investigation data are currently unavailable. As previously discussed, the biocompatibility of an implant material is highly dependent upon its corrosion resistance. Undoubtedly, Ti and Ti-based alloys are the preeminent materials in this respect. Like cobalt-based alloys, Ti-alloys have excellent corrosion resistance due to a passive oxide film. A significant distinction lies in the fact that, owing to the intrinsic properties of titanium, the matrix element, the corrosion resistance of Ti-alloys does not necessitate enhancement through alloying. The mechanical properties of Ti-based alloys are influenced by variations in the levels of interstitial and impurity elements. In another instance, the second-generation Ti-based alloy, known as the TNZT system, achieved the lowest elastic modulus among all metallic alloy biomaterials through the incorporation of novel alternative alloying elements [21]. It is noteworthy that, although the elastic modulus of Ti-alloys is lower than that of Co-based alloys, the stress shielding effect persists as a significant issue, as it can only be mitigated, not entirely eradicated. In terms of ultimate tensile stress, beta-titanium alloys exhibit values comparable to those of stainless steels, although lower than those of Co-based alloys. Hip joint simulation tests have demonstrated that Ti-alloys exhibit superior wear resistance compared to Co-alloys. However, externally applied stresses may compromise the stability of the oxide layer, resulting in the generation and subsequent release of hard oxide particles within the human body. This debris would subsequently further erode the oxide layer, leading to more severe surface damage to the implants. Consequently, Ti-based alloys are more suitably employed as components of modular constraints, rather than in articulating interfaces with other materials. Figure 2 shows the various prospects of hydrogel technology in the field of bio implants. A comparative analysis of metallic biomaterials, encompassing stainless steel, cobalt-based alloys, titanium, and Ti-based alloys, was summarized by Long and Rack. The information provided a clear perspective on why Ti and Ti-based alloys are considered the most promising metallic biomaterial to date [22].

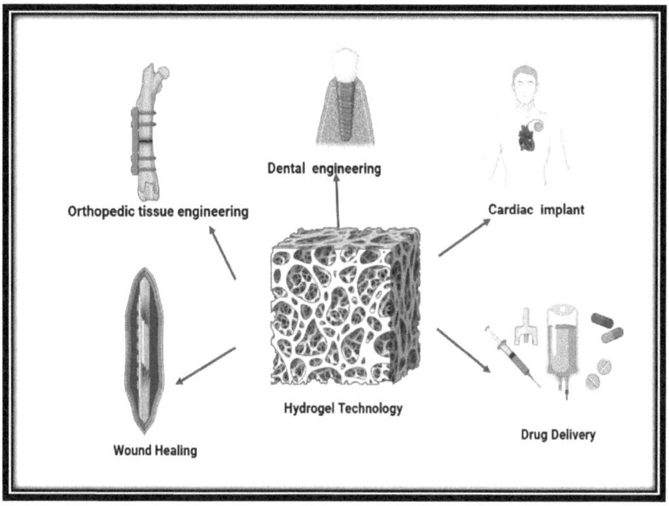

Fig. 2 Various prospects of hydrogel technology in bioimplant field

3.4 Types of Materials Used for Biological Implants or Have Different Medical

Throughout the twentieth century, materials selected for medical implants were frequently adopted from industrial sectors such as engineering and manufacturing. Surgeons usually chose materials that were already well-tested and trusted in other technical fields, assuming they would work similarly in the human body. However, the body reacts in unique ways—both inside the lab (in vitro) and inside the body itself (in vivo)—whenever any foreign material is introduced. It is a harsh and complex environment for any implant, and the materials need to go through strict testing to understand how they will behave biologically [23]. Looking at how implant materials have developed over the last 60 years, we can group their progress into three generations. The first-generation used materials that were chemically stable and did not interact much with the body—they were just meant to do their job without causing harm [24]. The second-generation introduced materials that could either dissolve safely in the body or interact positively with the surrounding tissues, such as bioactive ceramics and biodegradable polymers [25]. The third and current generation focuses on advanced materials designed to actively guide and influence the body's own cells, encouraging specific healing or regenerative responses at both the molecular and functional levels [26]. These generations reflect how our needs and understanding of materials have evolved over time, moving from simply being safe to actually helping the body heal better.

3.5 First Generation of Implants

The first generation of implants marked the beginning of using materials in the human body for medical support or replacement. These implants were made from inert materials such as stainless steel, titanium, alumina ceramics, and certain polymers. The term "inert" means that these materials did not actively interact with the body's tissues—they were chosen mainly for their mechanical strength, durability, and chemical stability, rather than for any biological compatibility[14, 22]. Their role was straightforward: to perform a function, such as supporting a broken bone or replacing a joint, without causing a harmful reaction in the body. The main goal of these implants was to restore mechanical function. For example, in hip replacements, a metal ball and stem made of stainless steel or titanium would be inserted into the thigh bone (femur) to replace the damaged joint. These parts did not bond with the bone or surrounding tissue, but they were designed to fit tightly and work reliably under pressure. Another example is the use of dental implants made from titanium, which were placed in the jawbone to hold artificial teeth [27, 28]. Although these implants did not promote healing directly, their strength and resistance to corrosion made them suitable for long-term use.

However, because the body recognizes these materials as foreign, it often responded by forming a thin layer of scar-like tissue around the implant. This response helped to isolate the implant but also prevented true integration with bone or tissue. As a result, while first-generation implants could last for many years, they sometimes failed due to wear, loosening, or inflammation over time. Still, these early materials laid the foundation for modern implant technology and inspired the search for materials that could not only replace damaged parts but also communicate with and support healing in the body [28].

3.6 Second Generation of Implants

The second generation of implants brought a big shift in how materials were used in the body. Unlike the earlier implants that were just meant to stay put and not cause harm, this new generation included materials that could actually interact with the body in helpful ways. These implants were either bioactive, meaning they could bond with living tissues, or biodegradable, meaning they would slowly break down and disappear after doing their job [29]. A good example of a bioactive material is hydroxyapatite (HA)—a substance similar to the mineral found in natural bone. When used to coat metal implants, hydroxyapatite helps bone cells grow and attach to the implant, making it more stable and better integrated with the body [30]. Another innovation was bioactive glass, which not only supports bone growth but also releases helpful ions like calcium and phosphate to speed up healing [26]. On the other hand, biodegradable polymers such as PLA (polylactic acid) and PGA (polyglycolic acid) were designed to be temporary. These materials were often used in things like dissolvable stitches or small bone pins, and once their job was done, they safely broke down into substances the body could absorb or flush out—so no second surgery was needed to remove them [31]. Overall, this generation of implants wasn't just about replacing damaged parts—it was about working with the body to heal itself. These smarter materials could guide tissue repair, encourage cells to grow, and even fade away when no longer needed. This leap in design laid the groundwork for even more advanced, interactive implants in the future.

3.7 Third Generation of Implants

The third generation of implants represents a huge leap forward in medical technology. Unlike earlier implants that either stayed in the body doing a mechanical job or helped tissue repair on the surface level, these newer implants are designed to work with the body on a much deeper level—even down to how our cells behave. These materials can actually send signals to cells, encouraging them to grow, repair, or even transform into the right type of tissue needed for healing [26]. One exciting example is in bone regeneration. Scientists have created biodegradable scaffolds—basically tiny frameworks made from special materials like calcium phosphate or advanced polymers—that can be loaded with stem cells or growth factors. When placed into a bone defect, these scaffolds guide the body's natural repair process, helping it build new bone from scratch. Over time, the scaffold itself breaks down and disappears, leaving behind healthy, functional tissue [32]. Another amazing development is the use of "smart" materials, like hydrogels or responsive polymers, which can react to changes in the body—such as temperature, pH, or enzymes. These materials can release drugs, change shape, or deliver specific signals depending on the environment they're in, offering a level of control that wasn't possible before [33]. Overall, third-generation implants are not just passive tools—they're active participants in the healing process. They reflect years of progress in both biology and materials science, with a clear goal: to help the body heal itself in the most natural, precise way possible.

4 Application of Bioimplants in Different Fields

4.1 Dental Tissue Engineering

Our teeth, intricate structures, are indispensable for the process of digestion and speech. The dental structure, including the teeth and their supporting periodontal tissues, are continually challenged by the hostile oral milieu. This triggers inflammation and destruction of the delicate dental structures, dentin, periodontal ligament (PDL), cementum, and alveolar bone. Indeed, dental caries and periodontal disease are two of the most prevalent chronic afflictions. A staggering 90% of American adults suffer from some form of tooth decay. In addition, a prevalence rate of 47% of periodontal disease was observed among the U.S. population aged 30 years and older. [34]

Bioimplants based tissue regeneration offers a revolutionary solution to repair and revitalize damaged dental structures. Bioimplants are the unsung heroes of tissue regeneration, carrying cells, providing structural support, and delivering vital bioactive molecules. These temporary scaffolds must be gentle on the body and naturally dissolve, making biocompatibility and biodegradability. Numerous types of biodegradable materials have been studied for their potential in dental tissue engineering applications [34, 35].

To mitigate the limitations associated with individual biomaterials, composite biomaterials that leverage the complementary properties of different components are increasingly utilize in dental tissue engineering. Composite bioimplants can more effectively replicate the composition of natural extracellular matrices, thereby promoting cell proliferation and tissue development. Given the composite nature of the periodontal extracellular matrix, comprising both soft and hard tissue components, the combination of polymeric and mineralized biomaterials is frequently employed to support periodontal tissue regeneration. Nevertheless, the majority of scaffold bioimplants are unable to provide the requisite biological cues to modulate cellular processes such as proliferation, chemotaxis, differentiation, and tissue development. To enhance the biological functionality of biomaterials, bioactive molecules are incorporated to regulate cellular processes and promote tissue formation. A significant proportion of bioactive molecules, such as VEGF, PDGF, BNPs, IGF, and TGF-beta, belong to the growth factor family and exert pleiotropic effects on cellular behavior. Furthermore, the incorporation of bioactive molecules, including platelet-rich plasma (PRP), platelet-rich fibrin (PRF), and enamel matrix proteins, has been utilized to augment dental tissue engineering. A primary limitations in the field of tissue engineering is the development of strategies for the spatiotemporal delivery of bioactive molecules to the defect site. Self-assembling peptides, these molecular architects, are revolutionizing dental regeneration by creating ECM- like structures that guide tissue repair [36]. These amazing amphiphilic peptides self-assembly like magic, forming hydrogels under the gentlest conditions. The introduction of oppositely charged ions or alteration of pH values initiates physical cross-linking, resulting in the formation of nanofibrous peptide hydrogels. Self-assembled peptides are the next-generation biomaterials, surpassing conventional materials with their custom-designed structures and endless possibilities for modification and improvement. Nevertheless, a significant limitation of peptide-based biomaterials is their relatively low mechanical strength, which an hinder their application in regenerative dentistry. Biomaterials employed for dental tissue engineering are fabricated into three-dimensional scaffolds possessing high porosity and sufficient mechanical stability to support cellular proliferation and differentiation. Interconnected porosity within the scaffold is essential to permit cellular infiltration and nutrient diffusion. The scaffold must degrade at the perfect pace, matching the rhythm of new tissue growth. Furthermore, the scaffold surface should possess suitable bio functional properties to facilitate cell-materials interactions. Based on their fabrication method, scaffolds can be classified as preformed scaffolds or injectable scaffolds. Preformed scaffolds are like ready-made puzzles, fitting perfectly into place, while injectable scaffolds are shape-shifting chameleons, adapting to the defect site. Injectable scaffolds are the future of regenerative medicine, offering minimally invasive delivery, shape-shifting abilities, and a welcoming environment for cell growth, surpassing traditional preformed scaffolds. Table 1 shows the summary of the commonly used biomaterial for dental applications. The complex anatomical features of dental tissues necessitate the use of injectable scaffolds, which can conform to irregular defect geometries and facilitate precise tissue regeneration.

4.2 Orthopedics

Orthopedic bioimplants play a crucial role in enhancing human quality of life by providing effective solutions for a wide range of musculoskeletal disorders and injuries. The nature of the interface between the bone and the implant significantly influences the process of osseointegration, which is essential for successful bone healing and implant stability. The appropriate surface properties of orthopedic bioimplants should be designed to modulate the differentiation of mesenchymal stem cells toward an osteogenic phenotype, thereby enhancing bone regeneration and implant integration. By modifying the surface properties of bioimplants, it is possible to control their rate of biodegradation, allowing for a tailored approach to meet specific clinical needs. Optimize the mechanical properties of the implant to ensure its long-term durability and functionality, while also promoting its biocompatibility and integration with the adjacent host tissue. Furthermore, surface modification techniques can be employed to impart antibacterial properties to the implant, thereby reducing the risk of postsurgical infections and promoting aseptic healing [41, 42]. The above-mentioned requirements for surface modifications can be applied to metallic materials, which possess inherent bulk properties suitable for orthopedic applications, to further enhance their biocompatibility, biodegradability, and antimicrobial properties. Advanced manufacturing technologies are used to produce metallic bioimplants for orthopedic applications. Bone fracture implants, which play a crucial role in promoting bone healing and restoring function, have emerged as a significant area of focus within the orthopedic bioimplant field. Among the various metallic biomaterials, stainless steel (SS 316L), titanium alloy (Ti–6Al–4V), and cobalt-chromium (Co–Cr) alloy have been extensively investigated for orthopedic applications due to their desirable bulk properties, including excellent mechanical strength, corrosion resistance, and biocompatibility. Nonmetallic materials such as alumina, nylon 6/6, and PMMA are alternative options for fracture fixation. By tailoring the mechanical properties of metallic biomaterials to better mimic those of bone, researchers aim to mitigate the adverse effects of stress shielding and promote optimal bone-implant integration, leading to improved clinical outcomes. Metallic biomaterials

Table 1 Summary of the commonly used biomaterials for dental tissue regeneration

Biomaterials	Target tissues	Characteristics	References
Hyaluronic acid	Pulp, PDL	Optimal biocompatibility, osmotic pressure control, tissue lubrication, enhanced healing, rapid degradation	[37]
Alginate	Pulp, alveolar bone	Ionic crosslinking, processability, low cell adhesivity, poor dimensional integrity	[38]
Chitosan	Pulp, alveolar bone, PDL	Exhibiting favorable biocompatibility and antibacterial efficacy, these materials necessitate integration with complementary biomaterials to optimize performance	[39]
Gelatin	Pulp, dentin, periodontal tissues	Hydrolysis product of collagen, water-soluble, readily modifiable to incorporate other bioactive molecules, possessing low mechanical strength	[40]
Collagen	Pulp, Periodontal ligament (PDL)	A primary component of the extracellular matrix, possessing excellent biocompatibility, these materials exhibit low mechanical properties and often necessitate combination with other biomaterials to augment their mechanical performance	[41]
Biphasic calcium phosphate (BCP)	Alveolar bone, cementum	A composite of hydroxyapatite and tricalcium phosphate, exhibiting a tunable degradation rate and osteoinductive properties	[42]
Bioactive glass (BG)	Bioactive glass (BG)	Exhibiting osteogenic and angiogenic activities, a tunable degradation rate, and low fracture resistance	[43]
Fibrin	Pulp, dentin	Superior biocompatibility, thrombin-inducible fibrin gel formation, growth factor and integrin binding motifs, rapid degradation kinetics	[44]
PLA, PLGA	Pulp, dentin, alveolar bone PDL	Exhibiting a controlled degradation rate, tunable mechanical strengths, and good biocompatibility, while lacking specific cell recognition sites	[43]
Hydroxyapatite (HA)	Alveolar bone, cementum	Possessing a chemical composition similar to that of bone, exhibiting osteoinductive properties, and capable of direct bonding with natural bone, while degrading at a slow rate	[45]

are ideal for orthopedic applications due to their excellent biocompatibility, hemocompatibility, and fatigue strength. However, these bioimplants are susceptible to wear and corrosion when exposed to the complex biological environment of blood and tissue, which can lead to the release of metal ions and debris, potentially causing adverse local and systemic effects. The corrosion and infection susceptibility of metallic bioimplants can result in the release of metal ions into the surrounding tissues, which can elicit immune responses. This may necessitate the administration of immunosuppressive therapy to mitigate these adverse effects, potentially leading to complications and increased healthcare costs. The limitations of metallic bioimplants include poor tissue integration, infection risk, and the possibility of revision surgery. To optimize the biological performance of orthopedic implants, it is essential to modify their surfaces with bioactive nanomaterials. These nanomaterials can create a favorable microenvironment for cell growth, differentiation, and tissue regeneration, ultimately improving implant integration and long-term clinical outcomes. Surface engineering with nanomaterials improves implant osseointegration. Surface treatment strategies for orthopedic implants have been commercialized: A variety of surface treatment techniques, including grit blasting, acid etching, anodization, laser texturing, chemical etching, and plasma treatment, have been employed to modify the surface properties of orthopedic metallic bioimplants, enhancing their biocompatibility, osseointegration, and overall clinical performance. A variety of surface treatments have been commercialized for orthopedic implants, including Ti Unite, SLA, microchanneled, grit-blasted, and Nano Tite surfaces. These treatments enhance implant osseointegration and clinical performance. Nanomaterial-based grit blasting and acid washing are used to modify implant surfaces. Strong acids, such as hydrofluoric, nitric, and sulfuric acid, are employed in acid-etching treatments to create micro- and nanoscale surface features on bioimplants. These surface modifications can promote cell-material interactions, stimulate bone growth, and enhance the long-term stability of the implant. A variety of surface modification techniques, such as dip coating, laser treatment, surface patterning, thermal spraying, biomimetic deposition, and sol–gel deposition, have been explored to improve the biocompatibility, osseointegration, and overall performance of metallic bioimplants. Many of these surface treatments have been clinically approved [38].

4.3 Cardiology

In this medical discipline, biodegradable stents and heart valves are of specific interest. A stent is a cylindrical, mesh-like device employed as an endovascular prosthesis to preserve the patency of dilated blood vessels. Stents were developed to address the acute complications of percutaneous transluminal coronary angioplasty, including dissection, thrombosis, and elastic recoil. Subsequently, stents became the cornerstone of percutaneous coronary intervention (PCI) owing to a reduction in restenosis rates and favorable clinical outcomes in follow-up studies. Three primary stent technologies are utilized in clinical practice: bare-metal stents

(BMS), drug-eluting stents (DES), and bioresorbable vascular scaffolds (BVS) [46, 47]. In response to the limitations of drug-eluting stents, particularly the potential for long-term adverse events associated with permanent metallic implants, bioresorbable vascular scaffolds were developed. Several theoretical advantages arose, such as the elimination of the metallic foreign body to mitigate inflammation and neo atherosclerosis, thereby restoring vasomotor function and facilitating future grafting or repeated revascularization procedures. Bioresorbable products offer the advantage of providing temporary structural support to vessels during the initial healing phase, followed by controlled degradation and release of the vessel. The poly-L-lactic acid-based Absorb scaffold was the first bioresorbable vascular scaffold to be widely evaluated in large-scale, multicenter clinical trials. However, studies demonstrated that, in comparison with market-leading metal stents, the utilization of Absorb BVS resulted in a heightened risk of stent thrombosis. Subsequent to the discontinuation of the Absorb BVS clinical trials, other bioresorbable scaffold platforms have garnered increased interest. Among these, the magnesium-based bioresorbable vascular scaffold, Magmaris, has gained significant attention, while other platforms are currently in development or early clinical evaluation. The development of novel biomaterials, such as desaminotyrosine polycarbonate, salicylates, ϵ-caprolactone, and poly(d,l-lactide-co-glycolide), is ongoing for these applications. The primary impetus for advanced heart valve tissue engineering is to provide solutions for the increasing number of patients requiring valve replacement procedures. A diverse range of biomaterials, encompassing biological, synthetic, and hybrid materials, can be employed to fabricate scaffolds that may promote host tissue regeneration postimplantation. These scaffolds can facilitate tissue engineering by providing a three-dimensional matrix for cellular attachment, proliferation, and differentiation. Autologous cells are often harvested from the patient and seeded onto the scaffold prior to implantation. These scaffolds are typically presented in the form of durable shapes or hydrogels. Porous scaffolds can be fabricated via 3D printing techniques utilizing various polymeric materials, including polyglycolic acid (PGA), polylactic acid (PLA), and collagen. The research aimed to identify a biomaterial suitable for 3D printing and subsequent cell seeding, meeting the necessary criteria for tissue engineering applications. Hydrogel scaffolds are synthesized by crosslinking hydrophilic polymers, often through free radical polymerization. Hydrogels offer the advantage of a high water content, which permits the facile penetration of nutrients. It was postulated that vascular prostheses could be engineered from biomaterials such as collagen, elastin, gelatin, fibrinogen, or silk fibroin, or their combinations with synthetic polymers [46].

5 Applications of Polymeric Hydrogels in Bioimplants

5.1 Drug Delivery Systems

Drug delivery systems or DDS, such as hydrogels and nanoparticles (NPs), have been used extensively to address problems like their limited ability to reach the target tissues and minimal bioavailability, and solubility of drugs. Additionally, by decreasing undesirable side effects generated by pharmaceuticals' uneven distribution throughout various tissues, the DDS increases the therapeutic effectiveness of medications and particularly for hydrophobic medications and macromolecule cargos, hydrogels provide significant benefits for drug delivery [48, 49]. A number of studies have been published in the scientific literature that address the immediate, long-term systemic, and local effects that occur after the implantation of polymers comprising lactide and glycolide. There are many orthopedic applications of lactide and glycolide, and the FDA-approved such copolymer is the poly lactide-co-glycolide copolymer or PLGA. Drug delivery techniques for controlled release have extensively used PLGA. Both local and systemic adverse consequences were not observed in microspheres containing bioactive substances, and hydrolytic chain breakage is the degradation mechanism in this instance. Interestingly, the response to the biodegradable polymer differs depending on the various tissues not only the nanoparticle size; as a result, the toxicological assessment is only based on the application [50]. PLGA with hydrogel of poly (ethylene glycol) diacrylate and hyaluronic acid combination for paclitaxel drug is applied effectively in lung cancer to inhibit tumor growth [48, 50]. Ethyl cellulose nanoparticles can be used to make nanoparticle loaded hydrogel for carboplatin drug delivery that can treat numerous cancers namely lung, colon, breast, prostate, etc. The carboplatin loaded ethyl cellulose nanoparticle are added to a chitosan-based thermosensitive hydrogel for effective drug delivery with minimal toxicity due to controlled release of the drug to the target site. It is observed from various studies that the efficacy and pharmacokinetics of nanoparticles loaded hydrogel (NLH) for the carboplatin drug delivery process is more than that of the only NPs with lesser systemic toxicity [50].

According to the study of Gao et al. [51], a thermosensitive hydrogel composed of the Pluronic F-127 polymer can be used effectively for intertumoral administration. Poly(ε-caprolactone)-poly(ethylene glycol)-poly(ε-caprolactone) nanoparticle with norcantharidin and doxorubicin are co-loaded within hydrogel of Pluronic F-127 and it increases solubility and overall administration of drug [51]. Another effective drug delivery system can be achieved from Pluronic F127, the anticancer activity of Deguelin drug can be uplifted if the NP coated deguelin is delivered to the bladder cancer site with Pluronic F127 hydrogel [48]. In case of breast cancer treatment, siRNA delivery using oligopeptide-terminated poly(β-aminoester) (pBAE) nanoparticles is effective but rapid degradation of NPs limits the application. pBAE with siRNA for breast cancer gene therapy loaded with polyamidoamine hydrogen and cross-linked dextran aldehyde can increase the stability of the gene-loaded drug [52]. The delivery of siRNA with NLH improved human breast cancer cells' capacity

Table 2 Name and function of various hydrogel [54]

Hydrogel product	Chemical used to make hydrogel	Use of the product
Biotene product	Carbomer and hydroxyethyl cellulose	To treat dryness of the mouth
Simpurity™ hydrogel	Olyethylene oxide (PEO), polyurethane, polyvinyl alcohol (PVA), acrylate	To heal wound
Nicorette® hydrogel	Hydroxypropyl methylcellulose	To help in nicotine quitting process
Neutrogena® Hydro Boost® hydrogel	Hyaluronic acid	As moisturizing agent
Suprasorb® G hydrogel	Acrylic polymers, polyethylene (PE), and phenoxyethanol	To heal wound

to silence genes because it possesses a better transfection activity than hydrogels containing free siRNA or siRNA with NPs in solution [48]. According to the research of Brachi et al., it was suggested that Poloxamer 407 thermosensitive hydrogel containing polyurethane (PUR) nanoparticles has been developed for delivery to glioblastoma tumors [53]. The bioavailability of the drug zinc phthalocyanine (ZnPC) is very low due to the hydrophobicity of the drug, and as a result, it shows a poor range of solubility. To promote suitable drug delivery as well as increase therapeutic efficacy a thermal-sensitive PCL-PTSUO-PEG hydrogel can be used and high loading capacity can be reached [49]. Oral moisturizing product Biotene contains carbomer and hydroxyethyl cellulose hydrogels. Hydrogels can retain water within the 3D structure and promote prolonged release of water to solve the dry mouth problem [9, 49]. Table 2 depicts the various hydrogel products and their chemical constituents with versatile application avenues.

5.2 *Tissue Engineering Scaffolds and Wound Healing Applications*

Scaffolds made of bioactive hydrogel are being researched extensively as a possible way to direct tissue regeneration. Hydrogel can also eliminate wound exudate, create a moist environment, and promote and direct tissue regeneration. It is one of the most often used sectors in wound dressing. The development of a composite collagen-based hydrogel scaffold with bFGF-loaded PLGA microspheres proved successful in accelerating the healing of wounds. In order to treat an infected and dry wound, researchers placed antibacterial silver or medications for example commercial silvasorb into hydrogel for wound healing. Hydrophilic domains of polymeric hydrogels support mechanical stability as well as provide proper nutrition for the development of the endogenous cell [49, 55]. Remarkably, the vascularization process might be aided by the application of hydrogels. A biomimetic PEG-based hydrogel could create complex networks with capillary-like structures that encourage the development of functioning blood vessels indicating the strong pro-angiogenesis potential for other tissue engineering uses. Many studies have revealed that the d-MAP scaffold shows a quicker rate of disintegration in vivo when incorporated within the hydrogel, which would be advantageous for IL-33 + type 2 myeloid cell activation. In addition, it triggers wound healing as it enhances tensile strength around the wound and promotes hair neogenesis [49]. Hydrogels made of two or more molecularly interlocked networks of polymers are known as interpenetrating network (IPN) hydrogels. The bridging of the polymer networks boosts the IPN hydrogel's structure. Their excellent mechanical qualities and biocompatibility facilitate cell proliferation and tissue regeneration, thus explaining the reason that they are widely utilized in tissue engineering. IPN hydrogels provide a moist and suitable environment for tissue repair, which has demonstrated potential in wound healing applications. Hydrogels based on neutral polymers are usually created via crosslinking polymerization of monomers like vinyl, methacrylates, or acrylates. The mechanical characteristics of neutral polymer-based hydrogels, including their stiffness or elasticity, can be modified to meet the needs of the desired tissue or application. Table 3 explores the various hydrogel types and its characteristic features in various application areas. Neutral hydrogels can serve as a scaffold for tissue regeneration and cell proliferation in tissue engineering by simulating the extracellular matrix just like a natural extracellular matrix and it can encourage healing and tissue regeneration, they can be applied to wounds to provide a moist, protected environment. Cationic and anionic polymer-based hydrogels have been used in tissue engineering and wound healing as cellular adhesion and tissue engineering are aided by interactions with negatively or positively charged biomolecules [54]. Alginate with RGD or laminin is proven a good hydrogel to resolve the problem of peri-implantitis as it is designed with Aggregatibacter actinomycetemcomitans [55, 56]. Alginate hydrogels functioned as an appropriate framework to promote dental regeneration and disperse growth factors, enhancing the natural ability of the tooth pulp to regenerate. Cyclodextrin hydrogel can act as dental tissue regeneration material because of low biotoxicity and good biocompatibility. Studies show that the use of gums in tissue engineering and the pharmaceutical business has increased dramatically. Its ideal characteristics, such as biological activity, biodegradability, and water solubility, make it a preferred scaffolding material. Gellan gum hydrogels that are cross-linked with carbodiimide induce the vacuolation of human tooth-derived stem cells in vitro. According to many studies, PEG hydrogel membranes were just as successful as conventional collagen membranes in addressing bone problems surrounding dental implants [55, 57].

Table 3 Hydrogel in biomedical application [58]

Applied area	Hydrogel type	Characteristic features	References
Drug delivery system (DDS)	Stimulating, collagen-based, peptide in nature, nanofiller-enhanced, and molecularly imprinted polymer (MIP) based	Wound healing, soft gel capsule	[59]
Ophthalmology	Methacrylate	Eye drop, contact lens	[60]
Cardiology	Enhanced with nanofiller, stimulating, oscillating in itself	Implantation	[61]
Dental	Collagen based and peptide in nature	Ointment, implantation, wound dressing	[58]
Cosmetic and reconstructive surgery	Hyaluronic acid, methacrylate	Implanting a transdermal part	[62]
Tissue culture	Collagen, gelatin	Primary support for cells to adhere and regeneration of tissue	[63]

References

1. Buenger D, Topuz F, Groll J (2012) Hydrogels in sensing applications. Prog Polym Sci 37(12):1678–1719. https://doi.org/10.1016/j.progpolymsci.2012.09.001
2. Dimatteo R, Darling NJ, Segura T (2018) In situ forming injectable hydrogels for drug delivery and wound repair. Adv Drug Deliv Rev 127:167–184. https://doi.org/10.1016/j.addr.2018.03.007
3. Sershen S, West J (2002) Implantable, polymeric systems for modulated drug delivery. Adv Drug Deliv Rev 54(9):1225–1235. https://doi.org/10.1016/S0169-409X(02)00090-X
4. Wahid F, Zhao X-J, Jia S-R, Bai H, Zhong C (2020) Nanocomposite hydrogels as multifunctional systems for biomedical applications: current state and perspectives. Compos B Eng 200:108208. https://doi.org/10.1016/j.compositesb.2020.108208
5. Mondal P, Chatterjee K (2022) Injectable and self-healing double network polysaccharide hydrogel as a minimally-invasive delivery platform. Carbohydr Polym 291:119585. https://doi.org/10.1016/j.carbpol.2022.119585
6. Cai J, Guo J, Wang S (2023) Application of polymer hydrogels in the prevention of postoperative adhesion: a review. Gels 9(2):98. https://doi.org/10.3390/gels9020098
7. Zhang W et al (2024) Hydrogel-based dressings designed to facilitate wound healing. Mater Adv 5(4):1364–1394. https://doi.org/10.1039/D3MA00682D
8. Günay KA et al (2023) Photo-expansion microscopy enables super-resolution imaging of cells embedded in 3D hydrogels. Nat Mater 22(6):777–785. https://doi.org/10.1038/s41563-023-01558-5
9. Jacob S, Nair AB, Shah J, Sreeharsha N, Gupta S, Shinu P (2021) Emerging role of hydrogels in drug delivery systems, tissue engineering and wound management. Pharmaceutics 13(3):357. https://doi.org/10.3390/pharmaceutics13030357
10. Winblade ND, Nikolic ID, Hoffman AS, Hubbell JA (2000) Blocking adhesion to cell and tissue surfaces by the chemisorption of a Poly-l-lysine-graft-(poly(ethylene glycol); phenylboronic acid) copolymer. Biomacromol 1(4):523–533. https://doi.org/10.1021/bm000040v

11. Xue X, Hu Y, Wang S, Chen X, Jiang Y, Su J (2022) Fabrication of physical and chemical crosslinked hydrogels for bone tissue engineering. Bioact Mater 12:327–339. https://doi.org/10.1016/j.bioactmat.2021.10.029
12. Manivasagam G, Dhinasekaran D, Rajamanickam A (2010) Biomedical implants: corrosion and its prevention—a review. Recent Patents Corros Sci 2(1):40–54. https://doi.org/10.2174/1877610801002010040
13. Luthringer BJC, Feyerabend F, Willumeit-Römer R (2014) Magnesium-based implants: a mini-review. Magnes Res 27(4):142–154. https://doi.org/10.1684/mrh.2015.0375
14. Geetha M, Singh AK, Asokamani R, Gogia AK (2009) Ti based biomaterials, the ultimate choice for orthopaedic implants—a review. Prog Mater Sci 54(3):397–425. https://doi.org/10.1016/j.pmatsci.2008.06.004
15. Yin Yee Chin P, Cheok Q, Glowacz A, Caesarendra W (2020) A review of in-vivo and in-vitro real-time corrosion monitoring systems of biodegradable metal implants. Appl Sci 10(9):3141. https://doi.org/10.3390/app10093141
16. Hussain O, Saleem S, Ahmad B (2019) Implant materials for knee and hip joint replacement: a review from the tribological perspective. IOP Conf Ser Mater Sci Eng 561(1):012007. https://doi.org/10.1088/1757-899X/561/1/012007
17. Spoială A, Ilie C-I, Ficai D, Ficai A (2023) Biomaterials, pp 97–129. https://doi.org/10.1007/978-3-031-35832-6_4
18. Öztürk O, Türkan U, Eroglu AE (2006) Metal ion release from nitrogen ion implanted CoCrMo orthopedic implant material. Surf Coat Technol 200(20–21):5687–5697. https://doi.org/10.1016/j.surfcoat.2005.08.113
19. Yaghoubi S, Schwietert CW, Mccue JP (2000) Biological roles of titanium. Biol Trace Elem Res 78(1–3):205–218. https://doi.org/10.1385/BTER:78:1-3:205
20. Okazaki Y (2002) Effect of friction on anodic polarization properties of metallic biomaterials. Biomaterials 23(9):2071–2077. https://doi.org/10.1016/S0142-9612(01)00337-4
21. Niinomi M (2008) Biologically and mechanically biocompatible titanium alloys. Mater Trans 49(10):2170–2178. https://doi.org/10.2320/matertrans.L-MRA2008828
22. Long M, Rack HJ (1998) Titanium alloys in total joint replacement—a materials science perspective. Biomaterials 19(18):1621–1639. https://doi.org/10.1016/S0142-9612(97)00146-4
23. Wagner WR (2020) The biomaterials literature In: Biomaterials science. Elsevier, pp 1533–1536. https://doi.org/10.1016/B978-0-12-816137-1.15004-4
24. Williams DF (2008) On the mechanisms of biocompatibility. Biomaterials 29(20):2941–2953. https://doi.org/10.1016/j.biomaterials.2008.04.023
25. Batool M, Abid MA, Javed T, Haider MN (2025) Applications of biodegradable polymers and ceramics for bone regeneration: a mini-review. Int J Polym Mater Polym Biomater 74(1):39–53. https://doi.org/10.1080/00914037.2024.2314601
26. Hench LL, Polak JM (2002) Third-generation biomedical materials. Science 295(5557):1014–1017. https://doi.org/10.1126/science.1067404
27. Sidambe A (2014) Biocompatibility of advanced manufactured titanium implants—a review. Materials 7(12):8168–8188. https://doi.org/10.3390/ma7128168
28. Bose S, Roy M, Bandyopadhyay A (2012) Recent advances in bone tissue engineering scaffolds. Trends Biotechnol 30(10):546–554. https://doi.org/10.1016/j.tibtech.2012.07.005
29. Alizadeh-Osgouei M, Li Y, Wen C (2019) A comprehensive review of biodegradable synthetic polymer-ceramic composites and their manufacture for biomedical applications. Bioact Mater 4:22–36. https://doi.org/10.1016/j.bioactmat.2018.11.003
30. LeGeros RZ (2002) Properties of osteoconductive biomaterials: calcium phosphates. Clin Orthop Relat Res 395:81–98. https://doi.org/10.1097/00003086-200202000-00009
31. Middleton JC, Tipton AJ (2000) Synthetic biodegradable polymers as orthopedic devices. Biomaterials 21(23):2335–2346. https://doi.org/10.1016/S0142-9612(00)00101-0
32. Langer R, Vacanti JP (1993) Tissue engineering. Science 260(5110):920–926. https://doi.org/10.1126/science.8493529
33. Place ES, George JH, Williams CK, Stevens MM (2009) Synthetic polymer scaffolds for tissue engineering. Chem Soc Rev 38(4):1139. https://doi.org/10.1039/b811392k

34. Jensen J et al (2015) Functionalization of polycaprolactone scaffolds with hyaluronic acid and β-TCP facilitates migration and osteogenic differentiation of human dental pulp stem cells in vitro. Tissue Eng Part A 21(3–4):729–739. https://doi.org/10.1089/ten.tea.2014.0177
35. Chen G et al (2015) Combination of aligned PLGA/Gelatin electrospun sheets, native dental pulp extracellular matrix and treated dentin matrix as substrates for tooth root regeneration. Biomaterials 52:56–70. https://doi.org/10.1016/j.biomaterials.2015.02.011
36. Sumayli A (2021) Recent trends on bioimplant materials: a review. Mater Today Proc 46:2726–2731. https://doi.org/10.1016/j.matpr.2021.02.395
37. Areevijit K, Dhanesuan N, Luckanagul JA, Rungsiyanont S (2021) Biocompatibility study of modified injectable hyaluronic acid hydrogel with mannitol/BSA to alveolar bone cells. J Biomater Appl 35(10):1294–1303. https://doi.org/10.1177/0885328220971746
38. Zhang R et al (2020) Alginate/laponite hydrogel microspheres co-encapsulating dental pulp stem cells and VEGF for endodontic regeneration. Acta Biomater 113:305–316. https://doi.org/10.1016/j.actbio.2020.07.012
39. Liao F et al (2010) A novel bioactive three-dimensional β-tricalcium phosphate/chitosan scaffold for periodontal tissue engineering. J Mater Sci Mater Med 21(2):489–496. https://doi.org/10.1007/s10856-009-3931-x
40. Li Q, Hu Z, Liang Y, Xu C, Hong Y, Liu X (2021) Multifunctional peptide-conjugated nanocarriers for pulp regeneration in a full-length human tooth root. Acta Biomater 127:252–265. https://doi.org/10.1016/j.actbio.2021.03.059
41. Lausch AJ, Chong LC, Uludag H, Sone ED (2018) Multiphasic collagen scaffolds for engineered tissue interfaces. Adv Funct Mater 28(48). https://doi.org/10.1002/adfm.201804730
42. Uchikawa E et al (2021) Tooth transplantation with a β-tricalcium phosphate scaffold accelerates bone formation and periodontal tissue regeneration. Oral Dis 27(5):1226–1237. https://doi.org/10.1111/odi.13634
43. Zhang J, Chen Y, Xu J, Wang J, Li C, Wang L (2018) Tissue engineering using 3D printed nanobioactive glass loaded with NELL1 gene for repairing alveolar bone defects. Regen Biomater 5(4):213–220. https://doi.org/10.1093/rb/rby015
44. Huang GT (2009) Pulp and dentin tissue engineering and regeneration: current progress. Regenerative Med 4(5):697–707. https://doi.org/10.2217/rme.09.45
45. Huang GT-J et al (2010) Stem/progenitor cell-mediated de novo regeneration of dental pulp with newly deposited continuous layer of dentin in an in vivo model. Tissue Eng Part A 16(2):605–615. https://doi.org/10.1089/ten.tea.2009.0518
46. Dobrzyńska-Mizera M et al (2024) Engineering of bioresorbable polymers for tissue engineering and drug delivery applications. Adv Healthc Mater 13(30). https://doi.org/10.1002/adhm.202401674
47. Wallbrecher R et al (2017) Membrane permeation of arginine-rich cell-penetrating peptides independent of transmembrane potential as a function of lipid composition and membrane fluidity. J Control Release 256:68–78. https://doi.org/10.1016/j.jconrel.2017.04.013
48. Nunes D, Andrade S, Ramalho MJ, Loureiro JA, Pereira MC (2022) Polymeric nanoparticles-loaded hydrogels for biomedical applications: a systematic review on in vivo findings. Polymers (Basel) 14(5):1010. https://doi.org/10.3390/polym14051010
49. Cao H, Duan L, Zhang Y, Cao J, Zhang K (2021) Current hydrogel advances in physicochemical and biological response-driven biomedical application diversity. Signal Transduct Target Ther 6(1):426. https://doi.org/10.1038/s41392-021-00830-x
50. Pappalardo D, Mathisen T, Finne-Wistrand A (2019) Biocompatibility of resorbable polymers: a historical perspective and framework for the future. Biomacromol 20(4):1465–1477. https://doi.org/10.1021/acs.biomac.9b00159
51. Gao B et al (2021) Intratumoral administration of thermosensitive hydrogel co-loaded with norcantharidin nanoparticles and doxorubicin for the treatment of hepatocellular carcinoma. Int J Nanomedicine 16:4073–4085. https://doi.org/10.2147/IJN.S308057
52. Segovia N, Pont M, Oliva N, Ramos V, Borrós S, Artzi N (2015) Hydrogel doped with nanoparticles for local sustained release of siRNA in breast cancer. Adv Healthc Mater 4(2):271–280. https://doi.org/10.1002/adhm.201400235

53. Brachi G et al (2020) Intratumoral injection of hydrogel-embedded nanoparticles enhances retention in glioblastoma. Nanoscale 12(46):23838–23850. https://doi.org/10.1039/D0NR05053A
54. Thang NH, Chien TB, Cuong DX (2023) Polymer-based hydrogels applied in drug delivery: an overview. Gels 9(7):523. https://doi.org/10.3390/gels9070523
55. Atia GAN et al (2023) New challenges and prospective applications of three-dimensional bioactive polymeric hydrogels in oral and craniofacial tissue engineering: a narrative review. Pharmaceuticals 16(5):702. https://doi.org/10.3390/ph16050702
56. Devillard R et al (2017) In vitro assessment of a collagen/alginate composite scaffold for regenerative endodontics. Int Endod J 50(1):48–57. https://doi.org/10.1111/iej.12591
57. Sarkar K et al (2023) Nanotherapeutic potential of antibacterial folic acid-functionalized nanoceria for wound-healing applications. Nanomedicine 18(2):109–123. https://doi.org/10.2217/nnm-2022-0233
58. El Sayed MM (2023) Production of polymer hydrogel composites and their applications. J Polym Environ 31(7):2855–2879. https://doi.org/10.1007/s10924-023-02796-z
59. Misal RS, Potphode VR, Mahajan VR (2017) Review on: new approaches in self micro-emulsifying drug delivery system. Res J Pharm Technol 10(4):1215. https://doi.org/10.5958/0974-360X.2017.00218.9
60. Xinming L et al (2008) Polymeric hydrogels for novel contact lens-based ophthalmic drug delivery systems: a review. Cont Lens Anterior Eye 31(2):57–64. https://doi.org/10.1016/j.clae.2007.09.002
61. Paul A et al (2014) Injectable graphene oxide/hydrogel-based angiogenic gene delivery system for vasculogenesis and cardiac repair. ACS Nano 8(8):8050–8062. https://doi.org/10.1021/nn5020787
62. Pei M, Jia X, Zhao X, Li J, Liu P (2018) Alginate-based cancer-associated, stimuli-driven and turn-on theranostic prodrug nanogel for cancer detection and treatment. Carbohydr Polym 183:131–139. https://doi.org/10.1016/j.carbpol.2017.12.013
63. Bullo S, Hussein MZ (2015) Inorganic nanolayers: structure, preparation, and biomedical applications. Int J Nanomed:5609. https://doi.org/10.2147/IJN.S72330

Chapter 6
PLA-Based Composites for Tissue Engineering

M. Vishnuvarthanan, Arbind Prasad, and V. Subha

1 Introduction to PLA-Based Composites in Tissue Engineering

Polylactic acid (PLA), a biodegradable and biocompatible polymer derived from renewable sources, has become a cornerstone material in tissue engineering [1]. Its favourable properties such as ease of processing, tunable degradation rate and mechanical strength, make it suitable for creating scaffolds and implants [2]. However, the inherent limitations of PLA, including its brittleness and lack of bioactivity, necessitate the development of composites. With the incorporation of various reinforcements such as bioceramics, natural fibres and nanomaterials, PLA-based composites exhibit enhanced properties tailored for specific tissue engineering applications [3]. These composites support cell adhesion, proliferation and tissue regeneration, positioning PLA as a promising candidate for advanced biomedical applications in regenerative medicine.

M. Vishnuvarthanan (✉)
Department of Mechanical Engineering, Kalasalingam Academy of Research and Education, Srivilliputhur, Tamil Nadu, India
e-mail: vishnuvarthanan.india@gmail.com

A. Prasad
Mechanical Engineering Department, Katihar Engineering College (Under Department of Science, Technology and Technical Education, Government of Bihar), Katihar, Bihar, India

V. Subha
Department of Chemical and Biological Engineering, University of Sheffield, Sheffield, UK

A. Prasad and V. Katiyar (eds.), *Resorbable Polymers for Bioimplants and Fixation Devices*, Materials Horizons: From Nature to Nanomaterials,
https://doi.org/10.1007/978-981-96-9306-1_6

1.1 Importance of Resorbable Polymers

Resorbable polymers have revolutionized the biomedical field by offering materials that naturally degrade within the body, eliminating the need for secondary surgeries to remove implants [4]. These polymers, which are designed to break down into biocompatible by-products, are particularly valuable in applications such as tissue engineering, drug delivery systems and bioimplants. Their ability to integrate seamlessly with biological systems while supporting tissue regeneration underscores their importance in modern medicine. One of the primary advantages of resorbable polymers is their tailored degradation rates, which can be engineered to match the healing process of the tissue they support [5]. This adaptability ensures the gradual transfer of load to regenerating tissues, preventing stress shielding and enhancing recovery outcomes. Polymers such as polylactic acid (PLA), polyglycolic acid (PGA) and their copolymers are widely used because of their predictable degradation, mechanical strength and proven biocompatibility [6]. In tissue engineering, resorbable polymers provide scaffolds that mimic the extracellular matrix, guiding cell attachment, proliferation and differentiation [7]. The versatility of these materials allows for modification through blending, copolymerization or reinforcement with bioactive materials, further improving their performance. Moreover, in drug delivery, these polymers act as carriers that degrade to release therapeutic agents in a controlled manner, enhancing treatment efficacy while minimizing side effects.

The importance of resorbable polymers extends to environmental sustainability as well. The derivation of these materials from renewable sources, coupled with their biodegradable nature, makes them eco-friendly alternatives to conventional materials in healthcare [8]. The challenges such as limited mechanical properties, complex manufacturing processes and continuous advancements in polymer science are addressing these limitations, unlocking new possibilities for their applications. Resorbable polymers represent a critical component of biomaterial innovation, offering unique advantages for tissue regeneration, medical device design and sustainable healthcare solutions.

1.2 Role of PLA in Biomedical Applications

Polylactic acid (PLA), a biodegradable polymer derived from renewable resources such as corn starch and sugarcane has emerged as a key material in biomedical applications [9]. Its biocompatibility, ease of processing and ability to degrade into non-toxic by-products, primarily lactic acid, make it particularly suitable for medical use [10]. Over the years, PLA has been extensively utilized in drug delivery systems, tissue engineering and bioimplants because of its versatile properties and sustainable origins. One of the most significant roles of PLA in biomedicine is its use in creating scaffolds for tissue engineering. These scaffolds provide structural support for cell attachment, proliferation and differentiation, mimicking the extracellular matrix [11].

The tunable degradation rate of PLA allows the scaffold to degrade as new tissue forms, reducing the need for surgical removal. Additionally, its mechanical properties can be enhanced by blending with other polymers or reinforcing with bioceramics or nanomaterials, making it adaptable for various tissue types, including bone, cartilage and skin.

In drug delivery, PLA serves as an excellent carrier for controlled release systems [12]. Its degradation behaviour can be engineered to release therapeutic agents over extended periods, improving treatment efficacy and patient compliance. PLA-based microspheres and nanoparticles are widely researched for their ability to encapsulate drugs, proteins and even genetic material [13]. PLA has also been applied in surgical sutures, fixation devices and resorbable implants, where its biocompatibility and mechanical stability are crucial. Products such as resorbable screws, plates and pins made from PLA are commonly used in orthopaedic and dental surgeries. Despite its many advantages, PLA faces challenges such as brittleness and slow degradation under physiological conditions. PLA plays a pivotal role in advancing biomedical technology, combining functionality with sustainability.

1.3 Objectives of PLA-Based Composite Development

The development of PLA-based composites aims to address the inherent limitations of pure polylactic acid (PLA) because of its biodegradable and biocompatible nature for advanced biomedical applications [14]. The brittleness, limited mechanical strength and lack of bioactivity of PLA restrict its effectiveness in demanding medical applications such as tissue engineering and bioimplants. To overcome these challenges, the primary objective of PLA-based composite development is to enhance its mechanical, thermal and structural properties. PLA composites exhibit improved strength, toughness and flexibility, making them suitable for applications requiring mechanical resilience by incorporating reinforcements such as bioceramics, natural fibres or nanoparticles. Another critical objective is to tailor the degradation behaviour of PLA composites. The degradation rate of PLA often needs to be synchronized with the tissue regeneration process in biomedical applications. The addition of bioactive reinforcements or copolymers allows the degradation profile to be adjusted to match the specific healing needs of different tissues. This ensures a controlled and gradual release of the composite's structural support as the tissue regenerates.

The improving of PLA's bioactivity is also a key goal of PLA-based composite development [15]. PLA lacks inherent bioactive properties, which are essential for promoting cell attachment, proliferation and differentiation. These composites can provide a more favourable environment for tissue regeneration with the incorporation of bioactive materials such as hydroxyapatite, bioactive glasses or natural extracts. Additionally, the development of PLA composites seeks to expand their functionality through surface modifications or the inclusion of therapeutic agents. This enables the composites to serve as drug delivery systems, antimicrobial scaffolds or carriers for growth factors. The objectives of PLA-based composite development focus on enhancing its properties, tailoring degradation rates, improving bioactivity and introducing multi-functionality, thereby making PLA composites versatile and effective for diverse biomedical applications.

2 Properties of the PLA Relevant to Tissue Engineering

Polylactic acid (PLA) is widely used in tissue engineering because of its biodegradability, biocompatibility and mechanical properties. It is derived from renewable sources and degrades into non-toxic lactic acid, ensuring safety in biomedical applications. Its high tensile strength and stiffness provide structural integrity, whereas its tunable degradation rate allows scaffolds to match tissue regeneration timelines [16]. The thermal stability and ease of processing of PLA enable the fabrication of complex structures such as scaffolds and implants. However, its brittleness and lack of inherent bioactivity require modification or reinforcement for enhanced performance, making it an adaptable and promising material for tissue engineering applications. The properties of PLA for tissue engineering are shown in Fig. 1.

2.1 *Biodegradability and Biocompatibility*

Polylactic acid (PLA) is widely recognized for its biodegradability and biocompatibility, two essential characteristics that make it a preferred material in biomedical applications [17]. Biodegradability refers to the ability of PLA to break down into non-toxic by-products through natural metabolic processes, eliminating the need for surgical removal of implants once they have served their purpose. This degradation occurs via hydrolysis, where water molecules gradually break the ester bonds in PLA, leading to its breakdown into lactic acid, which is then safely absorbed or excreted by the body. The degradation rate of PLA can be tailored by adjusting its molecular weight, crystallinity and copolymer composition, making it suitable for different tissue regeneration [18]. For example, in bone tissue engineering, PLA can degrade slowly to provide long-term mechanical support as the bone heals, whereas in soft tissue applications, it may degrade more rapidly to accommodate faster healing.

Fig. 1 Properties of PLA for tissue engineering

Biocompatibility is another crucial property of PLA. As a biocompatible polymer, PLA does not provoke a significant immune response when it is implanted in the body, making it well suited for medical and tissue engineering applications. The ability of PLA to integrate with living tissue without causing irritation or rejection ensures its widespread use in drug delivery systems, sutures and biodegradable implants. PLA can also be modified to enhance its surface properties, promoting cell adhesion, proliferation and differentiation, which are vital for tissue regeneration. PLA is a core component in scaffold fabrication for applications such as bone, cartilage and skin regeneration. The biodegradability and biocompatibility of PLA make it an excellent candidate for biomedical applications, offering safe, effective and sustainable solutions for tissue engineering and medical device development.

2.2 *Mechanical Properties and Thermal Stability*

Polylactic acid (PLA) possesses a combination of mechanical properties that make it suitable for various biomedical applications, although it also presents challenges that require modification for specific uses. PLA has relatively high tensile strength and stiffness compared with other biodegradable polymers, which provides structural integrity for applications such as scaffolds and implants in tissue engineering [19]. The mechanical strength of PLA allows it to support the body's load-bearing requirements, particularly in applications such as bone implants and surgical sutures. However, PLA is also known for its brittleness, especially at lower temperatures, which can limit its performance under dynamic loading conditions. To overcome this limitation, PLA is often blended with other materials or reinforced with nanoparticles, fibres or bioceramics to increase its flexibility and toughness without compromising its biodegradability [20].

Thermal stability is another critical property of PLA, especially when it is used in manufacturing processes such as extrusion, moulding or 3D printing, where high temperatures are involved. PLA exhibits good thermal stability up to a certain temperature, typically approximately 60–65 °C, beyond which it can start to lose its mechanical properties due to thermal degradation [21]. This temperature range is suitable for most tissue engineering applications, but it may limit the use of PLA in environments requiring increased heat resistance. To improve its thermal stability, PLA can be modified through copolymerization or blending with materials such as polyhydroxyalkanoates (PHAs) or bioactive glass. These modifications not only enhance the thermal resistance but also improve the overall mechanical properties of PLA. PLA offers desirable mechanical properties for biomedical applications, and its brittleness and limited thermal stability at high temperatures necessitate further modifications and composite development to optimize its performance for specific tissue engineering and medical device applications.

2.3 Challenges: Brittleness and Bioinertness

Despite its many advantages, polylactic acid (PLA) faces two significant challenges in biomedical applications: brittleness and bio inertness [22]. These limitations can delay its effectiveness in certain medical and tissue engineering applications, necessitating further material development and modification. Brittleness is one of the primary drawbacks of PLA, particularly in load-bearing applications. While PLA offers good tensile strength and stiffness, its tendency to fracture or fail under impact or stress limits its suitability for certain implants and scaffolds. This is particularly problematic in applications where flexibility and toughness are essential, such as in soft tissue engineering or dynamic orthopaedic implants. The brittleness of PLA is due to its crystalline structure and lack of molecular flexibility, which makes it less able to absorb and distribute mechanical stress [23]. To address this, PLA is often blended with other polymers, such as polycaprolactone (PCL), or reinforced with natural fibres, nanoparticles or bioceramics to increase its toughness and flexibility while maintaining its biodegradability.

Bio inertness is another challenge that restricts the biological performance of PLA in tissue engineering. PLA, in its pure form, lacks significant bioactivity, meaning that it does not actively promote cellular attachment, proliferation or differentiation, which are essential for tissue regeneration. This bio inertness can result in poor integration with surrounding tissues, reducing the efficacy of PLA-based scaffolds in supporting tissue growth [24]. To overcome this, PLA is often combined with biologically active materials such as hydroxyapatite, collagen or growth factors to enhance cell interactions and promote tissue regeneration. Additionally, surface modifications such as plasma treatment or coating with bioactive peptides are employed to improve the cellular compatibility. PLA has great potential in biomedical applications, and its brittleness and bio inertness limit its performance.

3 Synthesis and Processing of PLA

The synthesis and processing of polylactic acid (PLA) are crucial steps in optimizing its properties for various applications, especially in the biomedical and tissue engineering fields. PLA is synthesized primarily through the ring-opening polymerization of lactide, a cyclic dimer of lactic acid, using catalysts to control the molecular weight and polymer structure. Processing methods such as extrusion, injection moulding and 3D printing are employed to fabricate PLA-based products with specific shapes and mechanical properties. By adjusting the processing conditions and incorporating additives or reinforcing materials, the performance of PLA can be tailored for diverse applications, including scaffolds, drug delivery systems and implants.

3.1 Overview of PLA Synthesis Techniques

Polylactic acid (PLA) is synthesized through various techniques, primarily focussing on the polymerization of lactic acid or lactide. The most commonly used methods for PLA synthesis include direct condensation polymerization and ring-opening polymerization (ROP) of lactide. Each method has distinct advantages and limitations, which make them suitable for different applications. The methods for the synthesis of PLA are shown in Fig. 2.

1. **Direct Condensation Polymerization**: This method involves the direct polymerization of lactic acid by eliminating water molecules [25]. It is a simple and cost-effective approach often used for low molecular weight PLA. However, the reaction is limited by side reactions such as the formation of oligomers, which can reduce the overall yield and molecular weight of the polymer. The reaction typically requires high temperatures and catalysts to drive condensation.
2. **Ring-Opening Polymerization (ROP) of Lactide**: ROP is the most widely used technique for the industrial-scale synthesis of PLA, as it produces high molecular weight PLA with better control over the polymer structure [26]. This process involves the polymerization of lactide, the cyclic dimer of lactic acid, which can exist in different stereoisomeric forms (L-lactide, D-lactide and meso-lactide). Lactide is polymerized via metal-based catalysts such as tin octoate or aluminium alkoxides under controlled conditions. ROP allows precise control over the polymer's molecular weight, stereochemistry and crystallinity, which directly impacts its properties, such as mechanical strength and degradation rate.
3. **Solvent-based methods**: PLA can also be synthesized by dissolving lactic acid in solvents and using various polymerization techniques, such as bulk polymerization or solution polymerization [27]. These methods are typically used to increase control over the polymerization process and improve the uniformity of the product.

Each synthesis method impacts the final properties of PLA, including its molecular weight, crystallinity and degradation behaviour, making it essential to select the

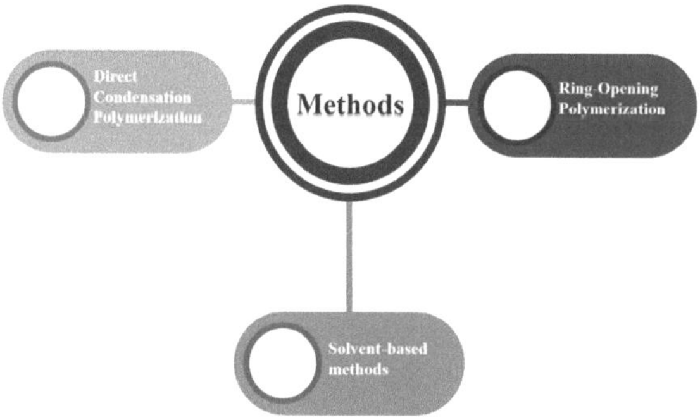

Fig. 2 Synthesis methods of PLA

appropriate technique on the basis of the intended application such as for biomedical implants, tissue scaffolds or drug delivery systems.

3.2 Polymer Processing Methods

Polymer processing is a crucial step in transforming raw polymer materials into useful products with desired shapes, sizes and properties. Various methods are employed to process polymers, depending on the type of polymer, the product's intended use and the required characteristics. Among the most common polymer processing techniques are extrusion, injection moulding, blow moulding, thermoforming and compression moulding. Each method has its own advantages and applications, and selecting the right method is essential for optimizing material properties for specific uses. The various polymer processing methods are shown in Fig. 3.

1. **Extrusion**: Extrusion is a widely used process for forming continuous shapes such as films, sheets, fibres and pipes [28]. In extrusion, polymer pellets or granules are fed into a heated barrel, where they are melted by mechanical shear and heat. The molten polymer is then forced through a die, which shapes it into the desired form, such as a film or rod. After exiting the die, the extruded material is cooled and solidified to retain its shape. Extrusion is especially suitable for producing large quantities of polymer products with consistent cross-sectional shapes. It can be used with various materials, including thermoplastics such as polylactic acid (PLA), polyethylene (PE) and polystyrene (PS). The cooling rate during extrusion can influence the crystallinity and mechanical properties of the polymer, which is especially important for applications in biomedical devices.
2. **Injection Moulding**: Injection moulding is a high-speed manufacturing process used to create complex parts with high precision [29]. In this method, polymer pellets are heated until they melt, and the molten polymer is injected under high pressure into a mould cavity. Once the material cools and solidifies, the moulded part is ejected from the mould. Injection moulding is used for the mass production of components with intricate designs, such as medical devices, consumer

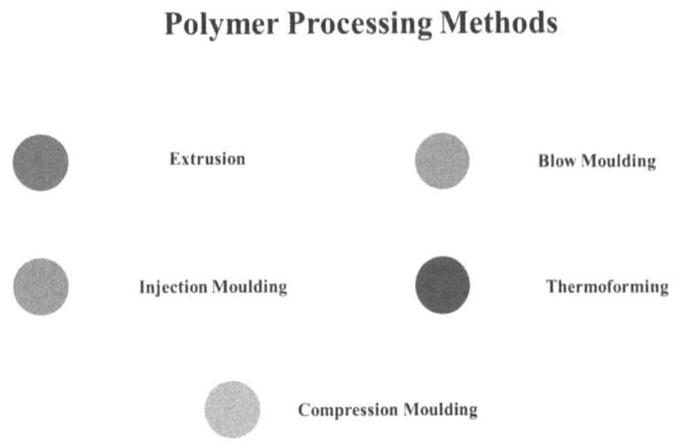

Fig. 3 Polymer processing methods

products and automotive parts. It offers excellent repeatability and dimensional accuracy, making it ideal for producing high-quality products. However, this method requires precise control over processing conditions such as temperature, pressure and cooling rates to avoid defects such as warping or residual stresses.

3. **Blow Moulding**: Blow moulding is commonly used to manufacture hollow plastic parts, such as bottles and containers. In this process, a polymer tube or mixture is extruded or injection moulded, and then air is blown into the tube to inflate it into a mould cavity [30]. The material expands to fit the mould, and after cooling, the hollow part is ejected. This technique is particularly useful for producing lightweight, high-volume containers and packaging products. Blow moulding offers good flexibility in terms of material choice and mould design and is often used in the production of bottles made from materials such as PET, HDPE and PVC. However, it is not suitable for parts with complex internal features or thin walls.
4. **Thermoforming**: Thermoforming is a process used to transform plastic sheets into specific shapes by heating them to a pliable temperature and then applying pressure to mould them [31]. The heated sheet is either vacuum-formed or draped over a mould and then cooled to retain the desired shape. Thermoforming is widely used for packaging products, disposable containers and automotive parts. It offers rapid production cycles and is cost-effective for low to medium production volumes. However, thermoforming is generally limited to products with relatively simple geometries, and the material must be uniformly heated for optimal results.
5. **Compression Moulding**: Compression moulding involves placing a premeasured amount of polymer into a heated mould cavity. The mould is then closed, and pressure is applied to the polymer to form the desired shape. Once cooled, the part is ejected from the mould. This method is commonly used for thermosetting polymers and composite materials [32]. It is often employed in the production of automotive parts, electrical components and industrial goods. Compression moulding offers high material utilization and can accommodate complex shapes, but it is generally slower than other methods, such as injection moulding.

3.3 Tailoring Properties Through Molecular Weight and Crystallinity

Molecular weight and crystallinity are two fundamental factors that significantly influence the properties of polymers, including their mechanical strength, thermal stability and degradation behaviour. Tailoring these properties through molecular weight control and crystallinity manipulation is essential for optimizing polymers such as polylactic acid (PLA) for various applications, including biomedical implants.

1. **Molecular Weight**: Molecular weight refers to the average mass of a polymer molecule, which is determined by the number of repeating units (monomers)

in the polymer chain. The molecular weight of a polymer is a key factor influencing its mechanical properties such as tensile strength, elasticity and viscosity. Polymers with high molecular weights generally exhibit superior mechanical strength, enhanced thermal stability and better resistance to environmental stress cracking [33]. In contrast, polymers with low molecular weights tend to be more flexible but less durable. In polymer processing, controlling the molecular weight is often achieved through adjusting the polymerization conditions, such as the reaction time, temperature and choice of catalyst. Higher molecular weights can improve the material's strength and resistance to wear, whereas lower molecular weights can enhance processing flowability. The molecular weight distribution also plays a significant role. A broad molecular weight distribution can impart toughness and impact resistance to a material, whereas a narrow distribution can enhance processing consistency and reproducibility.

2. **Crystallinity**: Crystallinity refers to the degree to which the polymer chains are organized into a regular, repeating pattern or crystal structure. The degree of crystallinity of a polymer affects its mechanical strength, rigidity and thermal properties. Highly crystalline polymers tend to be more rigid, with higher tensile strength and greater resistance to heat, whereas amorphous polymers are generally more flexible and transparent [34]. For PLA and other polymers, the crystallinity can be controlled by adjusting the processing conditions, such as the cooling rate, pressure and temperature. Slow cooling from the molten state allows for the formation of crystalline regions, whereas rapid cooling typically results in an amorphous structure. The crystallization rate can also be influenced by the presence of nucleating agents or plasticizers, which promote or hinder crystallization, respectively. The degree of crystallinity also impacts the polymer's degradation behaviour. More crystalline polymers generally degrade at a slower rate because of their tightly packed structure, whereas amorphous polymers degrade faster because their disordered structure is more amenable to hydrolytic degradation [35]. This is particularly relevant in biomedical applications where the degradation rate must match the healing process of tissues.
3. **Tailoring PLA for Biomedical Applications**: For PLA-based materials used in tissue engineering or drug delivery, tuning the molecular weight and crystallinity is essential for achieving the desired balance between the mechanical strength, degradation rate and biocompatibility [36]. For example, high molecular weight PLA with controlled crystallinity can be used for orthopaedic implants, providing the necessary strength while allowing gradual degradation as new tissue grows. Conversely, lower molecular weight PLA with reduced crystallinity may be ideal for drug delivery systems, ensuring a faster degradation rate that corresponds to controlled drug release. The properties of PLA and other polymers can be finely tuned to meet specific application requirements, by adjusting the molecular weight and crystallinity, allowing for enhanced performance in a wide range of industries especially in biomedicine.

4 Reinforcement for PLA-Based Composites

Reinforcements for PLA-based composites include the addition of materials to enhance the mechanical, thermal and functional properties of the polymer. Common reinforcements include natural fibres such as jute, flax, hemp and bamboo, as well as inorganic fillers such as glass fibres, carbon nanotubes and nanocellulose. These reinforcements improve the stiffness, strength, and impact resistance of PLA while maintaining its biodegradability. Natural fibre reinforcements also contribute to sustainability by reducing the reliance on petroleum-based materials. The integration of these reinforcements is achieved through various processing techniques, such as extrusion or injection moulding, tailoring PLA-based composites in biomedical fields.

4.1 Bio Ceramics and Bioactive Glasses

Bioceramics and bioactive glasses are advanced materials that are extensively used in the field of biomedical applications, particularly in tissue engineering, bone repair and dental implants [37]. These materials are known for their excellent biocompatibility, osteoconductivity and ability to integrate with surrounding tissues, making them ideal for use in resorbable bioimplants and fixation devices. Bioceramics are a group of ceramic materials that are specifically designed for medical and dental applications. They are typically inorganic, non-metallic materials that can be bioinert, bioactive or biodegradable depending on their composition and intended use. The most commonly used bioceramics include alumina (Al_2O_3), zirconia (ZrO_2) and calcium phosphate ceramics such as hydroxyapatite (HA) [38]. These materials are selected for their excellent mechanical properties, high wear resistance and biological stability.

Bioceramics such as hydroxyapatite (HA) closely resemble the mineral phase of bone, making them highly suitable for bone tissue engineering applications. When implanted, these ceramics promote bone growth and regeneration because of their similarity to the natural bone mineral content. Additionally, bioceramics have low toxicity and can be designed to degrade over time, facilitating tissue regeneration as the material is resorbed by the body. However, the brittleness of many bioceramics remains a challenge, especially for load-bearing applications, which often require the use of composites or reinforced structures.

Bioactive glasses are a class of glasses that exhibit a unique ability to bond with living bone and promote the growth of new bone tissue. The most widely studied bioactive glass is 45S5 bio glass which is composed of silica (SiO_2), calcium oxide (CaO), sodium oxide (Na_2O) and phosphorus pentoxide (P_2O_5) [39]. When bioactive glasses come into contact with body fluids, they form a hydroxyapatite layer on their surface which promotes bone cell attachment, growth and mineralization. The bioactivity of these glasses is attributed to their chemical composition, which encourages the formation of a surface layer that mimics natural bone minerals. This ability to bond to bone tissues makes bioactive glasses highly valuable in applications such as bone grafts, bone repair and drug delivery systems. Additionally, bioactive glasses can be tailored for different degradation rates to suit the specific needs of tissue regeneration, whether for rapid or slow healing. Bioactive glasses also have the advantage of being easily processed into various forms such as powders, fibres and scaffolds to create structures that can support cellular growth and function [40]. Owing to their versatility in processing, bioactive glasses are ideal for designing customized implants that are both effective and functional.

Both bioceramics and bioactive glasses are widely used in orthopaedic and dental implants, as well as in tissue engineering scaffolds. Their ability to promote bone healing and integration with natural tissues makes them an essential component in the development of resorbable bioimplants. However, challenges remain in terms of their mechanical strength, brittleness and slow degradation rates in some cases. Bioceramics and bioactive glasses are critical materials in the fields of regenerative medicine and tissue engineering. The ability of these materials to integrate with bone tissues and promote healing makes them invaluable for developing innovative, resorbable biomedical devices.

4.2 *Nanomaterials*

Nanomaterials have gained significant attention as reinforcements in polylactic acid (PLA)-based composites, particularly for tissue engineering applications. These materials provide enhanced mechanical, thermal and biological performance to the composites because of their unique properties at the nanoscale. Nanomaterials such as nanoparticles, nanofibers and nanocellulose are incorporated into PLA matrices to overcome the inherent limitations of PLA such as its low mechanical strength and brittle nature and to improve the overall performance of the material for biomedical applications [41].

1. **Nanocellulose**: Nanocellulose, derived from natural cellulose sources such as wood, cotton or agricultural waste is one of the most promising nanomaterials for reinforcing PLA. It exists in various forms such as cellulose nanocrystals (CNCs) and cellulose nanofibrils (CNFs) which have high surface areas and mechanical strengths [42]. When incorporated into PLA, nanocellulose significantly improved the tensile strength, stiffness and impact resistance of the composite while maintaining its biodegradability and biocompatibility. The addition of nanocellulose also enhances the porosity and surface roughness of PLA-based composites, which can support better cellular attachment and proliferation in tissue engineering applications. Moreover, nanocellulose can act as a delivery system for bioactive molecules, further enhancing the functional properties of the composite.

2. **Carbon Nanotubes (CNTs)**: Both single-walled (SWCNTs) and multi-walled (MWCNTs) carbon nanotubes (CNTs) constitute another class of nanomaterials commonly used to reinforce PLA. CNTs possess extraordinary mechanical strength, electrical conductivity and thermal stability, making them ideal for reinforcing PLA in demanding biomedical applications [43]. When incorporated into PLA, CNTs improve the composite's mechanical properties, including tensile strength, elongation at break and thermal stability. Additionally, CNTs can facilitate the alignment and orientation of the PLA chains, which enhances the structural integrity of the composite. In tissue engineering, CNT-reinforced PLA composites are useful for creating scaffolds with enhanced mechanical properties while promoting cellular growth and differentiation.
3. **Hydroxyapatite (HA) Nanoparticles**: HA, a naturally occurring mineral form of calcium apatite, is a key component of bone. The incorporation of HA nanoparticles into PLA-based composites for tissue engineering applications enhances the material's bioactivity and osteoconductivity. HA nanoparticles promote bone cell attachment and mineralization, making them particularly suitable for bone tissue regeneration [44]. The addition of HA to the PLA composites not only improved the mechanical properties but also accelerated the bone integration process by mimicking the mineralized phase of bone. The combination of HA nanoparticles with PLA offers a biodegradable, biocompatible material that can be used for bone scaffolds and other orthopaedic implants.
4. **Other Nanomaterials**: In addition to the commonly used nanomaterials mentioned above, other materials such as graphene, nanoclays and silica nanoparticles have also been explored as reinforcements for PLA composites. Graphene, known for its exceptional mechanical strength and electrical conductivity, when incorporated into PLA, improves the composite's mechanical properties and enables electrical stimulation for cell differentiation in tissue engineering [45]. Nanoclays and silica nanoparticles provide reinforcement and improve the composite's barrier properties, making them suitable for drug delivery applications. PLA-based composites reinforced with nanomaterials are increasingly being used in tissue engineering to create scaffolds that mimic the mechanical properties of natural tissues, particularly bone and cartilage. These composites offer better cell adhesion, proliferation and differentiation due to the increased surface area and bioactivity imparted by the nanomaterials. Additionally, nanomaterial incorporation allows for controlled degradation, ensuring that the scaffold resorbs at a rate compatible with tissue regeneration. This makes them ideal for creating temporary implants that provide mechanical support while allowing for tissue healing and integration.

These nanomaterials serve as excellent reinforcements for PLA-based composites, significantly improving their mechanical, thermal and biological properties. Their use in tissue engineering opens new possibilities for developing advanced, bioactive and biodegradable materials that can effectively support tissue regeneration and repair.

4.3 Natural and Synthetic Fibres

PLA-based composites are widely used in tissue engineering because of their biodegradability and biocompatibility. Both natural and synthetic fibres are incorporated as reinforcements to enhance the mechanical and structural properties of PLA for tissue regeneration applications. These fibres improve the strength, stiffness and durability of PLA, making the composites more suitable for a variety of biomedical uses, including bone scaffolds, wound healing and drug delivery systems.

4.3.1 Natural Fibres

Owing to their sustainability, cost-effectiveness and eco-friendly nature, natural fibres derived from plants, animals or minerals are increasingly being utilized as reinforcements in PLA-based composites. Some common natural fibres include jute, flax, hemp, bamboo and silk.

1. **Jute and flax**: These fibres are known for their high tensile strength and are widely used in PLA composites [46]. When incorporated into PLA, jute and flax improve the stiffness and mechanical properties of the composite while maintaining its biodegradability. These fibres also promote cellular adhesion and proliferation, making them suitable for tissue engineering applications, particularly for soft tissue regeneration.
2. **Hemp and bamboo**: Hemp and bamboo fibres are other popular choices because of their strength, low density and abundance [47]. They are lightweight and provide good reinforcement in PLA composites, making them ideal for applications in orthopaedics and bone tissue engineering. These fibres offer a renewable source of reinforcement contributing to the sustainability of PLA-based composites.
3. **Silk**: Silk fibres, especially those derived from silkworms, are known for their excellent mechanical properties, biocompatibility and biodegradability [48]. When used in PLA-based composites, silk fibres increase the composite's mechanical strength and improve cell growth, making them ideal for applications such as skin and bone tissue engineering.

Natural fibres are also biodegradable and encourage better tissue integration, which is vital in tissue engineering applications where the material needs to degrade as new tissue grows.

4.3.2 Synthetic Fibres

Synthetic fibres derived from petrochemical sources offer high mechanical strength, durability and resistance to environmental conditions. Some commonly used synthetic fibres include glass fibres, carbon fibres and aramid fibres.

1. **Glass fibres**: Glass fibres are commonly used to reinforce PLA in composites because they provide excellent tensile strength, impact resistance and dimensional stability [49]. Although they are not biodegradable, glass fibres are used in non-resorbable implants that provide long-term structural support in tissue engineering.
2. **Carbon fibres**: Carbon fibres are strong, lightweight and resistant to corrosion, making them ideal for PLA composites designed for load-bearing applications [50]. When incorporated into PLA, carbon fibres significantly increase the composite's mechanical strength and rigidity, making them suitable for use as scaffolds in bone tissue engineering.
3. **Aramid fibres**: Aramid fibres are known for their exceptional strength and resistance to abrasion and aramid fibres such as Kevlar fibres are used in PLA composites to increase their toughness and impact resistance [51]. These fibres can be applied in tissue engineering applications where high mechanical strength and stability are needed.

The incorporation of natural and synthetic fibres into PLA composites significantly improves their mechanical, thermal and biological properties, making them suitable for a wide range of tissue engineering applications. The bone scaffolds reinforced with natural fibres such as jute and flax exhibit enhanced mechanical properties and promote osteogenesis, whereas synthetic fibres such as carbon and glass provide additional strength for load-bearing applications. These fibre-reinforced PLA composites also offer controlled degradation rates, ensuring that the material resorbs in line with tissue regeneration. The use of both natural and synthetic fibres as reinforcements in PLA-based composites significantly enhances the properties of the material for tissue engineering. By combining the sustainability of natural fibres with the strength and durability of synthetic fibres, PLA composites can be optimized for various biomedical applications, leading to more effective and functional tissue engineering solutions.

5 Fabrication Techniques for PLA Composites

Fabrication techniques play an important role in tailoring the properties and applications of PLA-based composites, particularly for tissue engineering. These methods enable the incorporation of reinforcements such as fibres, nanoparticles or bioceramics into the PLA matrix, increasing its mechanical strength, thermal stability and biocompatibility. Common fabrication techniques include solvent casting, electrospinning, melt extrusion and 3D printing. Each method offers unique advantages, such as precise control over the material structure or efficient production of complex geometries.

5.1 Solvent Casting

Solvent casting is a widely used fabrication technique for creating PLA-based composites, particularly for tissue engineering applications [52]. This method involves dissolving PLA in an organic solvent and subsequently incorporating reinforcements such as nanoparticles, fibres or bioceramics to form a uniform composite solution. The solution is then cast into moulds or flat surfaces and allowed to dry, either under ambient conditions or controlled environments, to remove the solvent and produce a solid composite film or scaffold. One of the main advantages of solvent casting is its simplicity and ability to uniformly distribute reinforcements within the PLA matrix. This ensures homogeneity in the final composite material, which is crucial for consistent mechanical and biological properties. Additionally, solvent casting enables the incorporation of bioactive molecules such as growth factors or drugs, making it highly suitable for tissue engineering scaffolds that require controlled release or enhanced cellular interactions.

Solvent casting is particularly effective for producing thin films or membranes of PLA composites. These films are often used in applications such as wound healing, where they act as protective barriers and support cell adhesion and proliferation. The method can also create porous scaffolds by incorporating porogens such as salt or sugar into the solution, which are later leached out to form interconnected pores [53]. These porous structures are essential in tissue engineering, as they allow for nutrient diffusion, cell infiltration and tissue regeneration. However, solvent casting has limitations. The use of organic solvents such as chloroform or dichloromethane raises concerns about their toxicity and environmental impact. The residual solvent traces in the final composite can affect its biocompatibility. Additionally, the technique is generally limited to small-scale production and lacks precision in the fabrication of complex geometries or large scaffolds. Despite these challenges, solvent casting remains a versatile and cost-effective method for developing PLA-based composites. Its adaptability makes it a valuable tool in advancing biomaterials for regenerative medicine.

5.2 Melt Blending

Melt blending is a prominent fabrication technique used to produce PLA-based composites, particularly for tissue engineering applications. This method involves heating PLA above its melting point and mixing it with reinforcements such as fibres, nanoparticles or bioactive materials in a molten state. The resulting homogeneous blend is then processed into desired shapes via techniques such as extrusion, injection moulding or compression moulding. Melt blending is widely favoured for its solvent-free process, scalability and ability to enhance the mechanical and thermal properties of PLA [54]. One of the key advantages of melt blending is its environmental friendliness compared with solvent-based methods. Since no organic solvents are used, concerns about toxicity and residual solvent traces in the final composite are eliminated, making it more suitable for biomedical applications. This process is also versatile, accommodating a wide range of reinforcements, such as bioceramics, carbon-based nanomaterials or natural fibres, to create PLA composites with tailored properties for tissue engineering.

Melt blending significantly improves the mechanical strength and toughness of PLA-based composites because of its inherent brittleness. The addition of nanoparticles such as hydroxyapatite or carbon nanotubes enhances the structural integrity of the composites, making them suitable for use as bone tissue scaffolds. Similarly, the incorporation of biocompatible polymers such as polycaprolactone (PCL) or polyethylene glycol (PEG) during melt blending can improve the flexibility and biodegradability of PLA, ensuring better tissue integration and regeneration [55]. However, melt blending has several challenges. The high temperatures required can degrade sensitive bioactive molecules such as growth factors or proteins, limiting their incorporation. Additionally, achieving uniform dispersion of reinforcements, especially nanoscale materials, can be challenging because of agglomeration during the mixing process. The optimization of processing parameters such as temperature, shear rate and blending time is essential to ensure consistent composite quality. Despite these limitations, melt blending remains a highly efficient and scalable method for producing PLA-based composites tailored to tissue engineering applications. Its ability to produce mechanically robust, biocompatible and biodegradable composites makes it a preferred choice for developing scaffolds for bone regeneration, cartilage repair and other biomedical uses.

5.3 *Electrospinning for Scaffold Fabrication*

Electrospinning is a versatile and widely used technique for fabricating scaffolds from PLA-based composites, particularly in tissue engineering. This method uses an electrostatic field to produce fine, continuous fibres from a polymer solution or melt, which are collected to form a non-woven fibrous mat. The resulting scaffolds mimic the extracellular matrix (ECM) of natural tissues, offering a highly porous structure that supports cell adhesion, proliferation and tissue regeneration. The process involves dissolving PLA, which is often blended with other polymers or reinforced with nanoparticles in a suitable solvent to create a spinning solution. The solution is then loaded into a syringe connected to a high-voltage power supply. As the voltage is applied, a charged jet of the solution is ejected from the syringe needle, elongating and solidifying into thin fibres as the solvent evaporates. These fibres are deposited on a grounded collector, forming a scaffold.

Electrospinning allows precise control over the fibre diameter, morphology and alignment by adjusting parameters such as the solution concentration, applied voltage, flow rate and collection distance [56]. This flexibility enables the fabrication of scaffolds with tailored properties to meet the specific requirements of different tissue engineering applications, such as bone, cartilage and skin regeneration. The incorporation of reinforcements such as hydroxyapatite, bioactive glass or carbon nanotubes into PLA enhances the mechanical properties, bioactivity and compatibility of the scaffold. Natural polymers such as gelatin, chitosan or collagen are also blended with PLA to improve hydrophilicity and support cellular interactions. The electrospun PLA-based scaffolds can be further functionalized with growth factors, drugs or peptides to promote tissue repair and regeneration [57]. Despite its advantages, electrospinning has several challenges. The process is often limited to small-scale production, and solvent usage raises environmental and safety concerns. Additionally, achieving uniform dispersion of reinforcements in the polymer matrix can be challenging, affecting the mechanical and biological performance of the scaffold. Nevertheless, electrospinning remains a powerful technique for creating PLA-based composite scaffolds that replicate the structural and functional properties of the native ECM. Its adaptability and ability to produce highly porous, biocompatible materials make it a cornerstone in advancing tissue engineering and regenerative medicine applications.

5.4 3D Printing

3D printing, also known as additive manufacturing, has revolutionized scaffold fabrication for tissue engineering by enabling precise control over structure, geometry and composition. PLA-based composites have emerged as ideal materials for 3D printing because of their biodegradability, biocompatibility and ease of processing. This method creates scaffolds layer-by-layer via computer-aided design (CAD) models, allowing customization for specific tissue engineering applications. One of the key advantages of 3D printing is its ability to produce scaffolds with complex architectures and well-defined pore structures. This is critical in tissue engineering, where scaffolds must mimic the natural extracellular matrix (ECM) to support cell attachment, proliferation and nutrient transport [58]. The incorporation of reinforcements such as bioceramics, natural fibres or nanoparticles and PLA-based composites can be tailored to enhance their mechanical, thermal and biological properties. For example, hydroxyapatite-reinforced PLA scaffolds are commonly used for bone tissue engineering because of their enhanced stiffness and osteoconductivity. 3D printing techniques such as fused deposition modelling (FDM) and stereolithography (SLA) are widely employed for PLA-based composites. FDM involves extruding a melted polymer filament through a nozzle to form layers, whereas SLA uses light to cure a photopolymer resin. FDM is particularly popular for PLA-based scaffolds because it is cost-effective, scalable and compatible with various reinforcements. The precision of 3D printing allows for gradient structures, multi-material scaffolds and the incorporation of bioactive agents such as growth factors or drugs [59]. These features enable scaffolds to address specific tissue engineering challenges, such as promoting vascularization or controlled drug release. Additionally, patient-specific scaffolds can be fabricated via medical imaging data, offering personalized solutions for complex defects or injuries. However, challenges remain in the 3D printing of PLA-based composites. The realization of uniform dispersion of reinforcements can be difficult, and the relatively slow printing speed limits large-scale production. Moreover, the thermal degradation of sensitive bioactive materials during processing must be carefully managed. Despite these challenges, 3D printing is transforming the field of scaffold fabrication for tissue engineering. Its ability to produce tailored, functional scaffolds from PLA-based composites continues to drive advancements in regenerative medicine, offering new possibilities for repairing and regenerating damaged tissues.

6 Tissue-Specific Applications of PLA-Based Composites

PLA-based composites have gained significant attention in tissue engineering because of their versatility, biodegradability and biocompatibility. These materials can be tailored for specific tissue regeneration needs by incorporating reinforcements such as bioceramics, fibres or nanoparticles. For bone tissue engineering, PLA composites enhanced with hydroxyapatite provide excellent mechanical strength and osteoconductivity. In cartilage repair, elastic PLA blends mimic the native environment of chondrocytes. PLA-based scaffolds also support skin and vascular regeneration by providing a suitable framework for cell attachment and growth. PLA composites are promising choices for addressing the unique demands of various tissues in regenerative medicine because of their adaptability and tunable properties,

6.1 *Bone Tissue Engineering: Scaffolds and Implants*

Bone tissue engineering focuses on developing scaffolds and implants that mimic the structural and functional properties of native bone [60]. PLA-based composites have emerged as preferred materials in this field because of their biodegradability, biocompatibility and mechanical tunability. These composites can be customized to provide the necessary strength, porosity and bioactivity required for effective bone regeneration.

Scaffolds play a critical role in bone tissue engineering by acting as temporary frameworks that support cell attachment, proliferation and differentiation. PLA-based scaffolds are often reinforced with bioceramics such as hydroxyapatite (HA) or tricalcium phosphate (TCP) to enhance their osteoconductivity and mimic the mineralized matrix of bone. These reinforcements improve mechanical properties such as stiffness and compressive strength, ensuring that the scaffold can withstand physiological loads. Additionally, the porous structure of the PLA-based scaffolds facilitates nutrient and oxygen diffusion while allowing for vascularization, a key factor in successful bone repair. Implants made from PLA-based composites also show promise in orthopaedic applications. Their biodegradability eliminates the need for secondary surgeries to remove implants, reducing patient recovery time and associated healthcare costs. Through the incorporation of nanoparticles or bioactive agents, these implants can be designed to deliver therapeutic compounds such as growth factors or antibiotics directly to the injury site, promoting faster healing and preventing infections. Fabrication techniques such as 3D printing and electrospinning enable precise control over the architecture and properties of PLA-based scaffolds and implants [61]. These methods allow for the creation of gradient structures that closely replicate the hierarchical organization of bone tissue. Furthermore, patient-specific implants can be produced by imaging data, ensuring a perfect fit and better integration with the surrounding bone.

Despite their advantages, challenges remain in optimizing PLA-based composites for bone tissue engineering. The issues such as the brittleness of PLA, limited bioactivity and slow degradation rates need to be addressed [62]. PLA-based composites represent a promising solution for bone tissue engineering, offering customizable, biodegradable and bioactive materials for scaffolds and implants. The potential to improve patient outcomes continues to drive innovations in regenerative medicine.

6.2 Cartilage Regeneration: Load-Bearing Applications

Cartilage regeneration presents unique challenges because of its avascular nature, limited self-healing capacity and need to withstand mechanical loads [63]. PLA-based composites have emerged as promising materials for fabricating scaffolds and implants for cartilage repair because of their biodegradability, biocompatibility and mechanical tunability. With the incorporation of various reinforcements, these composites can be tailored to mimic the properties of native cartilage, facilitating effective regeneration in load-bearing applications.

Cartilage requires scaffolds that are flexible yet strong enough to endure repetitive stress. PLA, in its pure form, lacks sufficient elasticity, but blending it with natural polymers such as collagen, gelatin or chitosan can increase its flexibility and hydrophilicity. Additionally, reinforcing PLA with nanoparticles such as graphene oxide or carbon nanotubes can improve its mechanical properties and durability. These modifications enable the development of scaffolds that replicate the viscoelastic behaviour of cartilage, providing an ideal environment for cell attachment, proliferation and extracellular matrix (ECM) deposition [64]. For load-bearing applications, structural integrity is crucial. Advanced fabrication methods such as 3D printing allow for precise control over scaffold architecture, enabling the creation of structures with interconnected pores that support nutrient diffusion and cellular infiltration. Scaffolds with gradient properties can be designed to mimic the zonal organization of cartilage, ensuring both functional and structural compatibility.

Bioactive reinforcements such as bioactive glass or growth factors can be incorporated into PLA-based composites to promote chondrogenesis. These additions enhance the ability of the scaffold to guide stem cells or chondrocytes to regenerate cartilage tissue effectively. Additionally, drug-loaded PLA composites can provide sustained release of anti-inflammatory agents, reducing joint inflammation and enhancing the healing process. The achievement of right balance of mechanical strength, elasticity and degradation rate is complex. Moreover, long-term studies are needed to evaluate the integration of regenerated cartilage with surrounding tissues and scaffold performance under physiological conditions. PLA-based composites hold significant potential for cartilage regeneration, particularly in load-bearing applications.

6.3 Skin Tissue Engineering: Wound Healing Materials

Skin tissue engineering focuses on developing materials that promote wound healing, repair damaged tissues and restore skin integrity. PLA-based composites have gained significant attention in this field because of their biocompatibility, biodegradability and ability to be modified for specific applications. These materials provide an effective platform for designing scaffolds, films or hydrogels that support the healing process by mimicking the extracellular matrix (ECM) of the skin. PLA-based wound healing materials offer numerous advantages. The degradation product lactic acid is naturally metabolized in the body, ensuring minimal toxicity. These composites can be engineered to achieve controlled degradation rates, aligning with the wound healing timeline. Additionally, PLA-based scaffolds provide a structural framework for cell attachment, proliferation and migration, which are essential for tissue regeneration. The incorporation of reinforcements such as chitosan, collagen or alginate enhances the properties of PLA composites, improving their hydrophilicity, mechanical strength and antimicrobial activity. Chitosan, for example, provides antibacterial properties, helping to prevent infections in wounds. The addition of nanoparticles such as silver or zinc oxide can further enhance antimicrobial effectiveness while promoting angiogenesis and cell proliferation.

Advanced fabrication techniques, including electrospinning and 3D printing, enable the creation of PLA-based scaffolds with nanofibrous structures that closely resemble the natural ECM of the skin. These scaffolds allow for efficient nutrient exchange, oxygen diffusion and waste removal, which are critical for wound healing. Moreover, such structures support the sustained release of bioactive agents, including growth factors, anti-inflammatory drugs or stem cells, to accelerate tissue repair.

For burn wounds or chronic ulcers, PLA-based composites have demonstrated significant potential [65]. The ability of these materials to be customized for specific wound types makes them highly versatile. However, challenges such as optimizing mechanical flexibility for dynamic skin areas and ensuring consistent bioactivity remain under investigation. PLA-based composites offer promising solutions for wound healing materials in skin tissue engineering. Owing to their adaptability, biocompatibility and ability to incorporate bioactive components, these materials are invaluable tools for developing effective and innovative treatments for skin injuries.

7 Biofunctionalization of PLA-Based Composites

The biofunctionalization of PLA-based composites involves the incorporation of bioactive molecules or surface modifications to enhance their interaction with biological systems, particularly in tissue engineering applications. This process aims to improve cellular responses such as adhesion, proliferation and differentiation, making the composites more effective for regenerative medicine. Biofunctionalization can be achieved through the attachment of proteins, peptides or growth factors or the introduction of bioactive particles such as nanoparticles or bioactive ceramics. Biofunctionalization enhances tissue integration, promotes healing and ensures better performance in specific medical applications, such as bone, cartilage and skin regeneration by tailoring the surface properties of PLA-based composites.

7.1 Surface Modifications for Improved Cell Interaction

Surface modification is a crucial strategy for enhancing the biological performance of PLA-based composites in tissue engineering applications. While PLA is biocompatible, its surface characteristics are not always conducive to optimal cell adhesion, proliferation and differentiation. Therefore, modifying the surface of PLA composites can significantly improve their interaction with cells and tissues, enhancing their effectiveness as scaffolds or implants in regenerative medicine.

One of the most common approaches to surface modification is the introduction of functional groups such as amine, carboxyl or hydroxyl groups onto the surface of PLA [66]. These groups can be grafted onto PLA through chemical reactions, such as plasma treatment, UV irradiation or chemical cross-linking. The introduction of hydrophilic groups enhances the wettability of a material, which improves protein adsorption and facilitates cell attachment. Hydrophilicity is particularly important in promoting initial cell adhesion, as cells rely on the presence of extracellular matrix proteins to anchor to a surface. Another approach involves coating PLA-based composites with biologically active molecules such as collagen, fibronectin or laminin. These proteins, which are commonly found in the extracellular matrix, provide a favourable microenvironment for cell attachment and growth. The coating can be performed through physical adsorption or covalent bonding, ensuring that the bioactive molecules remain stable on the surface for extended periods. Additionally, peptides such as RGD (Arg-Gly-Asp) sequences, which are known to promote cell adhesion, can be attached to the PLA surface to further enhance cellular responses.

For more advanced biofunctionalization, nanoparticles and bioactive agents can be incorporated into the PLA composite during or after fabrication. Nanoparticles such as hydroxyapatite, silica or titanium dioxide can not only improve mechanical strength but also enhance bioactivity, providing sites for cell attachment and promoting osteoconductivity in bone tissue engineering. Similarly, growth factors such as bone morphogenetic proteins (BMPs) or vascular endothelial growth factors

(VEGFs) can be loaded onto the surface to promote cell differentiation, angiogenesis and tissue regeneration [67]. Electrospinning is another surface modification technique commonly used to create nanofibrous mats that mimic the natural extracellular matrix. These mats, when applied to PLA scaffolds, significantly improved the ability of the scaffold to support cell infiltration and growth. Electrospun PLA fibres also provide a larger surface area for cell adhesion, aiding in the formation of a more robust tissue construct.

Surface modifications are essential for enhancing the bioactivity and cellular interactions of PLA-based composites in tissue engineering. The surface-modified PLA composites hold great promise for advancing the development of functional and efficient tissue engineering scaffolds and implants by improving cell adhesion, differentiation and tissue integration.

7.2 *Integration of Growth Factors and Drug Delivery Systems*

The integration of growth factors and drug delivery systems into PLA-based composites represents a promising strategy to enhance tissue regeneration in biomedical applications. PLA, a biodegradable polymer, is an ideal candidate for controlled release systems, offering the ability to degrade over time while delivering bioactive molecules that promote cellular behaviour, tissue healing and regeneration. The performance of scaffolds in tissue engineering can be significantly improved by bio functionalizing PLA-based composites with growth factors and incorporating drug delivery capabilities. Growth factors such as bone morphogenetic proteins (BMPs), vascular endothelial growth factor (VEGF) and epidermal growth factor (EGF) play crucial roles in cellular processes such as differentiation, proliferation and angiogenesis. When integrated into PLA-based composites, these factors can guide the regeneration of specific tissues. For example, BMPs stimulate osteoblast differentiation in bone tissue engineering, whereas VEGF promotes angiogenesis in vascular applications. The controlled release of these growth factors from PLA scaffolds can be achieved through surface modification or encapsulation within micro- or nanoparticles. This ensures that the bioactive molecules are gradually released over time, mimicking the natural healing process and enhancing tissue formation.

Drug delivery systems embedded in PLA composites further enhance their functionality. These systems can be designed to deliver drugs or anti-inflammatory agents, promoting faster recovery and reducing the risk of infection or immune rejection [68]. For example, antibiotics such as gentamicin or anti-inflammatory drugs such as dexamethasone can be incorporated into PLA-based scaffolds to provide localized therapy at the site of injury or implantation. Drug release can be precisely controlled, ensuring a steady therapeutic dose over an extended period without causing systemic side effects by the use of nanocarriers or microparticles. Several techniques have been employed to integrate growth factors and drugs into PLA-based composites. Physical entrapment, electrospinning and chemical conjugation are commonly used methods. For physical entrapment, growth factors or drugs are simply mixed with PLA during processing, allowing for their gradual release as the material degrades. Electrospinning allows the incorporation of bioactive molecules into nanofibrous structures, which increases the surface area for cell interactions and promotes effective drug release [69]. Chemical conjugation, on the other hand, involves covalent bonding of growth factors or drugs to the PLA surface or within the polymer matrix, ensuring more controlled release and longer-term bioactivity. The integration of growth factors and drug delivery systems into PLA-based composites significantly enhances their effectiveness in tissue engineering applications. These composites not only support tissue regeneration but also improve healing outcomes by providing sustained release of bioactive molecules.

7.3 Antimicrobial and Anti-inflammatory Enhancements

The biofunctionalization of PLA-based composites with antimicrobial and anti-inflammatory agents is an essential strategy to improve their performance in tissue engineering applications. The incorporation of these enhancements not only promotes better healing but also reduces the risk of infections and inflammatory responses, which are common challenges in implantable biomaterials. Antimicrobial agents such as silver nanoparticles, copper or essential oils can be integrated into PLA-based composites to combat bacterial infections at the implantation site [70]. These agents exhibit broad-spectrum antimicrobial activity, preventing the growth of pathogens such as *Staphylococcus aureus* and *Escherichia coli,* which are often responsible for implant-related infections. PLA composites can maintain an effective concentration at the infection site, thereby minimizing the need for systemic antibiotics and reducing the risk of resistance development by releasing these antimicrobial agents in a controlled manner over time.

Anti-inflammatory agents such as corticosteroids or non-steroidal anti-inflammatory drugs (NSAIDs) are also critical for reducing inflammation at the implant site, promoting tissue regeneration [71]. Chronic inflammation can delay healing, impair tissue integration and lead to implant failure. By incorporating these agents into PLA composites, it is possible to modulate the inflammatory response, promoting a more favourable environment for cell proliferation and tissue growth. This controlled release helps to manage inflammation at the local level without causing systemic side effects, which is particularly beneficial in long-term implants. The incorporation of antimicrobial and anti-inflammatory agents can be achieved through various methods, such as physical entrapment, chemical grafting or the use of nanocarriers. NPs offer an efficient way to load and release these agents, as they provide a high surface area and can be easily functionalized to target specific tissues or cells. The incorporation of antimicrobial and anti-inflammatory agents into PLA-based composites enhances their therapeutic potential in tissue engineering. These biofunctionalized composites not only promote tissue regeneration but also reduce complications associated with infections and inflammation, leading to better clinical outcomes in regenerative medicine.

8 Challenges and Limitations in PLA-Based Tissue Engineering

PLA-based composites have shown great promise in tissue engineering, but several challenges and limitations must be addressed for their widespread clinical application. While PLA is biodegradable and biocompatible, its mechanical properties such as brittleness and limited strength can hinder its use in load-bearing applications. Additionally, the slow degradation rate of PLA may not always match the tissue regeneration process, leading to potential complications. Additionally, the hydrophobic nature of PLA limits cell adhesion, and the risk of inflammation or an immune response remains a concern.

8.1 Issues with Degradation Rates and By-Products

One of the key challenges in using PLA-based composites for tissue engineering is managing the degradation rate of PLA and the by-products produced during its breakdown. PLA is a biodegradable polymer, which makes it an attractive material for biomedical applications, particularly in tissue engineering, where the material needs to degrade as the tissue regenerates. However, the rate at which PLA degrades may not always match the regeneration process of the tissue, leading to several issues. The degradation of PLA occurs through hydrolytic cleavage of its ester bonds, which results in the formation of lactic acid as the primary by-product. While lactic acid naturally occurs in the body and can be metabolized, excessive accumulation of lactic acid in tissues can lead to localized acidic environments. This can have detrimental effects on surrounding cells, leading to inflammation or impaired tissue healing. In some cases, acidic by-products can even contribute to osteolysis in bone tissue engineering or interfere with the regeneration of other tissues.

The degradation rate of PLA is another critical concern. PLA degrades relatively slowly compared with other biodegradable materials such as polycaprolactone (PCL) or poly (lactic co-glycolic acid) (PLGA) [72] Slow degradation can be beneficial in certain applications and can also lead to situations where the material remains in place longer than necessary, potentially causing a mechanical mismatch between the degrading scaffold and the newly formed tissue. This mismatch can hinder proper tissue function and may lead to complications such as poor tissue integration, delayed regeneration or the formation of fibrous tissue. The slow degradation rate of PLA also affects its application in dynamic and load-bearing tissues. In bone tissue engineering, for example, PLA scaffolds may not degrade fast enough to provide the necessary support for new bone growth, which could compromise the function of the implant. Conversely, if the material degrades too quickly, it may result in insufficient mechanical support, leading to scaffold collapse before proper tissue regeneration has occurred. PLA-based composites offer significant promise in tissue engineering, and controlling their degradation rate and the by-products generated during degradation is a critical challenge.

8.2 Mechanical and Thermal Property Optimization

One of the key limitations of PLA-based composites in tissue engineering is their mechanical properties, particularly their brittleness and low tensile strength compared with those of natural tissues. PLA is a rigid polymer that may not provide the necessary mechanical support required for load-bearing applications such as bone regeneration or cartilage repair [73]. This limitation often leads to insufficient mechanical integrity, which can result in scaffold failure or inability to support tissue regeneration effectively. Furthermore, PLA has poor resistance to impact and may crack under mechanical stress, which can compromise its performance in long-term applications. The thermal properties of PLA also present challenges in tissue engineering. PLA has a relatively low glass transition temperature (Tg) and melting temperature (Tm), which makes it susceptible to deformation under physiological conditions or during the fabrication process. The low thermal stability limits its application under certain high-temperature conditions and could impact the structural integrity of the scaffold over time. For example, during manufacturing techniques such as 3D printing or extrusion, PLA-based composites may not maintain their shape if exposed to elevated temperatures.

Various strategies have been explored to overcome these mechanical and thermal limitations. One approach involves blending PLA with other polymers or incorporating reinforcement materials, such as natural fibres, bioceramics or nanoparticles. These reinforcements can significantly improve the tensile strength, stiffness and toughness of PLA-based composites. For example, combining PLA with fibres such as cellulose or chitosan can increase a material's strength while maintaining its biodegradability and biocompatibility. Additionally, the inclusion of ceramic particles or bioceramics such as hydroxyapatite can improve the mechanical strength and bioactivity of PLA, making it more suitable for bone tissue engineering. Another strategy is to modify the crystallinity and molecular weight of PLA. The crystallinity of PLA can be increased by controlling the polymerization process, leading to enhanced mechanical properties such as improved stiffness and strength. Higher molecular weight PLA also improves a material's mechanical performance and resistance to thermal degradation. Additionally, optimizing the cooling rate during processing can increase the crystallinity, thereby improving the thermal stability and increasing the resistance of PLA to deformation under physiological conditions. The thermal stability can also be improved by incorporating heat-resistant additives or blending PLA with high-performance thermoplastics such as polycarbonate or polyamide. These modifications help improve the overall thermal and mechanical stability of PLA-based composites, making them more suitable for demanding tissue engineering applications. The mechanical and thermal properties of PLA present challenges for tissue engineering applications, and various strategies, such as reinforcement with other materials, modification of crystallinity and optimization of molecular weight, can significantly increase its performance.

8.3 Scalability and Reproducibility in Production

One of the significant challenges in the development and application of PLA-based composites for tissue engineering is the scalability and reproducibility of their production. While laboratory-scale production of PLA scaffolds may yield promising results, translating these findings to large-scale manufacturing for clinical use remains a complex task. The ability to consistently reproduce high-quality scaffolds on a larger scale is critical for their widespread adoption in regenerative medicine. Scalability issues arise primarily from the need for precise control over the fabrication process. Methods such as solvent casting, electrospinning and 3D printing are commonly used to create PLA-based scaffolds, but scaling these techniques for mass production can be difficult [74]. For example, electrospinning, while ideal for creating fine, fibre-like scaffolds with high surface areas, can be time-consuming and difficult to scale up. The process requires careful control of parameters such as voltage, solution viscosity and environmental conditions. Small fluctuations in these parameters during scaling can lead to inconsistencies in fibre diameter, porosity or scaffold morphology, which can affect scaffold performance and suitability for tissue regeneration.

Similarly, 3D printing, which allows for precise control over scaffold architecture, may face challenges when scaling up. The high precision required for creating scaffolds with uniform pore sizes, interconnectivity and mechanical properties can be difficult to achieve in large volumes. Moreover, variations in printer settings, such as extrusion rates and temperature control, can lead to inconsistencies in the final product. The speed of production also becomes a limiting factor when transitioning from small- to large-scale production. Reproducibility is another challenge in the production of PLA-based composites for tissue engineering. The variability in raw material quality, including the PLA polymer itself and the reinforcing agents used, can introduce inconsistencies in the final scaffold properties. For example, variations in the molecular weight, crystallinity or purity of PLA can lead to differences in the degradation rate, mechanical strength and biocompatibility. Even minor changes in processing conditions such as temperature, solvent concentration or mixing time can result in different scaffold qualities, which can affect their performance in tissue engineering applications. The improvements in the standardization of manufacturing processes are needed to address these scalability and reproducibility issues. This includes better control over raw material sourcing, precise process monitoring and the development of automated systems for large-scale production. Research into continuous manufacturing techniques, such as roll-to-roll processing or extrusion-based systems, could help optimize production efficiency and consistency. Additionally, advances in process modelling and simulation can provide valuable insights into scaling up production while maintaining quality. PLA-based composites show great potential for tissue engineering, and challenges related to scalability and reproducibility must be addressed to ensure their viability for clinical use.

9 Regulatory and Clinical Considerations

Regulatory and clinical considerations for PLA-based composites in tissue engineering are crucial to ensure their safety, efficacy and approval for medical use. These materials must meet strict regulatory standards set by agencies such as the FDA and EMA, which evaluate their biocompatibility, degradation rates and potential cytotoxicity. The comprehensive preclinical and clinical studies are needed to assess their long-term performance and risks in human applications. Additionally, proper manufacturing practices must be followed to ensure product consistency and traceability. The successful commercialization of PLA-based composites in tissue engineering and biomedical applications is to be carried out by meeting these regulatory requirements.

9.1 FDA and ISO Standards for Resorbable Polymers

The regulatory landscape for resorbable polymers, such as PLA-based composites in tissue engineering, is governed by strict guidelines to ensure their safety, effectiveness and quality for medical applications. In the United States, the Food and Drug Administration (FDA) is the primary regulatory body overseeing the approval of medical devices and biomaterials [75]. For PLA-based composites, which are intended for use in tissue engineering, these materials are typically classified as medical devices or drug delivery systems. As such, they must comply with FDA regulations, including those outlined in the Code of Federal Regulations (CFR) Title 21 [76].

For devices that come into direct contact with tissues, FDA guidance requires that the materials meet stringent requirements for biocompatibility, non-toxicity and proper degradation profiles. Biocompatibility testing is critical and is generally evaluated through the ISO 10993 standard series [77]. ISO 10993-1:2018, which addresses the biological evaluation of medical devices, provides a framework for testing the interaction of PLA-based composites with living tissues, ensuring that they do not cause adverse reactions such as inflammation, cytotoxicity or genotoxicity. Depending on the intended use, additional testing for systemic toxicity, sensitization and irritation may be needed. Furthermore, the FDA has evaluated the degradation rate of PLA-based composites to ensure that their resorption aligns with the tissue regeneration process. The material must degrade in a controlled manner not too quickly, as this could compromise tissue healing, or too slowly, which could lead to chronic inflammation or other complications. The FDA may require comprehensive preclinical studies, including in vivo testing, to assess the mechanical integrity of a polymer over time and its ability to support cell attachment and its biological degradation products.

In addition to FDA standards, ISO standards also play a significant role in the development of PLA-based composites. ISO 13485:2016, which specifies the requirements for a quality management system for medical devices, ensures that manufacturers maintain consistent quality throughout production [78]. The adherence to ISO standards guarantees that PLA-based composites are manufactured under controlled conditions with a documented process that complies with regulatory requirements. Adherence to FDA and ISO standards is crucial for the approval of PLA-based composites for clinical use in tissue engineering. These standards help ensure that the materials are safe, effective and reliable for their intended applications, providing confidence in their use in human clinical settings. Additionally, manufacturers must conduct thorough testing, follow good manufacturing practices (GMPs) and ensure proper documentation to meet both regulatory and clinical requirements.

9.2 Preclinical and Clinical Evaluations of PLA Composites

Preclinical and clinical evaluations are essential steps in the regulatory process for PLA-based composites used in tissue engineering. These evaluations help ensure that the materials are safe, effective and appropriate for human use. In preclinical studies, PLA composites undergo rigorous testing in laboratory and animal models to assess their biological, mechanical and degradation properties before human trials can begin. Preclinical evaluation typically starts with biocompatibility testing, which is vital for understanding how PLA-based composites interact with living tissues. A series of tests are performed to evaluate cytotoxicity, genotoxicity, local tissue irritation and systemic toxicity using standards such as ISO 10993 [79]. These tests help determine whether the composites cause any harmful reactions when they are implanted in the body. Moreover, testing in animal models is conducted to assess the ability of a material to support tissue growth, integration and healing. The degradation behaviour of the PLA composites was also examined, as they should degrade in a controlled manner, aligning with the tissue healing process without causing adverse inflammatory reactions. In addition to biocompatibility, preclinical studies have focused on the mechanical properties of PLA-based composites. For tissue engineering applications, the material must have suitable mechanical strength, stiffness and elasticity to support the load-bearing requirements of different tissues. For example, scaffolds designed for bone tissue engineering must have a similar mechanical strength to that of natural bone to provide adequate support for cell attachment and tissue regeneration.

Once preclinical studies demonstrate that PLA-based composites are biocompatible and mechanically suitable, the next step is clinical evaluation. The clinical trials are conducted in phases, starting with small-scale phase I trials, which assess the safety, tolerability and preliminary efficacy of PLA composites in human subjects. These trials often involve a limited number of patients and focus on monitoring adverse reactions such as infection, inflammation or immune responses. If successful, phase II trials are expanded to include a larger patient group to further evaluate the efficacy and safety of the composite materials [80]. In phase III trials, larger, multi-centre studies have been conducted to confirm the long-term safety and effectiveness of PLA-based composites in treating specific conditions, such as bone defects, cartilage injuries or skin wounds [81]. These studies are critical for gathering enough data to support regulatory approval for market release. Throughout these phases, regulatory bodies such as the FDA require comprehensive documentation of all testing and clinical trial results, which demonstrates that PLA-based composites meet all necessary safety and performance standards. Only after successful completion of these evaluations can PLA-based composites be approved for widespread use in clinical practice. Preclinical and clinical evaluations are critical for the safe and effective integration of PLA-based composites into tissue engineering applications. These evaluations provide the necessary data to ensure that the materials meet regulatory standards and perform as intended in real-world medical settings.

9.3 Ethical Considerations in Biomedical Applications

Ethical considerations play a crucial role in the development and application of PLA-based composites in biomedical tissue engineering. As these materials are designed for direct use in the human body, it is vital that their development and application adhere to ethical standards, ensuring safety, efficacy and respect for patient rights. One of the primary ethical concerns is informed consent. In clinical trials, patients must be fully informed about the risks, benefits and potential outcomes of using PLA-based composites in their treatment. This involves providing clear, understandable information about the material's properties, potential side effects and unknowns related to long-term degradation and tissue integration. Informed consent ensures that participants voluntarily agree to take part in clinical trials with a comprehensive understanding of the potential risks, which is crucial for maintaining ethical standards in clinical research. Another ethical issue is the impact of material degradation. PLA-based composites are designed to degrade over time, as they promote tissue regeneration [82]. However, the degradation rate and the by-products of this process must be carefully considered, as premature degradation or the production of harmful by-products could cause complications in patients, such as inflammation or tissue damage. Ethical considerations demand that thorough preclinical and clinical evaluations be conducted to ensure that PLA composites degrade in a controlled manner and that any degradation products are biocompatible and non-toxic. Failure to adequately address this could result in unintended harm to patients, undermining their trust in these materials.

Patient safety and well-being are paramount when implementing PLA-based composites in tissue engineering [83]. As these materials are often used in complex, high-risk medical procedures such as bone and cartilage repair or skin regeneration, minimizing the risk of complications such as infection, immune rejection or implant failure is essential. The regular monitoring of patients during and after implantation is needed to detect any adverse effects early and ensure that appropriate medical intervention is available if needed. Furthermore, the equity and accessibility of PLA-based composite therapies should be considered. While these materials show great promise for tissue engineering, they must be made accessible to a wide range of patients, including those from disadvantaged or underserved populations. The ethical development and use of PLA composites in biomedical applications require ensuring that their benefits are not restricted to certain groups but are available to all who might benefit from them. Finally, animal testing is an important ethical consideration during the preclinical evaluation phase. Animal models provide crucial insights into the safety and efficacy of PLA-based composites, and the ethical principle of minimizing harm to animals must be followed. Alternatives to animal testing, such as in vitro models, should be considered whenever possible, and when animal testing is necessary, it must be conducted under humane conditions with minimal suffering. The ethical considerations surrounding PLA-based composites in tissue engineering are multi-faceted and involve patient safety, informed consent, environmental impact and the humane treatment of animals. Adherence to these ethical principles ensures that the application of PLA composites is aligned with broader societal values and that they are used responsibly in biomedical practice.

10 Future Perspectives and Research Directions

The future of PLA-based composites in tissue engineering is promising, with several innovations in composite design, emerging applications in regenerative medicine and a growing focus on sustainability and cost-effectiveness. Recent advances in composite design and fabrication techniques are opening new possibilities for improving the performance of PLA-based materials. One significant trend is the development of multi-functional composites, which combine PLA with a variety of reinforcements, such as bioceramics, nanomaterials and natural fibres. These materials are engineered to enhance mechanical properties, improve bioactivity and control degradation rates, making them better suited for specific tissue engineering applications ranging from bone and cartilage regeneration to skin repair. The incorporation of smart materials such as temperature-responsive polymers or biodegradable nanostructures allows PLA-based composites to adapt to the changing physiological environment, offering enhanced functionality in dynamic tissue repair processes [84].

In terms of fabrication, 3D printing technologies are revolutionizing the production of customized scaffolds for tissue engineering. PLA-based composites are increasingly used in additive manufacturing, allowing the creation of complex, patient-specific scaffolds with precise control over pore structure and architecture.

This method ensures better tissue integration and supports cell migration and proliferation, which are essential for the successful regeneration of functional tissue. The future of 3D printing for PLA composites lies in improving the resolution and speed of printing while also developing multi-material printing techniques that combine PLA with other bioactive substances. PLA-based composites play a central role in the growing field of regenerative medicine. The ability of these materials to mimic the mechanical and biochemical properties of natural tissues makes them ideal candidates for scaffolds and implants used in a variety of therapeutic applications. PLA composites are already being explored for bone tissue engineering, where their controlled degradation matches that of bone healing and remodelling. In cartilage regeneration, PLA-based composites are designed to replicate the mechanical properties of cartilage, providing support for load-bearing joints and promoting tissue regeneration. Moreover, PLA composites are being integrated with growth factor, stem cell and gene therapy approaches, enabling advanced regenerative therapies that not only replace damaged tissues but also stimulate tissue regeneration. The future of PLA-based composites in regenerative medicine lies in their integration with drug delivery systems, where controlled release mechanisms can enhance therapeutic outcomes, reduce side effects, and accelerate tissue healing. Sustainability is a critical consideration in the development of biomaterials. PLA is already recognized as an environmentally friendly polymer due to its biodegradable nature and renewable source. However, future research will focus on improving the sustainability of PLA composites by optimizing the use of renewable resources, reducing waste and ensuring minimal environmental impact during production. The development of biobased additives and reinforcing agents such as biodegradable nanocellulose or bioactive glass will contribute to the eco-friendliness of PLA composites. From a cost-effective standpoint, PLA-based composites need to remain affordable for widespread clinical use [85]. As PLA production technologies advance and economies of scale are realized, the cost of PLA and its composites is expected to decrease, making them more accessible for both clinical and commercial applications. Additionally, the integration of low-cost materials such as natural fibres can reduce production costs while enhancing the mechanical properties of PLA composites.

The future of PLA-based composites in tissue engineering holds tremendous potential. The innovations in composite design, advanced fabrication techniques and emerging applications in regenerative medicine will continue to drive the field forward. Moreover, a strong focus on sustainability and cost-effectiveness will ensure that PLA composites remain a viable solution for tissue engineering in the future, balancing clinical effectiveness with environmental responsibility.

References

1. Nath PC, Sharma R, Debnath S, Nayak PK, Roy R, Sharma M, Sridhar K (2024) Recent advances in production of sustainable and biodegradable polymers from agro-food waste: applications in tissue engineering and regenerative medicines. Int J Biol Macromol:129129

2. Donate R, Monzón M, Alemán-Domínguez ME (2020) Additive manufacturing of PLA-based scaffolds intended for bone regeneration and strategies to improve their biological properties. e-Polymers 20(1):571–599
3. Arif ZU, Khalid MY, Noroozi R, Sadeghianmaryan A, Jalalvand M, Hossain M (2022) Recent advances in 3D-printed polylactide and polycaprolactone-based biomaterials for tissue engineering applications. Int J Biol Macromol 218:930–968
4. Morsada Z, Hossain MM, Islam MT, Mobin MA, Saha S (2021) Recent progress in biodegradable and bioresorbable materials: from passive implants to active electronics. Appl Mater Today 25:101257
5. Iqbal N, Khan AS, Asif A, Yar M, Haycock JW, Rehman IU (2019) Recent concepts in biodegradable polymers for tissue engineering paradigms: a critical review. Int Mater Rev 64(2):91–126
6. Jem KJ, Tan B (2020) The development and challenges of poly (lactic acid) and poly (glycolic acid). Adv Indus Eng Polym Res 3(2):60–70
7. Asti A, Gioglio L (2014) Natural and synthetic biodegradable polymers: different scaffolds for cell expansion and tissue formation. Int J Artif Organs 37(3):187–205
8. Elfaleh I, Abbassi F, Habibi M, Ahmad F, Guedri M, Nasri M, Garnier C (2023) A comprehensive review of natural fibers and their composites: an eco-friendly alternative to conventional materials. Results Eng:101271
9. Singhvi MS, Zinjarde SS, Gokhale DV (2019) Polylactic acid: synthesis and biomedical applications. J Appl Microbiol 127(6):1612–1626
10. Li G, Zhao M, Xu F, Yang B, Li X, Meng X, Li Y et al (2020) Synthesis and biological application of polylactic acid. Molecules 25(21):5023
11. Santoro M, Shah SR, Walker JL, Mikos AG (2016) Poly (lactic acid) nanofibrous scaffolds for tissue engineering. Adv Drug Deliv Rev 107:206–212
12. Tyler B, Gullotti D, Mangraviti A, Utsuki T, Brem H (2016) Polylactic acid (PLA) controlled delivery carriers for biomedical applications. Adv Drug Deliv Rev 107:163–175
13. Lassalle V, Ferreira ML (2007) PLA nano-and microparticles for drug delivery: an overview of the methods of preparation. Macromol Biosci 7(6):767–783
14. Omigbodun FT, Oladapo BI, Osa-uwagboe N (2024) Exploring the frontier of Polylactic Acid/Hydroxyapatite composites in bone regeneration and their revolutionary biomedical applications–A review. J Reinforced Plastics Compos:07316844241278045
15. Murariu M, Dubois P (2016) PLA composites: from production to properties. Adv Drug Deliv Rev 107:17–46
16. Gangolphe L, Leon-Valdivieso CY, Nottelet B, Déjean S, Bethry A, Pinese C, Garric X (2021) Electrospun microstructured PLA-based scaffolds featuring relevant anisotropic, mechanical and degradation characteristics for soft tissue engineering. Mater Sci Eng, C 129:112339
17. Ebrahimi F, Ramezani Dana H (2022) Poly lactic acid (PLA) polymers: from properties to biomedical applications. Int J Polym Mater Polym Biomater 71(15):1117–1130
18. Ogueri KS, Jafari T, Escobar Ivirico JL, Laurencin CT (2019) Polymeric biomaterials for scaffold-based bone regenerative engineering. Regen Eng Transl Med 5:128–154
19. Zhang F, King MW (2020) Biodegradable polymers as the pivotal player in the design of tissue engineering scaffolds. Adv Healthcare Mater 9(13):1901358
20. Sun J, Shen J, Chen S, Cooper MA, Fu H, Wu D, Yang Z (2018) Nanofiller reinforced biodegradable PLA/PHA composites: current status and future trends. Polymers 10(5):505
21. Nagarajan V, Mohanty AK, Misra M (2016) Perspective on polylactic acid (PLA) based sustainable materials for durable applications: focus on toughness and heat resistance. ACS Sustain Chem Eng 4(6):2899–2916
22. Liu S, Qin S, He M, Zhou D, Qin Q, Wang H (2020) Current applications of poly (lactic acid) composites in tissue engineering and drug delivery. Compos B Eng 199:108238
23. Farah S, Anderson DG, Langer R (2016) Physical and mechanical properties of PLA, and their functions in widespread applications—a comprehensive review. Adv Drug Deliv Rev 107:367–392

24. Gritsch L, Conoscenti G, La Carrubba V, Nooeaid P, Boccaccini AR (2019) Polylactide-based materials science strategies to improve tissue-material interface without the use of growth factors or other biological molecules. Mater Sci Eng, C 94:1083–1101
25. Ajioka M, Enomoto K, Suzuki K, Yamaguchi A (1995) The basic properties of poly (lactic acid) produced by the direct condensation polymerization of lactic acid. J Environ Polym Degrad 3:225–234
26. Södergård A, Stolt M, Inkinen S (2022) Industrial production of high-molecular-weight poly (lactic acid). In: Poly (lactic acid) synthesis, structures, properties, processing, applications, and end of life, pp 29–44
27. Marques DA, Jarmelo S, Baptista CM, Gil MH (2010) Poly (lactic acid) synthesis in solution polymerization. Macromolec Symp 296(1):63–71
28. Rosato DV, Rosato DV, Rosato DV, Rosato DV (1990) Extrusion. In: Plastics processing data handbook, pp 91–169
29. Fu H, Xu H, Liu Y, Yang Z, Kormakov S, Wu D, Sun J (2020) Overview of injection molding technology for processing polymers and their composites. ES Mater Manuf 8(20):3–23
30. Mount III EM (2024) Extrusion processes. In: Applied plastics engineering handbook. William Andrew Publishing, pp 313–362
31. Throne J (2024) Thermoforming. In: Applied plastics engineering handbook. William Andrew Publishing, pp 449–480
32. Dumont P, Martoïa F, Orgéas L (2023) Compression moulding. In: Design and manufacture of structural composites. Woodhead Publishing, pp 273–300
33. Sethi S, Ray BC (2015) Environmental effects on fibre reinforced polymeric composites: evolving reasons and remarks on interfacial strength and stability. Adv Coll Interface Sci 217:43–67
34. Galeski A (2003) Strength and toughness of crystalline polymer systems. Prog Polym Sci 28(12):1643–1699
35. Kamaly N, Yameen B, Wu J, Farokhzad OC (2016) Degradable controlled-release polymers and polymeric nanoparticles: mechanisms of controlling drug release. Chem Rev 116(4):2602–2663
36. Chong WJ, Shen S, Li Y, Trinchi A, Simunec DP, Kyratzis IL, Wen C et al (2023) Biodegradable PLA-ZnO nanocomposite biomaterials with antibacterial properties, tissue engineering viability, and enhanced biocompatibility. Smart Mater Manuf 1:100004
37. Schatkoski VM, do Amaral Montanheiro TL, de Menezes BRC, Pereira RM, Rodrigues KF, Ribas RG, Thim GP et al (2021) Current advances concerning the most cited metal ions doped bioceramics and silicate-based bioactive glasses for bone tissue engineering. Ceram Int 47(3):2999–3012
38. Ferrage L, Bertrand G, Lenormand P, Grossin D, Ben-Nissan B (2017) A review of the additive manufacturing (3DP) of bioceramics: Alumina, zirconia (PSZ) and hydroxyapatite. J Aust Ceram Soc 53:11–20
39. Ibrahim NF, Mohamad H, Noor SNFM (2017) Characterization on melt-derived bioactive glass powder from SiO_2-CaO-Na_2O-P_2O_5 system. J Non-Cryst Solids 462:23–31
40. Kaur G, Pandey OP, Singh K, Homa D, Scott B, Pickrell G (2014) A review of bioactive glasses: their structure, properties, fabrication and apatite formation. J Biomed Mater Res Part A 102(1):254–274
41. Bikiaris ND, Koumentakou I, Samiotaki C, Meimaroglou D, Varytimidou D, Karatza A, Papageorgiou GZ (2023) Recent advances in the investigation of poly (lactic acid)(PLA) nanocomposites: incorporation of various nanofillers and their properties and applications. Polymers 15(5):1196
42. Dufresne A (2019) Nanocellulose processing properties and potential applications. Curr For Reports 5:76–89
43. Serrano MC, Gutiérrez MC, del Monte F (2014) Role of polymers in the design of 3D carbon nanotube-based scaffolds for biomedical applications. Prog Polym Sci 39(7):1448–1471
44. Zhou H, Lee J (2011) Nanoscale hydroxyapatite particles for bone tissue engineering. Acta Biomater 7(7):2769–2781

45. Shin SR, Li YC, Jang HL, Khoshakhlagh P, Akbari M, Nasajpour A, Khademhosseini A (2016) Graphene-based materials for tissue engineering. Adv Drug Deliv Rev 105:255–274
46. Sanivada UK, Mármol G, Brito FP, Fangueiro R (2020) PLA composites reinforced with flax and jute fibers—a review of recent trends, processing parameters and mechanical properties. Polymers 12(10):2373
47. Liu D, Song J, Anderson DP, Chang PR, Hua Y (2012) Bamboo fiber and its reinforced composites: structure and properties. Cellulose 19:1449–1480
48. Ho MP, Lau KT, Wang H, Bhattacharyya D (2011) Characteristics of a silk fibre reinforced biodegradable plastic. Compos B Eng 42(2):117–122
49. Wang G, Zhang D, Wan G, Li B, Zhao G (2019) Glass fiber reinforced PLA composite with enhanced mechanical properties, thermal behavior, and foaming ability. Polymer 181:121803
50. Ismail KBM, Kumar MA, Mahalingam S, Raj B, Kim J (2024) Carbon fiber-reinforced polymers for energy storage applications. J Energy Storage 84:110931
51. Al-Furjan MSH, Shan L, Shen X, Zarei MS, Hajmohammad MH, Kolahchi R (2022) A review on fabrication techniques and tensile properties of glass, carbon, and Kevlar fiber reinforced rolymer composites. J Market Res 19:2930–2959
52. Tajbakhsh Raeisdasteh Hokmabad V, Davaran S, Ramazani A, Salehi R (2017) Design and fabrication of porous biodegradable scaffolds: a strategy for tissue engineering. J Biomater Sci Polym Ed 28(16):1797–1825
53. Wang L, Gardner DJ, Wang J, Yang Y, Tekinalp HL, Tajvidi M, Anderson J (2020) Towards industrial-scale production of cellulose nanocomposites using melt processing: a critical review on structure-processing-property relationships. Compos B Eng 201:108297
54. Backes EH, Harb SV, Beatrice CAG, Shimomura KMB, Passador FR, Costa LC, Pessan LA (2022) Polycaprolactone usage in additive manufacturing strategies for tissue engineering applications: a review. J Biomed Mater Res B Appl Biomater 110(6):1479–1503
55. Ji D, Lin Y, Guo X, Ramasubramanian B, Wang R, Radacsi N, Ramakrishna S (2024) Electrospinning of nanofibres. Nat Rev Methods Primers 4(1):1
56. Nazemi MM, Khodabandeh A, Hadjizadeh A (2022) Near-field electrospinning: crucial parameters, challenges, and applications. ACS Appl Bio Mater 5(2):394–412
57. Kim TG, Shin H, Lim DW (2012) Biomimetic scaffolds for tissue engineering. Adv Func Mater 22(12):2446–2468
58. Nikolova MP, Chavali MS (2019) Recent advances in biomaterials for 3D scaffolds: A review. Bioactive Mater 4:271–292
59. Collins MN, Ren G, Young K, Pina S, Reis RL, Oliveira JM (2021) Scaffold fabrication technologies and structure/function properties in bone tissue engineering. Adv Func Mater 31(21):2010609
60. Zhou X, Zhou G, Junka R, Chang N, Anwar A, Wang H, Yu X (2021) Fabrication of polylactic acid (PLA)-based porous scaffold through the combination of traditional bio-fabrication and 3D printing technology for bone regeneration. Colloids Surf, B 197:111420
61. Corneillie S, Smet M (2015) PLA architectures: the role of branching. Polym Chem 6(6):850–867
62. Guo X, Xi L, Yu M, Fan Z, Wang W, Ju A, Ren W et al (2023) Regeneration of articular cartilage defects: therapeutic strategies and perspectives. J Tissue Eng 14:20417314231164764
63. Huang D, Huang Y, Xiao Y, Yang X, Lin H, Feng G, Zhang X et al (2019) Viscoelasticity in natural tissues and engineered scaffolds for tissue reconstruction. Acta Biomater 97:74–92
64. Ilomuanya MO, Okafor PS, Amajuoyi JN, Onyejekwe JC, Okubanjo OO, Adeosun SO, Silva BO (2020) Polylactic acid-based electrospun fiber and hyaluronic acid-valsartan hydrogel scaffold for chronic wound healing. Beni-Suef Univ J Basic Appl Sci 9:1–13
65. Baran EH, Erbil HY (2019) Surface modification of 3D printed PLA objects by fused deposition modeling: a review. Coll Interfaces 3(2):43
66. Vo TN, Kasper FK, Mikos AG (2012) Strategies for controlled delivery of growth factors and cells for bone regeneration. Adv Drug Deliv Rev 64(12):1292–1309
67. Bai Y, Yin G, Huang Z, Liao X, Chen X, Yao Y, Pu X (2013) Localized delivery of growth factors for angiogenesis and bone formation in tissue engineering. Int Immunopharmacol 16(2):214–223

68. Ding Y, Li W, Zhang F, Liu Z, Zanjanizadeh Ezazi N, Liu D, Santos HA (2019) Electrospun fibrous architectures for drug delivery, tissue engineering and cancer therapy. Adv Func Mater 29(2):1802852
69. Sebbar K, El Aabedy A, Ibnsouda Koraichi S, Ulag S, Gunduz O, Elabed S (2024) Greener approaches to combat biofilm's antimicrobial resistance on 3d-printed materials: a systematic review. Coatings 14(4):400
70. Luo JD, Miller C, Jirjis T, Nasir M, Sharma D (2018) The effect of non-steroidal anti-inflammatory drugs on the osteogenic activity in osseointegration: a systematic review. Int J Implant Dentistry 4:1–12
71. Bazgir M, Zhang W, Zhang X, Elies J, Saeinasab M, Coates P, Sefat F (2021) Degradation and characterisation of electrospun polycaprolactone (PCL) and poly (lactic-co-glycolic acid)(PLGA) scaffolds for vascular tissue engineering. Materials 14(17):4773
72. Bakhtiari H, Nouri A, Khakbiz M, Tolouei-Rad M (2023) Fatigue behaviour of load-bearing polymeric bone scaffolds: a review. Acta Biomaterialia
73. Milovanovic S, Pajnik J, Lukic I (2022) Tailoring of advanced poly (lactic acid)-based materials: a review. J Appl Polym Sci 139(12):51839
74. Amato SF (2015) Regulatory strategies for biomaterials and medical devices in the USA: classification, design, and risk analysis. In: Regulatory affairs for biomaterials and medical devices. Woodhead Publishing, pp 27–46
75. Elchidana P (2024) Regulation of pharmaceutical polymers. In: Polymers for pharmaceutical and biomedical applications. Elsevier, pp 425–450
76. Sharma A, Luthra G (2023) Significance of ISO 10993 standards in ensuring biocompatibility of medical devices: a review. J Pharm Res Int 35(8):23–34
77. Ghasemi-Mobarakeh L, Kolahreez D, Ramakrishna S, Williams D (2019) Key terminology in biomaterials and biocompatibility. Curr Opin Biomed Eng 10:45–50
78. De Jong WH, Carraway JW, Geertsma RE (2020) In vivo and in vitro testing for the biological safety evaluation of biomaterials and medical devices. In: Biocompatibility and performance of medical devices. Woodhead Publishing, pp 123–166
79. Rounsaville BJ, Carroll KM, Onken LS (2001) A stage model of behavioral therapies research: getting started and moving on from stage I. Clin Psychol Sci Pract 8(2):133
80. Bobo D, Robinson KJ, Islam J, Thurecht KJ, Corrie SR (2016) Nanoparticle-based medicines: a review of FDA-approved materials and clinical trials to date. Pharm Res 33:2373–2387
81. Malekpour K, Hazrati A, Khosrojerdi A, Roshangar L, Ahmadi M (2023) An overview to nanocellulose clinical application: biocompatibility and opportunities in disease treatment. Regen Therapy 24:630–641
82. Trucillo P (2024) Biomaterials for drug delivery and human applications. Materials 17(2):456
83. Percival KM, Paul V, Husseini GA (2024) Recent advancements in bone tissue engineering: integrating smart scaffold technologies and bio-responsive systems for enhanced regeneration. Int J Mol Sci 25(11):6012
84. Tümer EH, Erbil HY (2021) Extrusion-based 3D printing applications of PLA composites: a review
85. Ramezani Dana H, Ebrahimi F (2023) Synthesis, properties, and applications of polylactic acid-based polymers. Polym Eng Sci 63(1):22–43

Chapter 7
Rheological Studies of Resorbable Polymeric Nanocomposites

Medi Harshith Kumar, Atanu Kumar Paul, Arbind Prasad, and Gourhari Chakraborty

1 Introduction

Degradable polymers can break down in different environmental conditions, including in vivo applications, making them appealing in modern-day applications because of the reduction in solid waste after use [1–4]. Biodegradable polymers have emerged as the clear alternative to plastic products including packaging, medical, cutlery items, and conductive materials preparations. Degradable polymers can be used for more than just short-term purposes since their characteristics can be tailored to allow for regulated breakdown rates, which facilitate tissue regeneration and sustained drug administration over a range of time [5–12]. However, limitations of thermal stability, mechanical strength, poor processability, and less melt strength bring down the applications in the real fields. Nanofillers have superior mechanical, thermal, and electrical conductivities, making them suitable for polymer matrix reinforcement [13, 14]. Recent advancements in polymer technology have addressed these shortcomings by creating novel composites or providing cutting-edge technologies for processing polymeric materials. The selection of nanofillers depends on the application, the nature of the fillers, and the nature of interaction with the matrix. Degradable polymers are mostly bio-based and biocompatible, leading to sustainable product development. Biomaterials have superior applicability in biomedical applications because of their non-toxic, biodegradable, and in vivo desorption [10, 15–17]. Resorbable polymers have biodegradable or in vivo desorption abilities, making

M. H. Kumar · A. K. Paul · G. Chakraborty (✉)
Department of Chemical Engineering, National Institute of Technology Andhra Pradesh, Tadepalligudem, Andhra Pradesh, India
e-mail: gourhari115@gmail.com

A. Prasad
Mechanical Engineering Department, Katihar Engineering College (Under Department of Science, Technology and Technical Education, Government of Bihar), Katihar, Bihar, India

A. Prasad and V. Katiyar (eds.), *Resorbable Polymers for Bioimplants and Fixation Devices*, Materials Horizons: From Nature to Nanomaterials,
https://doi.org/10.1007/978-981-96-9306-1_7

them resorbable inside the body. Polylactic acid (PLA), gelatin, collagen, bacterial cellulose, etc., are examples of some useful resorbable polymers in tissue engineering, drug delivery, implants, sutures, etc., applications [18, 19]. Three primary methods are used to create biodegradable composites: solution processing, melt processing, and in situ polymerization. The fundamental processing technique for creating composites from less thermally stable and hydrophilic polymers is solution casting. Film casting, casting–leaching, electrospinning, extrusion-based 3D printing, and fridge-drying are some of the solution-casting methods utilized in modern research and technologies. Melt processing is the most widely utilized technique for polymeric product development for biomedical applications [20]. PLA, PGA, PHA, and copolymer-based resorbable composites are mostly fabricated into useable items using extrusion, melt drawing, injection molding, fused deposition modeling-based 3D printing, etc. [11]. Nanocomposite properties strongly lie on the molecular weight of the polymer, filler properties, orientation of the fillers, dispersion of fillers, and interfacial interactions of reinforcements and matrix. The selection of processing technique depends on the end use of the products. Nanocomposite processing routes have a prominent role in determining the structural, mechanical, and thermal properties of end products [2, 11].

Rheology investigation gives information about the flow behavior of the polymeric chains in the solution or the molten state. Solution viscosity measurement reveals information about the flowability, activation energy requirement for flow, and gelation point of hydrogels [21, 22]. Thus processability of the composites strongly relies on the viscosity-related information for solution processing. The properties of melt-processed composites are strongly dependent on the molecular weight and the nature of polymer chain folding [23]. Rheology investigation followed by interpretation of data of storage modulus, loss modulus, and complex viscosity can indicate the nature of the composites. Melt strength and flowability are the two properties that govern successful and economical melt casting of composites. Melt rheology investigation of composites helps in optimizing the precursor compositions and processing conditions to get processable melt strength for the targeted applications [21]. Rheology investigation reveals information about the filler dispersion in the matrix. The effectiveness of the use of compatibilizers, cross-linkers, chain extenders, and plasticizers in composite systems can be examined through rheology investigations. The impact of rheology investigation in the development of paints, blends, and coatings is significant [6, 23–26].

In this chapter, discussions about the utility of rheological investigation for resorbable polymeric nanocomposite development are included. Flow measurement instruments for the viscosity and rheology of polymeric solutions and melts are suitably incorporated. Recent investigations related to the utilization of flow measurements in developing resorbable polymeric composites are extensively discussed. A summary and outlook section is appended at the end of this chapter highlighting the findings of recent works.

2 Rheology and Resorbable Polymeric Composites

Rheology is a crucial technique to study different filler and polymer matrix interactions. It explains how fillers act in polymers, how the arrangement matters, the distribution of nanofillers in the polymer, and how these interfaces can alter the structure and properties of biodegradable composites (Fig. 1). To process specific polymer matrix/filler systems, rheological assessments are really important, and it helps in determining suitable processing parameters, such as filler weight ratios, temperature, and screw speed, which are crucial to obtain the final composite materials [27, 28].

Bioresorbable polymeric composites are gaining attention in medical applications due to their ability to degrade in the body. Polydioxanone (PDO), polylactic acid (PLA), and polycaprolactone (PCL) are among the polymers studied for their tunable degradation profiles and mechanical properties [29]. The rheological properties of these composites are influenced by factors such as nanoparticle content and polymer–particle interactions [30]. Various synthetic and natural bioresorbable polymers have been developed, with degradation rates controllable through molecular weight and crystallinity adjustments [31]. Block copolymers, such as PLLA/PEG and PDLA/PEG, can form hydrogels with unique rheological properties and drug release capabilities [32]. These materials show promise in tissue engineering applications, offering tailored degradation times to match tissue regeneration processes and controlled drug delivery potential [29, 32].

Rheological tests help to understand the interaction of the matrix with other substances and their reinforcement in the composite material [33, 34]. Controlled geometric arrangements, with oscillatory frequency and strain sweep tests, provide important information on the complex viscosity of the polymer, storage, and loss moduli, phase lag, and the naturally relaxed state of the polymer chains. It has been observed that the complex viscosity of biocomposites displays shear-thinning behavior in all frequency ranges as per the power law relationship [24, 25].

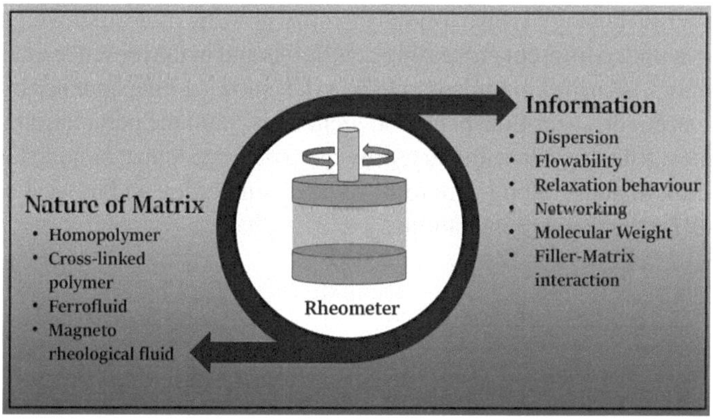

Fig. 1 Rheology analysis information

The polymers and additives interactions are dynamic in nature and are influenced by factors like miscibility, dispersion quality, etc. Cole–Cole plots (which plot imaginary viscosity concerning real viscosity over two flow modes) and Han plots (which plot storage and loss moduli) demonstrate these effects. Through this analysis, we can have a systematic understanding of the structure and function of the composite to direct the design requirement [6, 24]. By obtaining the crossover point where loss modulus (G'') equals storage modulus (G'), one can predict molecular weight and evaluate how processing affects its distribution from the rheological perspective. The nonlinear behavior in the low-frequency region of the curve shows the activity of G' and indicates a network formation in the composite during its molten state. Evaluation of the temperature-dependent behavior of G', G'', η, and related rheological parameters enables the estimation of flow activation energy in polymer matrix nanocomposites. Being familiar with melt rheology is useful for adjusting the material processing and improving the conditions for processing techniques as biodegradable polymers can contain fillers. This analysis will benefit more complicated systems with cross-linkers, plasticizers, and compatibilizers as it gives more advantages than normal polymer composites. Rheology helps clarify changes in major properties related to how blends function and the interaction of fillers in polymer blends [33, 35]. Rheological data can also provide major information about polymer chain relaxation and cross-linking levels in the materials. This information is really useful in systems for cross-linking using peroxide-type reactive extrusion or chemical linkers for gelling. Rheology measures the properties of materials to optimize manufacturing processes for obtaining the required desirable characteristics. This approach helps explore different ways in which viscosity varies in shear or the relationship between shear stress and shear rate for polymer solutions. For example, this is especially useful when the polymers that are being used are hydrophilic, such as cellulose, chitosan, and proteins all of which have high pervasiveness of gelling. The rheological studies are also indicative of the flowability and the gel points which simplifies the optimization of polymer applications and the preparation of the gels under the best possible conditions [36, 37].

Additionally, rheology is also used to quantify how the molecular weight of polymers changes under different processing conditions and in the presence of nanofillers, chain modifiers, or in the combined state [6, 25]. It shows the dependence of deformability, melt strength, shear flow behavior, yield stress, and the percolation thresholds in composites. Rheology, in summary, provides a deep systematic understanding of a material's functionality, along with tools for the formulation and process refinement of materials for particular applications.

3 Flow Behavior Measurements

Recent studies have explored the flow behavior of polymer nanocomposites, revealing complex interactions between nanoparticles and polymer matrices. Shear flow can enhance miscibility in polymer nanocomposites by altering polymer chain conformations [38]. The addition of small amounts of nanoparticles, such as nano-$CaCO_3$, can improve the melt flow behavior of polymers like poly-L-lactide [39]. However, nanoparticle surface polarity significantly influences composite flow behavior, with hydrophobic surfaces potentially reducing the maximum filler load due to increased particle–polymer interactions [40]. Interestingly, small, energetically neutral nanoparticles can act as plasticizers, reducing polymer melt viscosity regardless of chain length, an effect that persists for particles up to the size of the chain or entanglement mesh [41]. These findings provide valuable insights into the complex rheological behavior of polymer nanocomposites and offer a basis for predicting and optimizing their flow properties in various applications.

3.1 Rheology

A rheometer is a laboratory instrument used to measure and analyze characteristics such as flowability and flow nature of a liquid, slurry, or suspension under the effect of shear stress [42, 43]. It is utilized to investigate the primary chain relaxation behavior of polymer melts. A rheology study is helpful in cases where normal viscosity calculations cannot define the actual flow of some materials [44]. Rheometers can be classified into two categories depending upon the force applied to the fluid: rotational rheometer, which performs measurements under the control of shear stress or shear strain, and extensional rheometer, which functions under extensional stress or extensional strain [29, 44, 45].

There are two categories of oscillatory rheology investigations based on the type of inquiry: (i) strain sweep and (ii) frequency sweep [24, 32, 46, 47]. In the case of the strain sweep, angular frequency (ω) is kept constant and the impact of % strain change on the G', G'', and η is investigated, whereas for the second, % strain is kept constant and ω is varied with a defined frequency range. In the case of rotational rheology operation, measurements are conducted either at a definite torque within a time interval or for varying durations [24, 26].

Shearing Geometries (Fig. 2)

(a) *Concentric Cylinder (CC) Measuring Systems*

To perform tests on low-viscosity liquids generally, CC systems are used [26]. There is a restriction on the radius ratio of cup and bob, and it must not be more than 1.0487.

(b) *Cone/plate (CP) Measuring Systems*

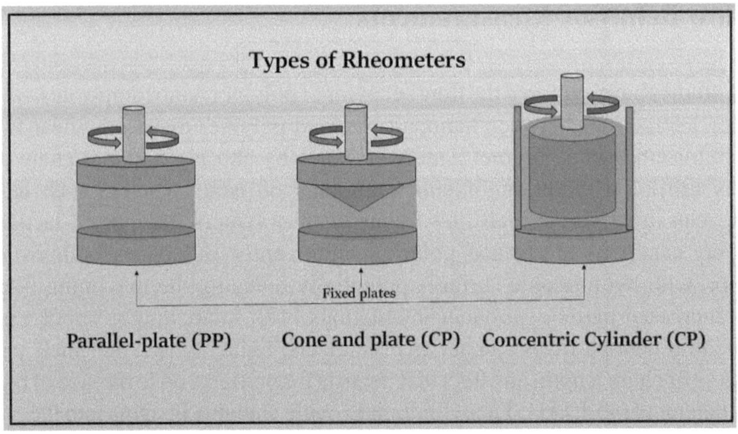

Fig. 2 Different types of rheometers

All kinds of fluids can benefit from CP systems; however, their applicability in dispersion studies depends on the size of the particles [47]. Cones having an angle greater than 4° are considered sub-standard, according to the ISO standard. The best CP systems to utilize for accurate measurements are those with truncated cone tips.

(c) *Parallel Plate (PP) Measuring Systems*

Polymeric melts, solutions, and gels are among the several polymeric systems that PP systems are primarily utilized for [6]. The ideal range for the distance between the plates is between 0.5 and 1.0 mm. The nature of the fluid, the sort of research, and other factors are determined.

3.2 *Viscosity*

One characteristic that indicates how much resistance a fluid offers to flow is its viscosity [47, 48]. Viscometer is an instrument that measures the viscosity of a material at laminar flow conditions; here, it refers to the resorbable polymeric nanocomposites. This resistance is due to the forces like surface drag. Different polymeric systems have different viscosity values based on their solvents, compositions, and different concentrations [26].

Given below are a few standard laboratory viscometers used for liquids.

Falling Sphere Viscometers

In this viscometer, the viscosity is determined by the motion of the particle through a stationary fluid governed by Stokes' law. A steel ball with a given size and density is let to move through the stationary liquid in a vertical capillary in a falling sphere viscometer [49]. The concept of terminal velocity is used to measure the viscosity. Depending on the different system conditions, different fluids are used [26].

For the industrial manufacture of composites, the melt flow index (MFI) is sometimes assessed. The MFI measurement is based on the principle that at a given temperature and under a predefined load, the amount of polymer passed through the capillary in ten minutes is measured. This preserves the consistency of the composite and the quality of the polymers in case of large-scale production. MFI is a representative of the chain mobility of the polymer melts, and it relies on the nature of the polymer, filler content, and the temperature at which it is taking place [49, 50].

U-Tube Viscometers

Wilhelm Ostwald invented the glass capillary viscometer, also known as the Ostwald viscometer. This device measures the duration it takes for a polymer solution to flow through a capillary tube compared to a solvent. This measurement helps determine the kinematic viscosity of the solution, and the device has to be calibrated to get accurate results.

U-tube viscometers are widely used for measuring fluid viscosity, but they have limitations. Caw and Wylie (1961) developed capillary viscometers with long-flared ends to minimize kinetic energy effects, improving accuracy in establishing viscosity scales [51]. Recent advancements include microfluidic capillary viscometers, which can measure viscosity using droplet spacing and require minimal sample volumes [52]. Sariyerli et al. (2018) compared capillary viscometers with the Stabinger viscometer SVM 3001, highlighting the importance of traceability in viscosity measurements [53]. To address the limitation of low shear rate measurements in traditional capillary tube viscometers, Cho et al. (1999) introduced a scanning capillary tube viscometer using a charge-coupled device sensor array, extending the measurable shear rate range [54]. These innovations have enhanced the capabilities of viscometers, allowing for more accurate and versatile measurements across various fluid types and viscosity ranges.

4 Rheology Studies of Resorbable Polymer and Nanocomposites (RNCs)

4.1 Gel/Hydrogel Rheology

Polylactic acid (PLA), polyglycolic acid (PGA), collagen, cellulose, gelatine, and CS are all examples of hydrogels and organogels that have high applications in commercial areas like tissue culture, drug delivery, adsorption, food, composites, etc. Cellulose, chitosan, gelatin, and modified cellulose are used to make gel composites for various applications. To find out the effectiveness and consistency of these gels, their viscosity is measured at different concentrations, pH levels, and shear rates (Table 1). Cellulose nanoparticle-based hydrogels have been synthesized by various research groups, and their gelation behavior has been studied. Different cross-linking methods are employed to create these gels, with the strength of the cross-links determining the gel's stability. Physical methods like using non-solvents or surfactants create weak cross-links, while chemical cross-linking and irradiation produce strong cross-links. The nature of the bonds and the stability of nanocomposite gels that are made from biodegradable polymers can be found by rheological studies.

The rheological and structural properties of cellulose nanocrystal (CNC) suspensions have been investigated using viscosity measurements with water as the medium. It was observed that the viscosity of CNC suspensions increases with concentration, likely due to increased collisions between CNCs. The viscosity also depends on CNC concentration, shear rate, and aspect ratio [61]. Wu et al. (2014) studied CNCs derived from switchgrass and cotton fibers for gel preparation and found that higher CNC content increases the storage modulus, loss modulus, and viscosity [55]. The gelation behavior was influenced by aspect ratio and CNC loading, with two critical gelation points identified [55, 61].

Similar studies were conducted on carboxymethyl cellulose gels in water with HCl and NaOH [36]. Gels with more than 5 wt% loading remained stable for 8 months, but 4 wt% gels lost about 30% of their strength in the same period. NaCl promoted cross-linking, increasing the storage and loss moduli, while NaOH reduced gel formation, and HCl enhanced it. Investigations into cellulose–aloe vera-based biogels for 3D printing found that high-viscosity gels (2800–4400 mPa·s) were suitable for extrusion, as confirmed using parallel plate and cone-and-plate rheometers [12, 36].

Cellulose-ether hydrogels for skin penetration showed pseudo-plastic behavior as per a study [62]. Ponsubha et al. (2020) synthesized gelatin, chitosan, and chondroitin sulfate-based 3D network gel using acetic acid solvent and gelling agents at 42 °C [56]. The viscosity of the gels was examined using a viscometer, and the

Table 1 Summary of viscosity studies of resorbable polymeric systems

Polymer system	Applications	Key findings	References
Cellulose	Hydrogel	Gel behavior depends on morphology and concentration of CNC	[55]
Gelatin–chitosan–chondroitin sulfate	Tissue culture	The highest viscosity reported at 5523 cp for chitosan-based gel	[56]
Gelatin–chitosan	Hydrogel	Chitosan addition has increased the rheological properties	[57]
Chitosan–gelatin	Drug delivery	Viscosity values indicated the thixotropic nature of the gels	[58]
Chitosan–gelatin	Drug delivery	Storage modulus increased for the cross-linked gels	[59]
Collagen–gelatin	Hydrogel	Time–temperature-dependent flow behavior observed	[60]

highest viscosity value was reported for chitosan hydrogel 5523 cP. Rheology analysis revealed that chitosan addition increased the linear region by 2.5% for oscillatory strain sweep investigation [56]. Protein-based hydrogel properties were investigated using rheology analysis. The albumin-based hydrogel's sol–gel transformation processes were investigated by frequency sweep analysis by Hájovská et al. (2020) [44]. Derkach et al. (2017) investigated the properties of chitosan–gelatin hydrogel under different pH conditions [57]. It was observed that in addition to chitosan from 0.1 to 0.6 wt% rheological properties of the gels increased to a large extent [57]. Rajput et al. (2023) synthesized biodegradable–biocompatible drug carriers using chitosan and gelatin [58]. It was observed that glutaraldehyde was effective in cross-linking these polymers. The thixotropic behavior of the gels was identified from the rheology investigation. The gels were observed to be effective carriers of the analgesic drug, aceclofenac [58]. Similarly, Omer et al. (2021) used chitosan–gelatin-based gels for oral drug delivery. Rheology investigations showed that the storage modulus increased for the cross-linked gels [59]. He et al. (2020) utilized rheology analysis for the examination of the flow behavior of collagen–gelatin gels. It was observed that gel properties are temperature and time-dependent [60]. Chitosan–collagen gel behavior was successfully examined by Zhang et al. (2021) using rheology analysis [63].

Rheology is critical for polymeric coatings, influencing spreadability, thickness, and uniformity. High apparent viscosity at low shear rates improves film formation during dip coating [64]. For edible coatings, incorporating pomegranate peel extract (PPE) into chitosan–gelatin mixtures increased viscosity, favoring applications in fruit coatings [65]. Chitin nanofibrils altered the rheological behavior of chitosan solutions, leading to solid-like properties and stable dynamic moduli [66]. Overall, it can be said that in the determination of the properties and applicability of chitosan, collagen, cellulose, gelatin, chondroitin sulfate, etc., rheology investigation of resorbable polymeric systems plays an important role. Gel usability in the internal and external human body also depends on the nature of the deformation of the gels which is reflected from the rheology investigations.

4.2 Thermoplastic Polymer Rheology Studies

Recent studies on the rheology of thermoplastic polymers reveal diverse behaviors influenced by their composition and processing conditions. Ghahramani et al. (2020) investigated thermoplastic vulcanizates (TPVs), highlighting the impact of cross-linked rubber content and curing levels on their complex rheological properties, including a novel observation of yield stress in extension [34]. Liu et al. (2019) developed a viscoelastic model for isothermal thermoplastic polymers, demonstrating a three-stage rheological characteristic that better aligns with experimental data than traditional models [67]. Mathew et al. (1993) focused on thermoplastic elastomers derived from NBR and PVC, finding that viscosity and flow behavior are significantly affected by the fusion extent of PVC particles [68]. Tajuddin et al. (2011)

characterized thermoplastic starches, revealing that flow properties are influenced by compounding parameters and melt temperatures, with wall-slip effects quantified through a modified Mooney procedure [45]. These studies collectively enhance the understanding of thermoplastic polymer rheology.

Thermoplastic resorbable polymers like PLA, PHB, PCL, PGA, PLGA, etc., have profound applicability in the development of resorbable transplants, scaffolds, drug carriers, sutures, etc. [20]. For polymers to be used in various industrial applications, different melt processing units are required for the development of composites and during casting of objects for specific applications. The efficiency of the resorbable polymeric nanocomposite production also relied on the proper picking of processing conditions. This remains true as major properties like melt strength, dispersion of filler, and deteriorations of molecular weight determine composite quality and process cost efficiency.

The flow behavior and nature of molten polymers can be understood by studying their rheological properties, which include storage modulus (G'), loss modulus (G''), and complex viscosity (η). Oscillatory rheology also is commonly used to study viscoelastic materials. This is done to monitor curing reactions in multi-component systems, particularly during the transition from a liquid to a solid state. Major parameters such as the start of the chemical reaction (t_0), the type and strength of hydrogel formation, the final complex dynamic viscosity (η*), and the gel point (t_{ael}) help to compare different network formulations and form new hybrid hydrogels [69].

As per a study by Shojaeiarani et al. (2019), focusing on the flow properties of melt PLA/CNC nanocomposites with varying CNC loadings, rheological tests were performed in oscillatory mode using a parallel plate geometry at 170 °C [70]. It was found that the addition of CNC increased the viscosity of the nanocomposites in both fabrication methods—film casting and spin coating. The storage modulus (G') and loss modulus (G'') also increased with higher CNC loading, which indicated the enhanced melt strength and greater chain entanglements within the PLA matrix [70]. Several other studies examined the effect of CNC reinforcement on PLA/PBAT blend properties [71, 72].

Chakraborty et al. (2023) investigated the effectivity of nanomaterial graphene (GR) distribution in PLA by frequency sweep rheology analysis between 0.1 and 500 s^{-1} frequency range under 5% strain rate [26]. It was observed that complex viscosity has two regions: One is solid-like and another Newtonian region with the increase of angular frequency ω. The flowability of melts was improved with the incorporation of GR in the PLA through extrusion [25].

Melt processing was conducted to investigate the dispersion of graphene in the PLA matrix, and before extrusion, a master batch of graphene was coated over the PLA pellets. The Han and Cole–Cole plots of masterbatch (MB)-coated nanocomposites showed increased miscibility and dispersion. The flowability of the composites was found to be improved by adding MB-coated graphene to the PLA matrix. Its homolymeric character was unaffected by the creation of the nanocomposite. These impacts are crucial considerations when producing PLA film on an industrial scale [6].

Also, a rheological analysis on polycaprolactone showed that the measurement results averaged over frequencies of 0.1 to 10 Hz and SF-01 had lower values for storage modulus (G'), loss modulus (G''), complex modulus (G^*), phase angle (δ), and complex viscosity (η^*) compared to the reference filler, PCL-01, within the same frequency range [73].

Viscosity measurements have long been utilized to determine the molecular weight of polymers, including biodegradable ones. Imaizumi et al. (2003) established relationships between intrinsic viscosity and molecular weight for biodegradable polyesters PBS and PBSA using SEC–LALLS and SEC–viscometry [74]. Dobrynin et al. (2023) proposed a universal approach for molecular weight determination from solution viscosity, applicable across a wide concentration range [75]. For biodegradable polymers like poly-D, L-lactide, and poly(glycolide-co-lactide), Rafler et al. (1990) determined viscosity–molecular weight relationships in various solvents using multiple characterization techniques [76]. Mahabadi (1984) introduced a novel method for estimating viscosity–average molecular weight from GPC data, particularly useful for polymers lacking known Mark–Houwink constants [77]. These studies demonstrate the versatility and importance of viscosity measurements in polymer characterization, offering various approaches to correlate viscosity with molecular weight for different polymer systems and solvents. Viscosity measurements were used to correlate the stereo-complexation of PLA fibers, calculate PLA molecular weight, and study the degradation of PLA/silk nanocomposites during recycling [78–81]. These rheological parameters provide insights into production consistency, material structure, flow behavior, synthesis conditions, and the efficiency of polymer-based solutions, gels, inks, and pastes.

4.3 Mathematical Modeling on Flow Behaviors of RNCs

Mathematical modeling of resorbable polymeric nanocomposites has been explored to understand their flow behaviors and degradation kinetics. Stephanou (2015) investigated how shear flow affects the phase behavior of polymer nanocomposites, finding that flow enhances miscibility [38]. Sevim and Pan (2018) developed a comprehensive model for hydrolytic degradation and erosion of biodegradable polymers, incorporating autocatalytic hydrolysis, oligomer production, and diffusion [82]. Liu et al. (2021) studied fluid flow effects on degradation kinetics, revealing that high porosity and dynamic conditions can slow down degradation, with a critical Reynolds number suppressing autocatalysis [83]. Sackett and Narasimhan (2011) reviewed mathematical modeling approaches for polymer erosion, categorizing them as phenomenological, probabilistic, or empirical [84]. These models consider various physical, chemical, and biological factors affecting polymer erosion, providing insights for rational design and optimization of biomaterials for drug delivery and tissue engineering applications.

Theoretical models are used to study molecular-level changes under different flow conditions, focusing on shear stress behavior at various strain rates. Dynamic

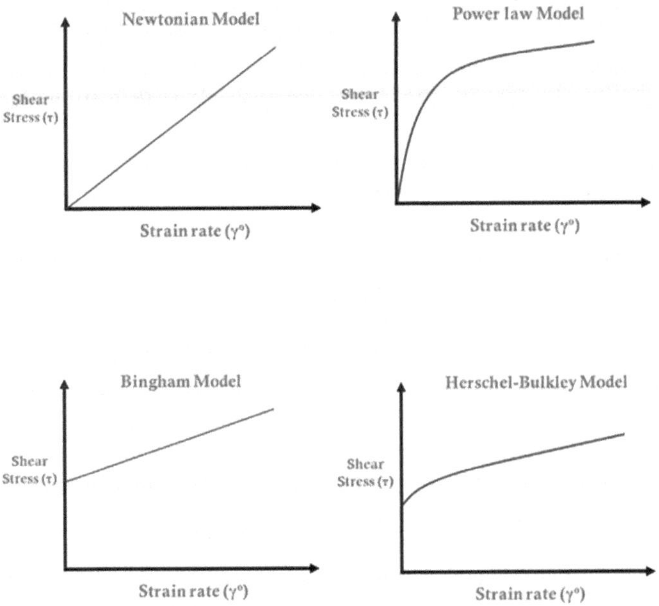

Fig. 3 Different flow models of biodegradable polymeric composites fluid

shear stress, indicating resistance to deformation in no-flow conditions, is calculated from these models. By combining theoretical models with experimental findings, the behavior of non-Newtonian flows can be predicted. Some models also relate viscosity to filler loading, while others are specifically applied to polymeric systems, including biodegradable nanocomposites (Fig. 3).

Power Law Model

The power law model has been widely applied to describe the flow behavior of resorbable polymeric nanocomposites. Studies have shown that these materials exhibit shear-thinning behavior, with the power law index (n) decreasing as filler concentration increases [39, 85]. A novel percolation-based theory has been developed to predict the power law index for highly filled composites, correlating n with filler concentration above the percolation threshold [86]. The addition of nanofillers can significantly affect rheological properties, with small amounts of nano-$CaCO_3$ improving the melt flow behavior of poly(L-lactic acid) composites [39]. Factors such as yield stress, relaxation time, and the width of the transition region between Newtonian and power law behaviors have been found to influence the loss factor (tan δ) in polymer nanocomposites [87]. These findings contribute to a better understanding of the flow characteristics of resorbable polymeric nanocomposites and their potential applications.

It is the most used model for the investigation of non-Newtonian fluid proposed by Ostwald and de Waele [88] and has a general form:

$$\eta = A\omega^n \quad (1)$$

Here, $A = A_o e\left[\frac{\Delta E}{R}\left(\frac{1}{T} - \frac{1}{T_0}\right)\right]$.

A and n are represented as preexponential factors and flow index.

Herschel–Bulkley (HB) Model

The Herschel–Bulkley (HB) model is widely used to describe non-Newtonian fluid behavior, particularly in blood flow and porous media applications. It incorporates yield stress and power law index parameters, making it more versatile than Bingham, power law, and Newtonian models [89]. The model has been applied to various scenarios, including blood flow through stenosed arteries [90] and fluid flow in propped fractures [91]. Numerical simulations and analytical calculations have been used to investigate HB fluid behavior in porous media, revealing the significant impact of yield stress on flow characteristics and pressure gradients [91]. Lattice Boltzmann methods have also been employed to model HB rheology in soft glassy materials, demonstrating a smooth transition between flowing and non-flowing states [92]. These studies highlight the HB model's importance in understanding complex fluid dynamics across diverse applications.

This model is used for polymeric systems where initial flow yield stress is required.

$$\tau = \tau_y + k\gamma^n$$

The viscosity can be calculated as:

$$\eta = \tau_0/\gamma + \dot{k}\gamma^n$$

Here, τ is the shear stress, and it is dependent on nanocomposite fluid composition and strain rate. τ_s is the yield stress, and it is a fluid internal property depending on the polymer as well as the loading of the composite. k is a measure of resistance toward flow, and n is the flow index. $n > 1$ indicates shear thickening behavior, i.e., increase in viscosity with strain rate which is normally not seen for biodegradable nanocomposite-based solutions and melts. It gets transformed into the **Bingham plastic model** when $n - 1$. In the most biodegradable nanocomposite-based systems, $n < 1$ indicates shear-thinning behavior. In case of $\tau_0 = 0$, it reduces to the power law model [93].

Casson Model

It is also similar to that of the HB model; only dependency of shear stress with strain rate is different while other terms have similar significance [93].

$$\tau^{1/2} = \tau_o^{1/2} + \eta_\infty^{1/2}(\gamma^o)^{1/2}$$

Mustafa and Khan (2015) examine the flow of Casson nanofluid over a nonlinearly stretching sheet under magnetic field effects, observing that temperature and nanoparticle volume fraction increase with the Casson fluid parameter [94]. Shahzadi and Ijaz (2019) studied Casson fluid flow in a curved annulus, noting increased stress formation and bolus variation compared to straight annulus models [95]. Zhirikova and Aloyev (2018) apply viscoelastic body models and fractal analysis to describe the flowability of polypropylene/carbon nanotube nanocomposites, demonstrating that yield stress is determined by elasticity modulus and structural state [96]. These studies collectively contribute to understanding the complex flow behaviors of polymeric nanocomposites and related materials under various conditions and in different geometries.

Cross-Model

The cross-model is a modified version of the power law model.

$$\eta = \eta_{\alpha} + \frac{\eta_0 - \eta_{\infty}}{1 + \left(\frac{\dot{\gamma}}{\gamma_c}\right)n}$$

Cross-equation was developed for non-Newtonian fluids considering different assumptions of pseudoplastic flow which is thought to be connected to the formation and breakage of structural linkages [97]. The cross-model is designed to fit the straight-line portion between η_0 and η_{∞}.

Burgos-Mármol et al. (2017) revealed that polymer chain stiffness affects nanoparticle mobility and dispersion, with a transition from isotropic to nematic phases influencing diffusion anisotropy and viscosity [98]. They observed a plasticizing effect at increased chain stiffness when shear is applied along the nematic axis. Zhirikova and Aloyev (2018) applied fractional derivatives and fractal analysis to describe the flowability of polypropylene/carbon nanotube nanocomposites, showing that yield stress depends on elasticity modulus and structural state [96]. These studies contribute to understanding the complex relationships between polymer structure, nanoparticle interactions, and flow behavior in nanocomposites.

Ellis Model

It is a three-parameter model and takes the form [99]:

$$\eta = \frac{\eta_0}{1 + \left(\frac{\tau}{\tau_{\frac{1}{2}}}\right)\alpha - 1}$$

Non-Newtonian viscosity is a function of the absolute value of shear stress tensor τ. Zero-shear viscosity is η_0, and α is obtained from the slope of $\log\left(\frac{\eta}{\eta_0} - 1\right)$ against $\log\left(\frac{\tau}{\tau_{\frac{1}{2}}}\right)$. In certain cases, the power law fails to recognize zero shear viscosity, and it predicts zero shear viscosity as infinity and infinite shear viscosity as zero [99–101].

Depending on the composition and nature of rheological fluids, different models were utilized by different groups. Such as Adhikary et al. (2021) utilized the HB model for the prediction of theoretical constants for hydroxyapatite–alginate–chitosan-based hydrogel. It was recorded that in incorporation 0.4% hydroxyapatite in system flow index (n) was 9.2 and consistency index (k) was 5.4 with negative yield stress [102]. Eslami et al. (2007) proposed a mesoscopic rheological model for polymer/layered silicate nanocomposites, which predicts responses to transient and steady shear flows [103].

There are some viscosity models also used in the case of polymeric suspensions such as [104] follows:

$$\textbf{Einstein Model} \quad \frac{\eta}{\eta_o} = 1 + 2.5\phi$$

$$\textbf{Brinkman Model} \quad \frac{\eta}{\eta_o} = \frac{1}{(1-\phi)^{2.5}}$$

η is the viscosity of the composite, η_o is the viscosity of the composite suspension without filler, and φ is the wt% of the filler within the suspension.

There are some other models also available for the prediction of polymer flow property based on the nature of the polymer and the aim of the investigation. These mechanical models and viscosity models also can be utilized for the prediction of magnetorheological behaviors of magnetorheological fluids (MRF).

5 Summary and Outlook

This chapter highlights the significance of biodegradable polymers and their composites, focusing on rheology as a crucial tool for understanding their viscoelastic properties. It discusses various instruments used to measure rheological properties, literature reviews of different biocomposites, and the growing importance of magnetorheological fluids. The chapter also discusses different rheological models for obtaining viscosity profiles and curve fitting parameters, emphasizing the importance of understanding the structure–property dependencies in polymer biocomposites.

The text discusses the importance of rheological investigation in the fabrication of biodegradable composites, focusing on filler–matrix interaction and processability. It covers various instruments used for determining the flow nature of these composites, including viscometers and rheometer instruments. The text also covers various rheological aspects of biodegradable composites, including solution rheology of polymers and nanocomposites, melt rheology, and advanced magnetorheological fluids. It also discusses general theoretical models for predicting flow patterns and their interdependence with rheology parameters. The text also discusses how data can be viewed from these models.

Future polymeric materials are focusing on biodegradable polymers for environmental sustainability. To achieve precise and cost-effective production, rheology parameters, particularly melt rheology, are crucial. Solution processing like gel drawing, coating, and electrospinning depend on solution flowability. Further investigation into improving properties is needed, and more applications for biodegradable nanocomposite-based magnetorheological fluids are needed for sustainable ferro-fluid development.

References

1. Auras R, Harte B, Selke S (2004) An overview of polylactides as packaging materials. Macromol Biosci 4:835–864. https://doi.org/10.1002/mabi.200400043
2. Babu RP, O'Connor K, Seeram R (2013) Current progress on bio-based polymers and their future trends. Prog Biomater 2:8. https://doi.org/10.1186/2194-0517-2-8
3. Lasprilla AJR, Martinez GAR, Lunelli BH, Jardini AL, Filho RM (2012) Poly-lactic acid synthesis for application in biomedical devices—a review. Biotechnol Adv 30:321–328. https://www.sciencedirect.com/science/article/pii/S0734975011000954?casa_token=i1YwQJe3j60AAAAA:LjzcAGK2rQtBxdAOZFK887fCNbiZBy5R0_eb1P9ZsUDT-bXzTYDiebTR_Bfkkv-VSt9ZQeIO6A
4. Savioli Lopes M, Jardini AL, Maciel Filho R (2012) Poly (lactic acid) production for tissue engineering applications. Procedia Eng 42:1402–1413. https://doi.org/10.1016/j.proeng.2012.07.534
5. Chakraborty G, Valapa RB, Pugazhenthi G, Katiyar V (2018) Investigating the properties of poly (lactic acid)/exfoliated graphene based nanocomposites fabricated by versatile coating approach. Int J Biol Macromol 113:1080–1091. https://doi.org/10.1016/j.ijbiomac.2018.03.037
6. Chakraborty G, Gupta A, Pugazhenthi G, Katiyar V (2018) Facile dispersion of exfoliated graphene/PLA nanocomposites via in situ polycondensation with a melt extrusion process and its rheological studies. J Appl Polym Sci 135. https://doi.org/10.1002/app.46476
7. Chakraborty G, Pugazhenthi G, Katiyar V (2019) Exfoliated graphene-dispersed poly (lactic acid)-based nanocomposite sensors for ethanol detection. Polym Bull 76:2367–2386. https://doi.org/10.1007/s00289-018-2494-8
8. Chakraborty G, Katiyar V, Pugazhenthi G (2021) Improvisation of polylactic acid (PLA)/exfoliated graphene (GR) nanocomposite for detection of metal ions (Cu^{2+}). Compos Sci Technol 213. https://doi.org/10.1016/j.compscitech.2021.108877
9. Hazarika D, Chakraborty G, Kumar A, Katiyar V (2023) Role of silk nanocrystal (SNC)-ZnO as an antibacterial nucleating nanohybrid for a patterned mimic poly(lactic acid) based nanofabric. Int J Biol Macromol 242. https://doi.org/10.1016/j.ijbiomac.2023.125126
10. Chakraborty G, Prasad A (2024) Bio-based materials: origin, synthesis, and properties. In: Sustainable bio-based composites: biomedical and engineering applications. De Gruyter, pp 1–24. https://doi.org/10.1515/9783111321530-001
11. Chakraborty G, Prasad A, Kumar A (2022) Processing of Biodegradable Composites. In: Biodegradable composites for packaging applications. CRC Press, Boca Raton, pp 33–48. https://doi.org/10.1201/9781003227908-3
12. Chakraborty G, Paul AK (2023) 3D printing in biomedical applications. In: Advanced materials and manufacturing techniques for biomedical applications. Wiley, pp 275–301. https://doi.org/10.1002/9781394166985.ch11
13. Chakraborty G, Padmashree R, Prasad A (2023) Recent advancement of surface modification techniques of 2-D nanomaterials. Mater Sci Eng B 297. https://doi.org/10.1016/j.mseb.2023.116817
14. Chakraborty G, Dhar P, Katiyar V, Pugazhenthi G (2020) Applicability of Fe-CNC/GR/PLA composite as potential sensor for biomolecules. J Mater Sci Mater Electron 31:5984–5999. https://doi.org/10.1007/s10854-020-03036-z

15. Das P, Chakraborty G (2024) Biodegradable polymers-based proton exchange membrane for fuel cells. In: Sustainable bio-based composites: biomedical and engineering applications. De Gruyter, pp 183–192. https://doi.org/10.1515/9783111321530-009
16. Chakraborty G, Das P, Prasad A (2024) Bio-based materials for adsorption and catalysis. In: Sustainable bio-based composites: biomedical and engineering applications. De Gruyter, pp 333–344. https://doi.org/10.1515/9783111321530-016
17. Prasad A, Paul AK (2024). Biodegradable waste processing for sustainable developments. https://doi.org/10.1201/9781003502012
18. Prasad A, Chakraborty G, Kumar A (2022) Bio-based environmentally benign polymeric resorbable materials for orthopedic fixation applications. In: Advanced materials for biomedical applications. CRC Press, Boca Raton, pp 251–266. https://doi.org/10.1201/9781003344810-15
19. Sringari SS, Chakraborty G, Prasad A (2022) Biodegradable and biocompatible polymeric nanocomposites for tissue engineering applications. In: Nanomanufacturing and nanomaterials design. CRC Press, Boca Raton, pp 205–225. https://doi.org/10.1201/9781003220602-13
20. Paul AK, Borkotoky SS, Prasad A (2023) Nanohydroxyapatite-based composite materials and processing. In: Advanced materials and manufacturing techniques for biomedical applications, pp 157–183. https://doi.org/10.1002/9781394166985.ch7
21. Kumar Paul A, Das S (2024) Biolubricants and its application in engineering. In: Sustainable bio-based composites: biomedical and engineering applications, pp 271–296. https://doi.org/10.1515/9783111321530-013
22. Paul AK, Prasad A, Mahto B, Kumar A, Singh P (2024) Introduction to bioenergy and waste to energy. In: Biodegradable waste processing for sustainable developments, pp 1–33. https://doi.org/10.1201/9781003502012-1
23. Paul AK (2022) Migration studies of biodegradable composites. In: Biodegradable composites for packaging applications, pp 193–212. https://doi.org/10.1201/9781003227908-11
24. Bhattacharjee SK, Chakraborty G, Kashyap SP, Gupta R, Katiyar V (2021) Study of the thermal, mechanical and melt rheological properties of rice straw filled poly (Butylene Succinate) bio-composites through reactive extrusion process. J Polym Environ 29:1477–1488. https://doi.org/10.1007/s10924-020-01973-8
25. Chakraborty G, Bhattacharjee SK, Katiyar V (2022) Rheological studies of biodegradable composites. In: Biodegradable composites for packaging applications. CRC Press, Boca Raton, pp 237–255. https://doi.org/10.1201/9781003227908-13
26. Chakraborty G, Bhattacharjee S, Katiyar V, Pugazhenthi G (2023) Melt rheology analysis through experimental and constitutional mechanical models of exfoliated graphene based polylactic acid (PLA) nanocomposites. J Polym Res 30. https://doi.org/10.1007/s10965-022-03353-3
27. Tesfaye M, Patwa R, Dhar P, Katiyar V (2017) Nanosilk-grafted poly(lactic acid) films: Influence of cross-linking on rheology and thermal stability. ACS Omega 2:7071–7084. https://doi.org/10.1021/acsomega.7b01005
28. Pal AK, Bhattacharjee SK, Gaur SS, Pal A, Katiyar V (2018) Chemomechanical, morphological, and rheological studies of chitosan-graft-lactic acid oligomer reinforced poly(lactic acid) bionanocomposite films. J Appl Polym Sci 135. https://doi.org/10.1002/app.45546
29. Fitzpatrick DP, Taghinezhad SF, Mansourieh M (2024) Rheological and thermal evaluation of polydioxanone and bioresorbable polymers for the optimization of downstream hot melt extrusion. AIP Conf Proc 3158. https://doi.org/10.1063/5.0204403
30. Sarvestani AS, Jabbari E (2006) Modelling and experimental investigation of rheological properties of injectable poly(lactide ethylene oxide fumarate)/hydroxyapatite nanocomposites. Biomacromol 7:1573–1580. https://doi.org/10.1021/bm050958s
31. Hajebi S, Nasr M, Rabiee N, Bagherzadeh M, Ahmadi S, Rabiee M, Tahriri M, Tayebi L, Hamblin MR (2021) Bioresorbable composite polymeric materials for tissue engineering applications. Int J Polym Mater Polym Biomater 70:926–940. https://doi.org/10.1080/00914037.2020.1765365

32. Li S, El Ghzaoui A, Dewinck E (2005) Rheology and drug release properties of bioresorbable hydrogels prepared from polylactide/poly(ethylene glycol) block copolymers. Macromol Symp 222:23–36. https://doi.org/10.1002/masy.200550403
33. D'Anna A, Arrigo R, Frache A (2022) Rheology, morphology and thermal properties of a PLA/PHB/clay blend nanocomposite: the influence of process parameters. J Polym Environ 30:102–113. https://doi.org/10.1007/s10924-021-02186-3
34. Ghahramani N, Iyer KA, Doufas AK, Hatzikiriakos SG (2020) Rheology of thermoplastic vulcanizates (TPVs). J Rheol (N. Y. N. Y) 64:1325–1341. https://doi.org/10.1122/8.0000108
35. Wang X, Peng S, Chen H, Yu X, Zhao X (2019) Mechanical properties, rheological behaviors, and phase morphologies of high-toughness PLA/PBAT blends by in-situ reactive compatibilization. Compos Part B Eng 173. https://doi.org/10.1016/j.compositesb.2019.107028
36. Lopez CG, Richtering W (2021) Oscillatory rheology of carboxymethyl cellulose gels: influence of concentration and pH. Carbohydr Polym 267. https://doi.org/10.1016/j.carbpol.2021.118117
37. Holder AJ, Badiei N, Hawkins K, Wright C, Williams PR, Curtis DJ (2018) Control of collagen gel mechanical properties through manipulation of gelation conditions near the sol-gel transition. Soft Matter 14:574–580. https://doi.org/10.1039/c7sm01933e
38. Stephanou PS (2015) How the flow affects the phase behaviour and microstructure of polymer nanocomposites. J Chem Phys 142. https://doi.org/10.1063/1.4907363
39. Liang JZ, Tang CY, Zhou L, Tsui CP, Li FJ (2012) Melt flow behavior in capillary extrusion of nanosized calcium carbonate-filled poly(L-lactic acid) biocomposites. Polym Eng Sci 52:1839–1844. https://doi.org/10.1002/pen.23130
40. Hanemann T (2013) Nanoparticle surface polarity influence on the flow behavior of polymer matrix composites. Polym Compos 34:1425–1432. https://doi.org/10.1002/pc.22428
41. Kalathi JT, Grest GS, Kumar SK (2012) Universal viscosity behavior of polymer nanocomposites. Phys Rev Lett 109. https://doi.org/10.1103/PhysRevLett.109.198301
42. Srivastava G, Paul AK, Goud VV (2018) Optimization of non-catalytic transesterification of microalgae oil to biodiesel under supercritical methanol condition. Energy Convers Manag 156:269–278. https://doi.org/10.1016/j.enconman.2017.10.093
43. Paul AK, Achar SK, Dasari SR, Borugadda VB, Goud VV (2017) Analysis of thermal, oxidative and cold flow properties of methyl and ethyl esters prepared from soybean and mustard oils. J Therm Anal Calorim 130:1501–1511. https://doi.org/10.1007/s10973-017-6424-z
44. Hájovská P, Chytil M, Kalina M (2020) Rheological study of albumin and hyaluronan-albumin hydrogels: Effect of concentration, ionic strength, pH and molecular weight. Int J Biol Macromol 161:738–745. https://doi.org/10.1016/j.ijbiomac.2020.06.063
45. Tajuddin S, Xie F, Nicholson TM, Liu P, Halley PJ (2011) Rheological properties of thermoplastic starch studied by multipass rheometer. Carbohydr Polym 83:914–919. https://doi.org/10.1016/j.carbpol.2010.08.073
46. Paul AK, Borugadda VB, Goud VV (2021) In-situ epoxidation of waste cooking oil and its methyl esters for lubricant applications: characterization and rheology. Lubricants 9:1–14. https://doi.org/10.3390/lubricants9030027
47. Paul AK, Borugadda VB, Reshad AS, Bhalerao MS, Tiwari P, Goud VV (2021) Comparative study of physicochemical and rheological property of waste cooking oil, castor oil, rubber seed oil, their methyl esters and blends with mineral diesel fuel. Mater Sci Energy Technol 4:148–155. https://doi.org/10.1016/j.mset.2021.03.004
48. Gope S, Samyor D, Paul AK, Das AB (2016) Effect of alcohol-acid modification on physicochemical, rheological and morphological properties of glutinous rice starch. Int J Biol Macromol 93:860–867. https://doi.org/10.1016/j.ijbiomac.2016.09.057
49. Patramanis-Thalassinakis C, Karavelas PS, Kominis IK (2023) A magnetic falling-sphere viscometer. J Appl Phys 134. https://doi.org/10.1063/5.0160731
50. Lloyd Sutterby J (1973) Falling sphere viscometer. J Phys E 6:1001–1005. https://doi.org/10.1088/0022-3735/6/10/016

51. Caw WA, Wylie RG (1961) Capillary viscometers of negligible kinetic energy effect. Br J Appl Phys 12:94–98. https://doi.org/10.1088/0508-3443/12/3/303
52. Delamarre MF, Keyzer A, Shippy SA (2015) Development of a simple droplet-based microfluidic capillary viscometer for low-viscosity newtonian fluids. Anal Chem 87:4649–4657. https://doi.org/10.1021/acs.analchem.5b00677
53. Sariyerli GS, Sakarya O, Akcadag UY (2018) Comparison tests for the determination of the viscosity values of reference liquids by capillary viscometers and stabinger viscometer SVM 3001. Int J Metrol Qual Eng 9. https://doi.org/10.1051/ijmqe/2018004
54. Cho YI, Kim WT, Kensey KR (1999) A new scanning capillary tube viscometer. Rev Sci Instrum 70:2421–2423. https://doi.org/10.1063/1.1149771
55. Wu Q, Meng Y, Wang S, Li Y, Fu S, Ma L, Harper D (2014) Rheological behavior of cellulose nanocrystal suspension: influence of concentration and aspect ratio. J Appl Polym Sci 131. https://doi.org/10.1002/app.40525
56. Ponsubha S, Jaiswal AK (2020) Effect of interpolymer complex formation between chondroitin sulfate and chitosan-gelatin hydrogel on physico-chemical and rheological properties, Carbohydr Polym 238. https://doi.org/10.1016/j.carbpol.2020.116179
57. Derkach SR, Voron'ko NG, Sokolan NI (2017) The rheology of hydrogels based on chitosan–gelatin (bio)polyelectrolyte complexes. J Dispers Sci Technol 38:1427–1434. https://doi.org/10.1080/01932691.2016.1250218
58. Rajput IB, Tareen FK, Khan AU, Ahmed N, Khan MFA, Shah KU, Rahdar A, Díez-Pascual AM (2023) Fabrication and in vitro evaluation of chitosan-gelatin based aceclofenac loaded scaffold. Int J Biol Macromol 224:223–232. https://doi.org/10.1016/j.ijbiomac.2022.10.118
59. Omer AM, Sadik WAA, El-Demerdash AGM, Hassan HS (2021) Formulation of pH-sensitive aminated chitosan–gelatin crosslinked hydrogel for oral drug delivery. J Saudi Chem Soc 25. https://doi.org/10.1016/j.jscs.2021.101384
60. He L, Li S, Xu C, Wei B, Zhang J, Xu Y, Zhu B, Cao Y, Wu X, Xiong Z, Huang R, Yang J, Wang H (2020) A new method of gelatin modified collagen and viscoelastic study of gelatin-collagen composite hydrogel. Macromol Res 28:861–868. https://doi.org/10.1007/s13233-020-8103-3
61. Li MC, Wu Q, Song K, Lee S, Qing Y, Wu Y (2015) Cellulose nanoparticles: structure-morphology-rheology relationships. ACS Sustain. Chem. Eng. 3:821–832. https://doi.org/10.1021/acssuschemeng.5b00144
62. Binder L, Mazál J, Petz R, Klang V, Valenta C (2019) The role of viscosity on skin penetration from cellulose ether-based hydrogels. Ski Res Technol 25:725–734. https://doi.org/10.1111/srt.12709
63. Zheng T, Tang P, Shen L, Bu H, Li G (2021) Rheological behavior of collagen/chitosan blended solutions. J Appl Polym Sci 138. https://doi.org/10.1002/app.50840
64. García MA, Pinotti A, Martino MN, Zaritzky NE (2009) Characterization of starch and composite edible films and coatings, edible film. Coatings Food Appl. 169–209. https://doi.org/10.1007/978-0-387-92824-1_6
65. Bertolo MRV, Martins VCA, Horn MM, Brenelli LB, Plepis AMG (2020) Rheological and antioxidant properties of chitosan/gelatin-based materials functionalized by pomegranate peel extract. Carbohydr Polym 228. https://doi.org/10.1016/j.carbpol.2019.115386
66. Mikešová J, Hašek J, Tishchenko G, Morganti P (2014) Rheological study of chitosan acetate solutions containing chitin nanofibrils. Carbohydr Polym 112:753–757. https://doi.org/10.1016/j.carbpol.2014.06.043
67. Liu DL, Zhou F, Fang K (2019) A theoretical study on transitional shear flow behavior of the compressible and isothermal thermoplastic polymer. Chin J Polym Sci English Ed. 37:518–526. https://doi.org/10.1007/s10118-019-2214-3
68. Mathew I, George KE, Francis DJ (1993) Studies on the rheological behavior of thermoplastic elastomers derived from NBR and PVC using a brabender plasticorder. Int J Polym Mater Polym Biomater 21:189–197. https://doi.org/10.1080/00914039308012046
69. Illner S, Sahmel O, Siewert S, Eickner T, Grabow N (2017) Rheological analysis of hybrid hydrogels during polymerization processes. Curr Dir Biomed Eng 3:699–702. https://doi.org/10.1515/cdbme-2017-0148

70. Shojaeiarani J, Bajwa DS, Stark NM, Bajwa SG (2019) Rheological properties of cellulose nanocrystals engineered polylactic acid nanocomposites. Compos Part B Eng 161:483–489. https://doi.org/10.1016/j.compositesb.2018.12.128
71. Mohammadi M, Heuzey MC, Carreau PJ, Taguet A (2021) Interfacial localization of CNCs in PLA/PBAT blends and its effect on rheological, thermal, and mechanical properties. Polymer (Guildf) 233. https://doi.org/10.1016/j.polymer.2021.124229
72. Sarul DS, Arslan D, Vatansever E, Kahraman Y, Durmus A, Salehiyan R, Nofar M (2022) Effect of mixing strategy on the structure-properties of the pla/pbat blends incorporated with cnc. J Renew Mater 10:149–164. https://doi.org/10.32604/jrm.2022.017003
73. Kim JS, In CH, Park NJ, Kim BJ, Yoon HS (2020) Comparative study of rheological properties and preclinical data of porous polycaprolactone microsphere dermal fillers. J Cosmet Dermatol 19:596–604. https://doi.org/10.1111/jocd.13076
74. Imaizumi M, Okino Y, Yoshikawa K, Koyama K (2003) Relationship between intrinsic viscosity and molecular weight (MW) of PBS and PBSA determined by SEC-LALLS and SEC-viscometry. Seikei-Kakou 15:419–423. https://doi.org/10.4325/seikeikakou.15.419
75. Dobrynin AV, Sayko R, Colby RH (2023) Viscosity of polymer solutions and molecular weight characterization. ACS Macro Lett 12:773–779. https://doi.org/10.1021/acsmacrolett.3c00219
76. Rafler G, Dahlmann J, Wiener K (1990) Biologisch abbaubare polymere. 1. Mitt. Viskositäts-molmassen-beziehung für poly-d,l-lactid und poly(glycolid(50)-co-lactid(50)). Acta Polym 41:328–333. https://doi.org/10.1002/actp.1990.010410604
77. Mahabadi HK (1984) New method for estimation of viscosity-average molecular weight of polymers from Gpc. J Polym Sci Part A-2 Polym Phys 22:449–457. https://doi.org/10.1002/pol.1984.180220307
78. Srinivas V, Bertella F, van Hooy-Corstjens CSJ, van Leeuwen B, Craenmehr EGM, Cavallo D, Rastogi S, Harings JAW (2020) Interfacial stereocomplexation in heterogeneous polymer powder formulations for reinforcing (laser) sintered welds. Addit Manuf 36. https://doi.org/10.1016/j.addma.2020.101665
79. Beltrán FR, Gaspar G, Dadras Chomachayi M, Jalali-Arani A, Lozano-Pérez AA, Cenis JL, de la Orden MU, Pérez E, Martínez Urreaga JM (2021) Influence of addition of organic fillers on the properties of mechanically recycled PLA. Environ Sci Pollut Res 28:24291–24304. https://doi.org/10.1007/s11356-020-08025-7
80. Tesfaye M, Patwa R, Gupta A, Kashyap MJ, Katiyar V (2017) Recycling of poly (lactic acid)/silk based bionanocomposites films and its influence on thermal stability, crystallization kinetics, solution and melt rheology. Int J Biol Macromol 101:580–594. https://doi.org/10.1016/j.ijbiomac.2017.03.085
81. Sangwan P, Way C, Wu DY (2009) New insight into biodegradation of polylactide (PLA)/clay nanocomposites using molecular ecological techniques. Macromol Biosci 9:677–686. https://doi.org/10.1002/mabi.200800276
82. Sevim K, Pan J (2018) A model for hydrolytic degradation and erosion of biodegradable polymers. Acta Biomater 66:192–199. https://doi.org/10.1016/j.actbio.2017.11.023
83. Liu Z, Zhang H, Lai H (2021) Fluid flow effects on the degradation kinetics of bioresorbable polymers. Comput Methods Biomech Biomed Engin 24:1073–1084. https://doi.org/10.1080/10255842.2020.1867115
84. Sackett CK, Narasimhan B (2011) Mathematical modeling of polymer erosion: consequences for drug delivery. Int J Pharm 418:104–114. https://doi.org/10.1016/j.ijpharm.2010.11.048
85. Kalambur S, Rizvi SSH (2006) Rheological behavior of starch-polycaprolactone (PCL) nanocomposite melts synthesized by reactive extrusion. Polym Eng Sci 46:650–658. https://doi.org/10.1002/pen.20508
86. Campbell GA, Zak ME, Radhakrishnan JS (2018) Development of a predictive power law relationship for polymer composites based on newtonial carrier concentrated slurries. Polym Compos 39:1172–1191. https://doi.org/10.1002/pc.24047
87. Zare Y, Rhee KY (2019) Prediction of loss factor (tan δ)for polymer nanocomposites as a function of yield tress, relaxation time and the width of transition region between Newtonian

and power-law behaviors. J Mech Behav Biomed Mater 96:136–143. https://doi.org/10.1016/j.jmbbm.2019.04.045
88. Power law (Ostwald-deWaele model). Encycl Dict Polym:781–781. https://doi.org/10.1007/978-0-387-30160-0_9193
89. Vajravelu K, Sreenadh S, Devaki P, Prasad KV (2011) Mathematical model for a Herschel-Bulkley fluid flow in an elastic tube. Cent Eur J Phys 9:1357–1365. https://doi.org/10.2478/s11534-011-0034-3
90. Kumar S, Garg NR, Gupta A (2015) Herschel Bulkley model for blood flow through an arterial segment with stenosis, pp 93–100
91. Ouyang L, Zhu D, Hill AD (2013) Theoretical and numerical simulation of Herschel-Bulkley fluid flow in propped fractures. https://doi.org/10.2118/163841-ms
92. Benzi R, Bernaschi M, Sbragaglia M, Succi S (2010) Herschel-Bulkley rheology from Lattice kinetic theory of soft glassy materials. Epl 91. https://doi.org/10.1209/0295-5075/91/14003
93. Lv H, Chen R, Zhang S (2018) Comparative experimental study on constitutive mechanical models of magnetorheological fluids. Smart Mater Struct 27. https://doi.org/10.1088/1361-665X/aae5e6
94. Mustafa M, Khan JA (2015) Model for flow of Casson nanofluid past a non-linearly stretching sheet considering magnetic field effects. AIP Adv 5. https://doi.org/10.1063/1.4927449
95. Shahzadi I, Ijaz S (2019) On model of hybrid Casson nanomaterial considering endoscopy in a curved annulas: a comparative study. Phys Scr 94. https://doi.org/10.1088/1402-4896/ab34bb
96. Zhirikova ZM, Aloyev VZ (2018) Application of model of the viscoelastic body and the fractal analysis for the description of process of flowability of polymeric nanocomposites. Mater Sci Forum 935 MSF:150–154. https://doi.org/10.4028/www.scientific.net/MSF.935.150
97. Xie J, Jin YC (2016) Parameter determination for the cross rheology equation and its application to modeling non-Newtonian flows using the WC-MPS method. Eng Appl Comput Fluid Mech 10:111–129. https://doi.org/10.1080/19942060.2015.1104267
98. Burgos-Mármol JJ, Álvarez-Machancoses Ó, Patti A (2017) Modeling the effect of polymer chain stiffness on the behavior of polymer nanocomposites. J Phys Chem B 121:6245–6256. https://doi.org/10.1021/acs.jpcb.7b02502
99. Raju KVSN, Krishna D, Devi GR, Reddy PJ, Yaseen M (1993) Assessment of applicability of carreau, ellis, and cross models to the viscosity data of resin solutions. J Appl Polym Sci 48:2101–2112. https://doi.org/10.1002/app.1993.070481205
100. Reddy MM, Gupta RK, Bhattacharya SN, Parthasarathy R (2007) Structure-property relationship of melt intercalated maleated polyethylene nanocomposites. Korea Aust Rheol J 19:133–139
101. Esposito Cordone C, Mensitieri G, Maffezzoli A (2009) Analysis of the structure and mass transport properties of nanocomposite polyurethane. Polym Eng Sci 49:1708–1718. https://doi.org/10.1002/pen.21410
102. Adhikari J, Perwez MS, Das A, Saha P (2021) Development of hydroxyapatite reinforced alginate–chitosan based printable biomaterial-ink. Nano-Struct Nano-Objects 25. https://doi.org/10.1016/j.nanoso.2020.100630
103. Eslami H, Grmela M, Bousmina M (2007) A mesoscopic rheological model of polymer/layered silicate nanocomposites. J Rheol (N. Y. N. Y) 51:1189–1222. https://doi.org/10.1122/1.2790461
104. Genc S, Derin B (2014) Synthesis and rheology of ferrofluids: a review. Curr Opin Chem Eng 3:118–124. https://doi.org/10.1016/j.coche.2013.12.006

Chapter 8
Materials Design and Migration Studies for Bone Tissue Engineering

Sunkari Dinesh, Subhasree Panda, and S. K. Khadheer Pasha

1 Fundamental Background

Tissue engineering (TE) is a branch of biomedical systemization that integrates cells, engineering techniques, material approaches, and suitable physicochemical and biochemical elements to repair, sustain, enhance, or substitute many kinds of biological tissues. TE often entails the use of cells positioned on tissue scaffolds to create new functional tissue for medicinal purposes, whilst it is not confined to applications using cells and scaffolds alone. Initially classified as a sub-discipline of biomaterial, it has grown in broadness and worth leading to recognition as a standalone field [1]. Although definitions of TE include a wide array of uses, the phrase is mostly linked to the restoration or substitute of specific tissues or entire organs, namely bone, blood vessels, bladder, cartilage, skin, and muscle. The associated tissues need precise mechanical and configurational qualities for optimal operation [2]. The word has also been used to describe endeavours aimed at executing certain biochemical processes using cells inside a synthetically constructed hold-up system (for example a man-made pancreas or a bioartificial liver). The phrase regenerative medicine is sometimes used interchangeably with TE; however, practitioners of regenerative medicine prioritize the use of stem cells or progenitor cells for tissue production.

S. Dinesh · S. K. Khadheer Pasha (✉)
Functional Nanomaterials and Polymer Nanocomposite Laboratory, Department of Physics, VIT-AP University, Amaravati, Guntur, Andhra Pradesh, India
e-mail: khadheerbasha@gmail.com

S. Panda
Department of Chemical Engineering, Yeungnam University, Gyeongsan 38541, Republic of Korea

A. Prasad and V. Katiyar (eds.), *Resorbable Polymers for Bioimplants and Fixation Devices*, Materials Horizons: From Nature to Nanomaterials,
https://doi.org/10.1007/978-981-96-9306-1_8

2 Introduction

Bone comprises 99% of the body's total calcium, with 85% derived from an inorganic material phase hydroxyapatite (Hap), with a chemical formula $Ca_{10}(PO_4)_6(OH_2)$. It has been studied that bone consists of cells such as osteoblasts (which synthesize new bone and repair existing bone), osteocytes (which enable communication amongst osteocytes and osteoblasts), and osteoclasts (which release the bone resorption enzymes), along with calcium [3].

Bones are the second most frequently transplanted tissues, and a survey indicates that the market value is reckoned to attain a significantly increased amount by the year 2027. In instances when the bones' natural healing is inadequate, other methods are required to facilitate the repair of major abnormalities. The traditional 'gold standard' method for this treatment included the use of autografts (bone tissue harvested from the patient) and allografts (bone tissue harvested from the donor) [3]. Nonetheless, the drawbacks linked to these techniques, such as afterwards problems, stress shielding and immunological response, served as a catalyst for the development of superior alternatives. This resulted in the employment of three-dimensional (3D) porous scaffolds in bone tissue engineering (BTE) that offers a 3D circumstance for cellular migration and proliferation, aiding eventual bone repair.

Bone has two components: the spongy cancellous bone and the denser cortical bone. It is comprised of organic components (collagen) and an inorganic mineral phase (Hap), that imparts durability to the overall structure [3, 4]. The human body has around 270 bones, some of which blend, resulting in 206 bones in adulthood. The features of bones are greatly influenced by their age and anatomical position in the internal environment of the body. It is categorized as an anisotropic material for this reason. The modulus of bone is typically in the range of 1–20 GPa. The implanted scaffold or graft must possess a modulus comparable to that of the target bone; otherwise, it may result in bone withering to stress shielding [5].

3 Scaffold

In BTE, the objective is to create unique biomimetic scaffolds as alternatives to conventional bone transplants. BTE calls for three primary components for successful bone regeneration: cells capable of generating a functional matrix, bioactive molecules including growth aspects that energize the development of specific tissues, and scaffolds that offer a 3D matrix for cellular adhesion, proliferation, and mechanical assistance. A bone tissue scaffold is a porous, biodegradable material that provides mechanical support and a surface for cells to grow on whilst damaged bone regenerates.

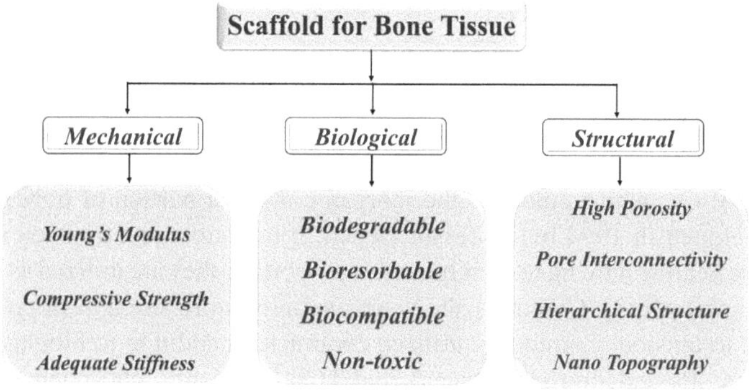

Fig. 1 Attributes of an optimal scaffold for osseous tissue

3.1 *Characteristics and Criteria of an Optimal Bone Scaffold Framework*

A scaffold operates as an extracellular matrix, providing rigidity against external stresses whilst aiming to align with the native tissue moduli to avert failure from stress shielding. At present, researchers' main intention is to attain a compacting strength between 2 and 12 MPa and a modulus of ~ 5 GPa, corresponding to the durability of cancellous bone [3, 6]. Furthermore, it is designed to exhibit regulated biodegradability, conceding the scaffold to disintegrate as the tissue moderately regenerates, eliminating the need for surgical removal [7]. This feature reduces possible post-operative problems such as shock and infection. The scaffold should ideally not cause the host body to mount an immune defence, mandating its biocompatibility [8].

An ideal scaffold is equipped with four fundamental properties: the capacity to relate to the design of the deciding sized defect, known as form; the ability to provide mechanical stability at the defect site, referred to as function; the need for secure attachment to prevent undesirable movement, termed fixation; and the capability to supply nutrients for cells, growth factors, and proteins, thereby promoting bone synthesis [9]. Figure 1 depicts the ideal properties sought in a scaffold.

3.2 *Scaffold Materials*

The materials for scaffold fabrication ought to be diligently chosen to accomplish the specifications of efficient diagnostics research. They are compelled to exhibit bioactive, enough robust strength in vivo and be biocompatible. Numerous materials have been examined for several years within the three primary stratifications: ceramics, polymers, and metals.

3.2.1 Ceramics

Ceramics serve as implants, often exhibiting bioactivity and releasing calcium ions, which enhance bioactivity and promote bone production [10]. The initial bioactive ceramic biomaterial employed by bioglass 45S5 with 45 wt% of SiO_2 and a 5:1 ratio of CaO to P_2O_5, which emulates the inorganic mineral portion of bone, is termed Hap. Developed in 1994 by Professor Hench, it has subsequently been utilized in bone tissue healing. Owing to their bioactive properties, they are utilized as both coatings and implants [3]. Ceramics show substantially more strength in compression compared to tension. Commonly utilized ceramics, in addition to bioglass, encompass Hap, CaPs (calcium phosphates), and TCP (tricalcium phosphate) [3]. These materials possess exceptional porosity and a highly coordinated architecture that facilitates bone and vascular inward growth whilst promoting nutrition transfer to the cells, along with displaying no negative side effects.

3.2.2 Metals

Historically, metals have persisted amongst the first biomaterials employed for biomedical applications mainly on account of their anti-corrosion capability and tensile strength. One drawback of making use of metals is the concern that harmful ions might seep from them, leading to inflammatory and allergy responses in the body. Furthermore, metals have a greater modulus, leading to progressive bone atrophy due to the stress-shielding effect [11]. Haugen et al. [3] demonstrated that titanium and titanium dioxide (TiO_2) utilized in rabbit tibia exhibited significant osseointegration and osteoconductivity. To address the constraints of metals, Hengel et al. [12] proposed the application of silver nanoparticle oxide (Ag NPO) sheathing to inhibit contamination, whilst He et al. [13] recommended augmenting porosity to $\approx$ 65% to decrease the implant's modulus by $\approx$ 0.18 GPa, aligning it more closely with the inherent bone. Chen et al. [3] discovered that when porosity surpasses 50%, the cohesion and multiplication of mesenchymal stem cells (MSCs) are boosted. In addition to titanium, different metals and metal amalgams such as calcium–magnesium (Ca–Mg) and strontium–magnesium (Sr–Mg) have been utilized in stents and orthopaedic prosthetics [14]. Magnesium–tungsten (Mg–W) and iron–magnesium (Fe–Mg) alloys were employed in the production procedure, evincing mechanical flexibility and favourable biological characteristics such as cytocompatibility and cell survival [3].

3.2.3 Polymers

Natural polymers have the additional benefit of enhanced biocompatibility and biodegradability. Chitosan is peculiarly appealing for implants due to its remarkable antibacterial qualities [15]. Chitosan can be used to fabricate implants, but it is not ideal for pharmaceutical delivery applications due to its low mechanical strength. [16]. Silk is a natural polymer characterized by its strength, elasticity, and strength-to-density ratio that exceeds that of steel by a factor of ten. The integration of carbonate-containing apatite (CAP) into silk polymer enhanced their porous architecture whilst maintaining their compressive strength [3]. There is a movement towards sourcing collagen from human origins via recombinant systems, as opposed to animal-derived collagen, which presents challenges such as elevated costs, disease transmission, and physicochemical degradation [3].

Synthetic polymers provide advantages by enabling the researcher to temper biological and mechanical aspects, together with regulating the detriment rate, through the manipulation of functional groups attached to the chain of the polymer and the monomer numbers utilized in the polymer fabrication. Some commonly deployed synthetic polymers are polycaprolactone (PCL), polyglycolic acid (PGA), polylactic glycolic acid (PLGA), and polylactic acid (PLA) [17]. Using a physical barrier that resembles a membrane to prevent connective tissue ingrowth and promote bone production is a recent development in directed bone regeneration [18]. Research conducted by Chuenjitkuntaworn et al. [3] showed that after 48 h of cell culture, PCL/HA scaffolds showed better cell proliferation than pure PCL scaffolds.

4 Bone Scaffold Design

Bone scaffold depictions might be categorized in two distinct manners. One categorization is based on overall designs, whilst the other is based on unit cell configurations. Figure 2 presents the categorization method for diverse scaffold designs.

4.1 Whole Design

Whole designs might be classified into three distinct categories: gradient design, uniform design, and topology optimization-based design. As implied by its name, porosity stays constant throughout. Porosity is a vital determinant of a scaffold's biological and robustness aspects. Young's modulus may be reduced to the level of natural bone with a concurrent augmented porosity or by the incorporation of augmentations [3]. Kadkhodapour et al. [19] observed a similar phenomenon, noting that the targeted Young's modulus was achieved even at porosities of 90%. The varying moduli at different spatial positions in natural bone prompted the creation of a scaffold with mechanical features that vary throughout its length [20]. Nune et al. [21] revealed that alkaline phosphatase activity and calcium concentration were closely correlated with porosity, hence boosting the stem cells' osteogenic differentiation by this design. Han et al. [22] developed a progressive operatively graded scaffold for bone regeneration, establishing that the mechanical characteristics, such as modulus

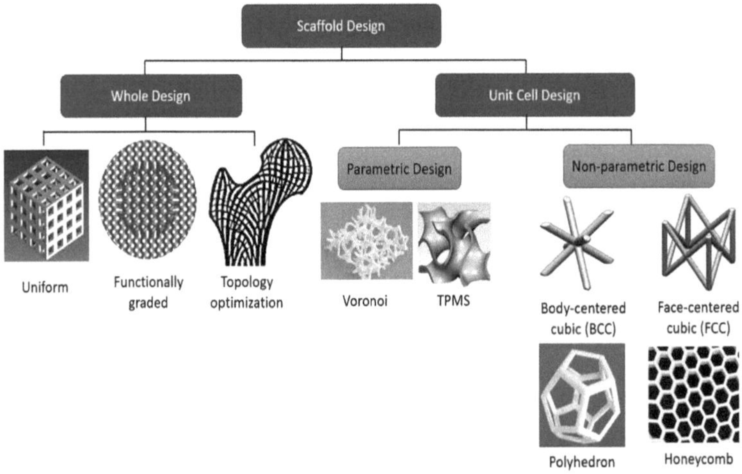

Fig. 2 Schematic illustrating the classes of scaffold designs according to the categorization scheme. Reproduced with permission from [3], copyright 2021. Elsevier

and yield strength, could be customized to attain preferred values. Topological optimization is an algorithmic method aimed at minimizing material usage in the making of a structure for a particular purpose. Common optimization strategies employed in the topology optimization of scaffolds comprise of solid isotropic material with penalization (SIMP) and the evolutionary structural optimization (ESO) approach [23]. In a topologically configured design, areas with tiny apertures are regarded as solid, whereas areas with big apertures are deemed voids [24]. Panesar et al. [25] chose the topology optimization methodology as the most efficient technique for scaffold design, owing to its superior resistance to loading anomalies. The printed structure often lacks precision contrasted to the original design attributable to the execution restrictions of printings. This enabled the application of topology fine-tuning methods to construct structures that align more closely with the intended design.

4.2 Unit Cell Designs

Unit cell architectures are further categorized as parametric and non-parametric frameworks. Parametric designs tend to be formalized than non-parametric patterns since they all use the same algorithms to be created, although the non-parametric designs share certain Basic Geometry. A high strength-to-weight ratio and outstanding insulating properties are two domains where these designs have demonstrated tremendous potential.

4.2.1 Parametric Designs

Structures generated by algorithms are termed parametric design. These standardized methods enable the generation of structures that are readily replicable. We might alter the construction by adjusting the parameters as needed.

Voronoi

Voronoi layouts are inclined by nature. They are constituted by a collection of points inside a designated volumetric pattern which might be either randomly arranged or systematically organized based on the application. The seed locations are eventually encircled by established geometric configurations that partition the architectural volume into subspaces. The convergence of any two spatial aspects delineates the demarcation line or plane of the subspaces. The ensuing configuration is referred to as the Voronoi design.

Voronoi framework is favoured for its attempt to replicate natural bone, seeking to provide an efficient porosity structure for BTE applications [26]. The Voronoi models are assembled by employing isotropic-linked porous models that convey numerous flaws including weak repeatability, excessive energy consumption, and prolonged cycle duration. A top-down design was offered that addressed non-uniformities by adjusting porosity, pore diameter, and structure breadth to replicate bone [27]. Numerous investigations have signified its high load-bearing capacity. Fantini et al. [28] fabricated scaffolds adopting Voronoi layouts, which showed a biomimetic morphology suitable for TE applications. Gomez et al. [29] engineered a specific scaffold that fitted the geometry of trabecular bone with a Voronoi pattern, yielding considerable osteoconductivity. They further noticed that isotropy is precisely linked to the quantity of seed items in the Voronoi structure.

Triply Periodic Minimal Surfaces (TPMSs)

A TPMS is defined as a minimum surface that is periodic in three distinct directions. Multiple techniques exist for the generation of TPMS. The coordinates of the TPMS can potentially be effectively determined using the Enneper–Weierstrass parameterization [30]. These are as detailed below.

$$u = re\left(e^{i\beta} \int_{\omega_0}^{\omega} \left(1 - \gamma^2\right) G(\gamma) d\gamma\right)$$

$$v = re\left(e^{i\beta} \int_{\omega_0}^{\omega} i\left(1 + \gamma^2\right) G(\gamma) d\gamma\right)$$

$$w = re(e^{i\beta} \int_{\omega_0}^{\omega} 2\gamma G(\gamma) d\tau)$$

Here, $i = \sqrt{-1}$; γ represents a complex variable, β denotes Bonnet angle, and re signifies the real component of the complex variable. The parameter ω evinces a direct correlation with the period of its associate direction. This approach is limited to producing only a few kinds of TPMS. TPMSs are unit cell layouts characterized by a mean curvature approaching zero. These are mainly characterized by combinations of trigonometric functions which are given in Table 1.

TPMSs offer greater transport of growth stimulants, oxygen, and sustenance to cells in conjunction with the waste materials' elimination, owing to their notably enhanced SA:V proportion and optimized permeability relative to lattice arrangements [36]. Their peculiar structure leads to reduced stress concentration, hence boosting mechanical aspects [37]. Numerous research evaluated the manufacturing capacity, mechanical characteristics, and bone ingrowth of various TPMS layouts, concluding that gyroid skeletal TPMS proved the leading and foremost scalable designed domain. Nevertheless, panel TPMS exhibited superior strength to withstand compression and impact resistance compared to skeletal TPMS. TPMS design is capable of being developed either by augmenting the trivial surface of a confined area, producing laminar structures, or by considering the volume that the laminar surface contains to be a solid, ensuing in skeletal structures [3].

4.2.2 Non-parametric Designs

Bone scaffolds frequently have four of the several non-parametric designs: BCC, FCC or diamond, polyhedron, and honeycomb. From these classifications, it is clear that non-parametric design is more concerned with a specific contour or framework, which might be flat, like a honeycomb or 3D inherently.

Body-Centred Cubic (BCC)

The ease of the BCC structure makes it a perfect target for simple production by means of linking the centre of the hexahedron with eight vertices. Unlike a solid framework, the modulus of a BCC-built scaffold might be lowered by $\approx 80\%$ to better match the natural bone's modulus [38]. This design leads to bone ingrowth and cell proliferation. Nonetheless, its compressive strength was deemed insufficient, exhibiting suboptimal surface area for cell adhesion and anisotropy. Consequently, several researchers enhanced this design by including vertical stiffeners to achieve improved compressive force tolerance [39]. Jamshidinia et al. [3] noted that in a BCC dental embedded configuration produced using electron beam machining (EBM), fractures started at the corners of a unit cell measuring 2 mm. The layout of scaffolding considerably influences a structure's collapse pattern. BCC design is frequently used because of its enhanced control over porosity.

Table 1 Trigonometric equations of TPMS unit cell designs

Serial number	Name of the design	Trigonometric equations	References
1	Primitive TPMS	$\cos(u) + \cos(v) + \cos(w) = k$	[31]
2	Neovius TPMS	$3\{\cos(u) + \cos(v) + \cos(w)\} + 4\cos(u)\cos(v)\cos(w) = k$	[32]
3	Gyroid TPMS	$\cos(u)\sin(v) + \cos(v)\sin(w) + \cos(w)\sin(u) = k$	[33]
4	I-graph-wrapped package (IWP) TPMS	$2\{\cos(u)\cos(v) + \cos(v)\cos(w) + \cos(w)\cos(u)\}$ $- \{\cos(2u) + \cos(2v) + \cos(2w)\} = k$	[34]
5	Diamond TPMS	$\sin(u)\sin(v)\sin(w) + \sin(u)\cos(v)\cos(w) + \cos(u)\sin(v)\cos(w)$ $+ \cos(u)\cos(v)\sin(w) = k$	[35]

Face-Centred Cubic (FCC)

Numerous diamond structures would be achieved by altering parameters such as θ (the angle subtended by the plane and struts), L (strut length), and α (unit cell length). The framework of the diamond/FCC scaffold unveiled Young's modulus comparable to that of unrefined bone, and both strength and Young's modulus were determined in an inverse aligned to the subtended angle [40]. Zhang et al. [41] observed a correlation between the porosity and the mechanical properties of the scaffold. Porosities at a reduced level of 66% exhibited mechanical qualities comparable to those of real bone tissue. FCC designs demonstrated the ability to produce targeted porosities whilst maintaining enough interconnectivity amongst the pores, achieving a maximum of 96% [3]. The mechanical characteristics of porous structures persisted despite considerable deterioration in FCC-designed scaffolds composed of magnesium and iron [42]. The diamond/FCC configuration demonstrates a more consistent stress allocation in struts in comparison with octahedral designs or cubic [43].

Polyhedron

Polyhedral architectures comprise the RD (rhombic dodecahedron), TC (truncated cube), octet, and RB (rhombic cube octahedron). The RD exhibited symmetry, isotropy, comparable compressive strength to natural bone and demonstrated superior mass transport capabilities for nutrients and oxygen for cells [44]. Rodríguez-Montaño et al. [45] showed in their investigation that the TC design exhibited significant bone growth under low stress conditions (less than 0.5 MPa). Conversely, at high stress conditions (exceeding 0.5 MPa), the hexahedron framework demonstrated the greatest degree of bone production. The rhombocuboctahedron unit cell was deemed optimal for medium loads [3].

Honeycomb

The honeycomb framework is beneficial for the sake of its characteristics of minimal weight, superior rigidity, and significant porosity. Its characteristics, including Young's modulus, are adjustable by the regulation of porosity. Moreover, a honeycomb shape facilitates a better homogeneous dissemination of pressure and promotes bone ingrowth. Silva et al. [3] constructed a composite scaffold of poly (d, l-lactic acid) (PDLLA) and graphene/multi-walled carbon nanotube oxides (MWCNTO-GO, 50% w/w) with a honeycomb configuration. Studies carried out in vitro on the scaffold showed that the cultivated cells improved osteoblast activity, which in turn led to a greater formation of lamellar bone. Favi et al. [3] demonstrated that a honeycomb scaffold with a pore size of 300 μm, composed of bacterial cellulose mineralized with nano-HaP, displayed cell adhesion, vitality, regulated disintegration, and enhanced mechanical capabilities.

5 Migration Studies in BTE

BTE employs embedded biomaterials as scaffolds for stem cells and osteoprogenitor cells to transiently preserve the activity, enlargement, and structure of bone tissue over the course of regeneration. The contemporary approach to bone tissue regeneration includes tissue-engineered products (TEP) created by seeding scaffolds with patient-derived osteoprogenitor cells in vitro, referred to as live autologous bone grafts or constructs [46]. Biomaterials utilized in BTE as scaffolds must solely lend an interconnected, microporous, and 3D structure beneficial to the proliferation, growth, and differentiation of osteoprogenitor cells when seeded with patient cell [47]. Additionally, TEPs are also made by saturating scaffolds with cytokines or growth factors like platelet-derived growth factor (PDGF), vascular endothelial growth factor (VEGF), or bone morphogenetic proteins (BMPs), in order to enhance the angiogenesis process and the osteoinductive traits of the material, thus expediting the osseointegration process post-implantation [48]. The main aspects of biointeraction between bone scaffolds and the cells are shown in Fig. 3. It is essential to mention that TEPs with stem cells enable upgraded bone repair by reducing the formation of fibrosis, promoting angiogenesis, and accelerating regeneration at the implantation site. Nevertheless, stem cell-based TEPs are still a relatively young field that requires a deeper comprehension of the biointeractions between eukaryotic cells and biomaterials.

Migration studies in BTE focus on understanding and optimizing how cells move within engineered scaffolds to promote uniform bone regeneration. Effective migration of osteoprogenitor cells, MSCs, and vascular cells is critical for scaffold integration, new tissue formation, and vascularization. Figure 4 shows the main aspects of migration studies in BTE.

Chemical gradients, often created by growth factors like BMPs (bone morphogenetic proteins), VEGF, and SDF-1 (stromal cell-derived factor), can attract stem cells to specific regions in the scaffold. Controlled release of these factors within the scaffold can establish gradients that guide cell movement and differentiation. Incorporating growth factors within the scaffold in degradable microspheres or through surface modifications can create localized signals that attract and guide cells. Cells

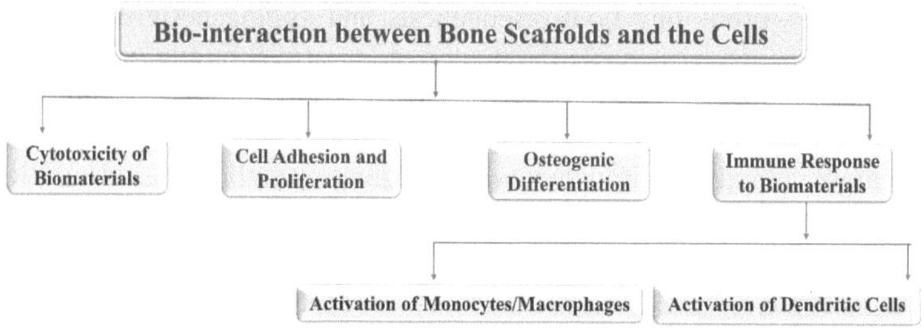

Fig. 3 Schematic showing the biointeraction between the cells and bone scaffolds

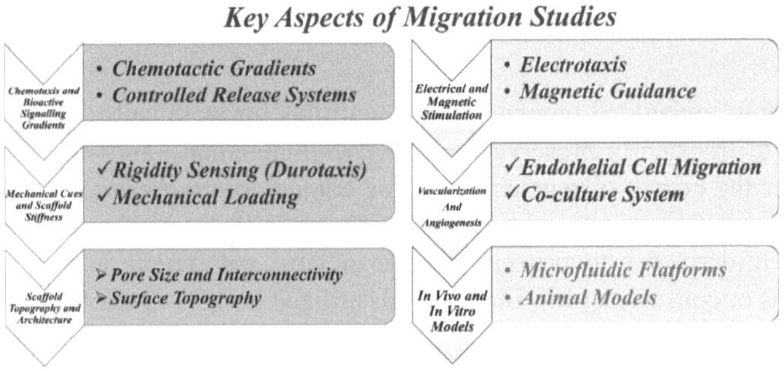

Fig. 4 Illustration of main characteristics of migration studies

respond to the stiffness of their environment, a process known as durotaxis. Gradients in scaffold stiffness can direct cells towards bone-like regions with optimal mechanical properties for bone formation. Applying dynamic loading or mechanical stress to the scaffold can encourage cell migration and alignment. This mimics the natural forces experienced by bone, promoting a structure conducive to bone growth. Scaffolds with well-designed pore structures and sizes (typically 100–500 μm for bone tissue) support nutrient diffusion, oxygen flow, and cellular movement. A highly interconnected structure enables cells to move freely and spread uniformly. Nano- and microscale features on the scaffold surface can impact how cells adhere, migrate, and differentiate. Rougher textures, for instance, can promote osteogenic differentiation whilst supporting better cell grip for migration.

Applying an electric field can influence cell migration by promoting electrotaxis. Cells like MSCs have shown improved migration and differentiation in response to electrical fields, enhancing osteogenic capacity within the scaffold. Magnetically responsive scaffolds, created by incorporating magnetic nanoparticles, can also guide cell movement when unshielded to an outer magnetic field, enhancing the efficiency of cell migration and scaffold colonization. Effective BTE requires blood vessel formation to deliver nutrients to the growing tissue. VEGF and other angiogenic factors stimulate endothelial cell migration, which is necessary for new blood vessel formation and improves overall bone scaffold integration. Co-culturing osteoblasts with endothelial cells encourages a coordinated migration and interaction between these cell types, promoting both osteogenesis and angiogenesis within the scaffold. Microfluidic devices enable the precise study of cell migration in response to various factors, allowing researchers to create gradients and monitor cell movement under controlled conditions. Implanting scaffolds in vivo helps assess cell migration in a more complex environment with real-time signalling, vascularization, and tissue formation. By fine-tuning these conditions, researchers can create scaffolds that better imitate the natural bone surroundings, leading to faster and more potent bone regeneration.

6 Interactions at the Scaffold Interface

Interactions at scaffold interfaces encompass the complex interplay between a scaffold's surface and its surrounding environment, which is crucial for fields like TE and drug delivery. Scaffold interfaces are engineered with specific biochemical and physical properties that influence how they interact with cells, biological molecules, or therapeutic agents. These interactions are mainly categorized into physical and chemical types. Physical interactions involve the scaffold's structural features such as porosity, surface roughness, and mechanical properties that determine cell adhesion, migration, and proliferation. For example, a porous scaffold allows for nutrient and oxygen transport, which is essential for cell survival and tissue growth. Additionally, the topography of the scaffold can direct cellular alignment and differentiation, especially in applications aimed at regenerating muscle or nerve tissues. Chemical interactions, on the other hand, involve functional groups and surface modifications that enhance biocompatibility and cell signalling. Functionalizing scaffold surfaces with biomolecules, such as peptides, growth factors, or antibodies, can attract specific cell types and stimulate desired cellular responses. Chemical modifications, such as coating the scaffold with hydrophilic or hydrophobic molecules, influence protein adsorption, which is a primary mediator of cell–scaffold interactions.

Understanding and tailoring these interactions at the scaffold interface is fundamental in creating successful biomaterials that mimic the natural extracellular matrix and promote desired outcomes in regenerative medicine. By customizing scaffold properties, researchers can create environments that assist tissue formation, fine-tune immune resolutions, or deliver therapeutic agents effectively, advancing both clinical and research applications.

6.1 *Cytotoxicity of Biomaterials*

Biomaterials designed for bone scaffolds must meet specific biocompatibility and safety requirements, particularly in minimizing cytotoxicity. Cytotoxicity specifies the potential of a material to cause cell damage or death upon interaction. When biomaterials are introduced into a biological environment, their cytotoxicity can affect cell viability, proliferation, and differentiation, crucial aspects for successful bone tissue engineering.

6.1.1 Cytotoxicity and Its Impact on Bone Scaffolds

The cytotoxicity of biomaterials intended for bone scaffolds is typically gauged using in vitro assays with osteoblast-like cells or stem cells, assessing cell viability, metabolic activity, and membrane integrity. Standard assays, such as 3-(4,5-dimethylthiazol-2-yl)-2,5-diphenyltetrazolium bromide (MTT), lactate dehydrogenase (LDH) release, or live/dead staining, provide initial insights into cytotoxicity (ISO 10993-5 standards for in vitro cytotoxicity testing); materials like HaP, TCP, and bioactive glasses generally exhibit low cytotoxicity because of their chemical homogeneity to the mineral phase of bone [49]. However, synthetic polymers, composites, and metal-based scaffolds may release ions or degrade into products that could exhibit cytotoxic effects. For example, metals such as titanium and its blends are frequently employed owing to their mechanical strength and biological compatibility. However, if improperly processed or if degradation occurs, metal ions can be released, potentially causing oxidative stress in cells.

6.1.2 Cyclodextrins (Cy) Mechanisms in Biointeractions

Cyclodextrins are a family of cyclic oligosaccharides consisting of α-D-glucose units connected by α (1 $\rightarrow$ 4) glycosidic bonds. They have a hydrophilic outer layer and a hydrophobic inner void, which makes them useful for forming inclusion complexes with various molecules. Cyclodextrins are primarily classified according to the quantity of glucose units they comprise (Table 2). Mainly, the roles of cyclodextrins include scaffold functionalization, enhancing osteogenesis, cholesterol, and lipid modulation, improving angiogenesis and gene and protein delivery [50–53]. Several mechanisms contribute to biomaterial-induced cytotoxicity, including oxidative stress, inflammatory response, and immune cell activation. Oxidative stress results when materials, such as certain nanoparticles or metal ions, generate reactive oxygen species (ROS) upon degradation. ROS can damage cellular components, leading to apoptosis or necrosis in bone-related cells. Furthermore, biopolymers like PLGA used in bone scaffolds degrade into acidic by-products, which may lower the pH around the scaffold and create a cytotoxic environment for cells [54] (Table 2).

6.1.3 Strategies to Minimize Cytotoxicity

To mitigate cytotoxic effects, biomaterial modifications have been explored. Coating scaffolds with bioactive molecules, optimizing porosity, or blending polymers with bioactive ceramics can reduce degradation-induced toxicity whilst promoting osteogenesis. For example, incorporating growth factors or peptides on scaffold surfaces can enhance cell adhesion and proliferation, supporting scaffold integration without increasing cytotoxicity. The cytotoxicity remains a pivotal concern in BTE. Through in vivo and in vitro research, explorers continue to refine materials and scaffold designs to ensure biocompatibility and optimize biointeractions between cells and scaffolds. Developing materials that support cellular functions without triggering adverse effects is critical for advancing successful clinical applications of bone scaffolds.

Table 2 Types of cyclodextrins and their applications

Type of cyclodextrin	Structure	Cavity diameter	Applications	References
Alpha-Cyclodextrin (α-CD)	Consists of six glucose units	4.7–5.3 Å	Used in food, pharmaceuticals, and cosmetics for solubilizing smaller molecules like ethanol or flavour compounds	[55]
Beta-Cyclodextrin (β-CD)	Consists of seven glucose units	6.0–6.5 Å	Commonly used in drug delivery, food preservation, and removing cholesterol or odours due to its optimal cavity size for many small molecules	[56]
Gamma-Cyclodextrin (γ-CD)	Consists of eight glucose units	7.5–8.3 Å	Used for encapsulating larger molecules such as vitamins or fatty acids and has the highest solubility amongst natural cyclodextrins	[57]

6.2 *Cell Adhesion and Proliferation*

Cell adhesion and proliferation in bone scaffolds are key factors in BTE, as these processes enable cells to attach, grow, and ultimately regenerate damaged bone. Recent advances emphasize the design of biomimetic scaffolds that replicate natural bone properties to optimize these functions. Various materials like calcium phosphates ($Ca_3(PO_4)_2$), polymers, and composites are popular choices because they mimic the bone matrix and support osteointegration. Calcium-based materials, including $Ca_3(PO_4)_2$ and calcium silicate (Ca_2O_4Si), have gained prominence due to their biocompatibility and ability to encourage osteoblast attachment and proliferation, essential for bone regeneration. These materials, often modified to improve their porosity and surface characteristics, support efficient cell adhesion by enhancing surface roughness and promoting protein absorption, crucial for initial cell attachment. The scaffold's microstructure, particularly nanofibrous designs, has been shown to significantly improve cellular responses by providing a 3D matrix that cells find favourable for attachment and growth [58].

Innovations also include bioactive molecules, such as flaxseed extracts, that have been integrated into nanofibrous scaffolds to further boost bioactivity without compromising scaffold stability. This approach aids in maintaining a conducive environment for osteoblast adhesion and proliferation, as well as the gradual release of osteoinductive factors that promote new bone tissue formation. As research advances, optimizing these composite materials and scaffold designs promises to enhance clinical applications by creating more durable, compatible scaffolds that align closely with natural bone properties [58].

6.3 Osteogenic Differentiation for Bone Scaffolds

Osteogenic differentiation is a critical process in BTE, aimed at developing effective bone scaffolds that can support new bone formation and repair. This process involves the transformation of stem or progenitor cells into osteoblasts, the cells reliable for bone construction. The progress in biomaterials, scaffold design, and biochemical stimuli have significantly enhanced the osteogenic potential of bone scaffolds. Biocompatible materials can be processed to create porous structures that facilitate cell attachment, proliferation, and differentiation. For example, electrospinning techniques have been employed to create nanofibrous scaffolds that bear a striking resemblance to the fibrous structure of natural bone, enhancing the scaffold's mechanical properties and biological performance [59]. Modernistic investigations have confirmed the successful integration of bioactive components, for example, $Ca_3(PO_4)_2$ into scaffolds to promote mineralization. In situ mineralization techniques utilize enzymes like alkaline phosphatase to catalyse mineral formation within the scaffold's structure, effectively enhancing osteogenic differentiation [60]. This approach simplifies the TE process by enabling mineralization to occur directly within the physiological environment.

6.3.1 The Role of Stem Cells

The selection of appropriate cell types is crucial for effective osteogenic differentiation. MSCs are often utilized thanks to their proficiency in differentiating into multiple cell lineages, including osteoblasts. When cultured on specially designed scaffolds, MSCs respond to mechanical and biochemical cues, leading to enhanced osteogenic differentiation. Factors such as the surface properties of the scaffold, the appearance of osteogenic growth factors (e.g. bone morphogenetic proteins), and the scaffold's physical structure significantly influence the differentiation process [59].

6.3.2 Scaffolds' Design Considerations

Scaffold design must consider porosity, mechanical strength, and biodegradability. The porosity of the scaffold is vital for nutrient transport and waste removal, whilst mechanical strength is necessary to withstand physiological loads during bone healing. New developments have produced intelligent scaffolds that can regulate the release of growth factors, further enhancing osteogenic differentiation. Additionally, modifications to scaffold surfaces, such as functionalization with peptides or proteins, have shown promise in promoting cell adhesion and differentiation. The evaluation of osteogenic differentiation in scaffolds often involves both in vitro and in vivo studies. In vitro inspections typically assess cell viability, escalation, and differentiation markers using strategies to illustrate enzyme-linked immunosorbent assay (ELISA) and quantitative polymerase chain reaction (qPCR). In vivo studies are essential for understanding the scaffold's integration with host tissues and the functional outcomes of bone healing.

6.4 Immune Response of Biomaterials for Scaffolds

The biomaterials' compatibility is essential for their biological purpose and use, as the host defence mechanism is the primary responder to shock or biomaterial implantation surgical procedure [61]. The use of scaffolds invariably elicits a foreign object response that could hinder tissue resumption and lead to fibrosis or scarring. The immune response to tissue damage can facilitate the recuperation too. The process of the host defence mechanism is complex due to the recruitment of several cell types. Consequently, techniques such as scaffold feature modification, stem cell treatment, growth factor application, and drug delivery systems have been employed to enhance scaffold efficiency [62].

6.4.1 Immunological Cell Interconnections

Post-implantation surgery, the recipient's body undergoes a series of immunological responses, both chronic and acute. Both the ingrained and modifiable immune systems contribute to the response [63]. The innate immune system comprises various resistance-related cells, combining lymphocytes (innate lymphoid cells, NK cells, γδT cells), mononuclear phagocyte cells (macrophages and monocytes), and polymorphonuclear cells (basophils, eosinophils, granulocytes), whereas adaptive immunity consists of T and B lymphocytes [64, 65] shown in Fig. 5.

The innate immune system participates in the preliminary stage of the immune response, which could be initiated within hours and last for days, triggered by inflammatory cytokines and extracellular matrix elements such as hyaluronic acid, laminin, and collagen [67]. Neutrophils, which may secrete growth factors such as VEGF to facilitate angiogenesis and increase proliferation, are also adept at clearing dead cells and debris whilst releasing chemokines to attract macrophages [68]. This method is harmful to healing due to extended activation and presence.

The compliant immune response system is initiated within one week. T cells are essential for both pro-regenerative and anti-regenerative responses, as well as the inflammatory cascade. CD4+ (T helper 1, Th1) and CD8+ (cytotoxic) T cells impair bone and cartilage regeneration by secreting TNF-α and IFN-γ, which impede rupture restoration and osteogenesis. Research indicates that the active depletion of CD8+ T cells accelerates the bone-mending operation. Tregs and γδT cells secrete essential

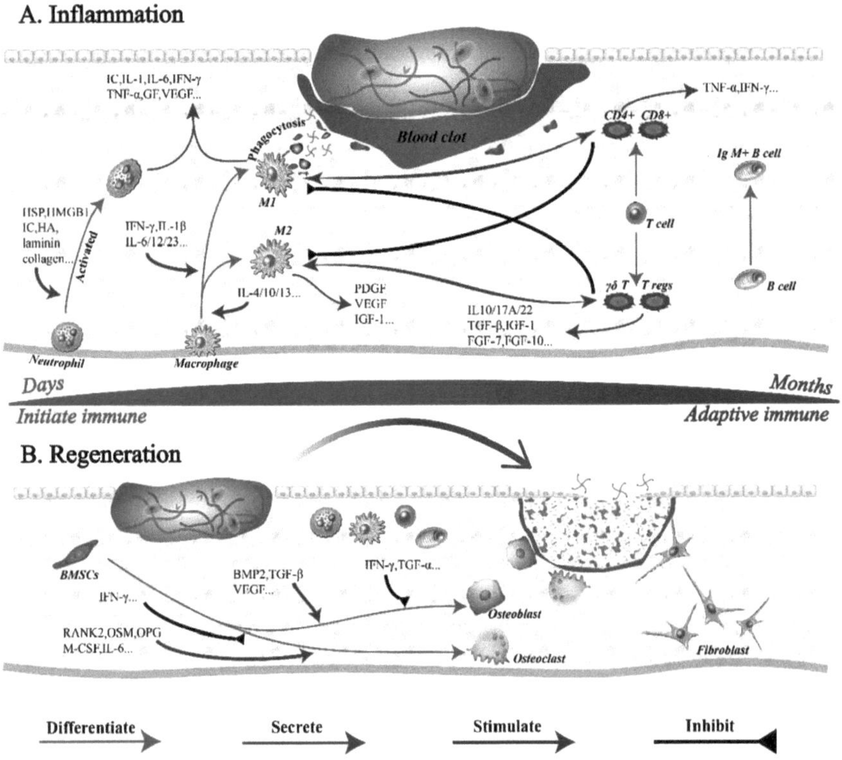

Fig. 5 A. Significant inflammatory connections amongst immunological cells, biomolecules, and implanted scaffolds and B. the differentiation and role of BMSCs in tissue regeneration. Reproduced with permission from [66], copyright 2019, Elsevier

cytokines and growth factors for tissue healing, including KGF-1 (FGF-7), KGF-2 (FGF-10), TGF-β, IGF-1, IL-10, IL-22, and IL-17A, which function as inflammatory inhibitors and possess the ability to modulate macrophage polarization. B cells are essential for TE too.

Stem cells exhibit immunosuppressive properties and the capacity for tissue regeneration. MSCs can constrain the formation of inflammatory cells, comprising dendritic cells, T cells, NK cells, B cells, and monocytes, through mechanisms involving the expression of major histocompatibility complex I/II and Fas ligand. Studies have shown that the existence of MSCs dramatically reduces pro-inflammatory cytokines such as TNF-α, IFN-γ, IL-6, IL-10, and IL-1β [69]. Chen et al. [66] examined osteonecrosis of the femoral head (ONFH) in cartilage and identified the up-regulation of chondrogenesis-related genes, featuring COL1A1, OGN, ASPN, MMP13, and BMP2. Liu et al. [70] included thermosensitive polypeptide hydrogels with the stem cells and evaluated their efficacy in the reclamation of hyaline-like cartilage and the decrease of fibrous tissue development in vivo. MSCs are intricately linked to macrophage polarization since they can induce the generation of M2 macrophages. M1 phenotype macrophages facilitate the development of mesenchymal stem cells into osteoblasts for bone and cartilage regeneration [71]. Immune-related cellular interactions can modulate osteogenesis and osteoclastogenesis throughout the bone and cartilage healing process. MSCs can transform into osteoblasts, which favourably contribute to the process of bone rebuilding. Cytokines such as bone morphogenetic protein 2 (BMP2), transforming growth factor beta (TGF-β), and VEGF can promote differentiation, but elevated levels of tumour necrosis factor-alpha (TNF-α) and interferon-gamma (IFN-γ) induce inflammatory performance and inhibit this process [66].

6.4.2 Modulation of Immune System to Strengthen Tissue Reconstitution

Scaffolds often demonstrate diverse immune responses and foreign body reactions attributable to their biological, physical, and chemical characteristics [72]. The physical features of particle configurations and scales are essential for the phagocytosis process. Materials over 100 μm necessitate the involvement of numerous monocytes and macrophages for phagocytosis, potentially leading to the release of proteolytic enzymes and the generation of reactive oxygen species. Surface topography might also affect the inflammatory response [73]. Materials like PEO (polyethylene oxide), PEG (polyethylene glycol), and PCL (polycaprolactone) have less cellular adhesion, potentially improving biocompatibility by reducing the build-up of pro-inflammatory substances. PEG is often regarded as a 'stealth' polymer that can attenuate the host immunological response. However, therapy with PEGylated medicines may induce an immunogenic response against PEG, hence diminishing therapeutic effectiveness. In the research conducted by Li and his colleagues [74], the customization of mesoporous silica micro-rod (MSR) scaffolds with PEG-RGD markedly reduced inflammation, which was not unique to RGD. In a separate study, pre-coating modified CD47 PVC (polyvinyl chloride) demonstrated decreased surface neutrophil recruitment and adherence [66]. The textured surfaces of materials demonstrate a quicker response to oxygen species compared to smooth scaffolds, indicating an amplified immunogenic reactivity. Various naturally sourced biomaterials, in particular, high molecular weight (HMW) hyaluronic acid and chitosan, possess radical oxygen species-scavenging capabilities, hence imparting anti-inflammatory effects [66]. The enhanced scaffold surface's hydrophilicity or liquid affinity promotes osteoblast cell attachment and the adherence of pro-inflammatory cells, like neutrophils. Moreover, + vely charged surfaces such as $-NH_2$ have superior cell adhesion capacity compared to negatively charged surfaces. Chemical functional groups like $-CH_3$ and –OH elicit a pronounced inflammatory response, whereas –COOH groups display diminished reactions.

The anti-inflammatory compounds' containment, or pro-inflammatory molecules in some instances, demonstrates advantages in repairing the integrity of injured tissue [66]. Common analgesics, including aspirin, pentoxifylline, and ibuprofen, might diminish the overexpression of TNF-α, which may trigger pathological osteoclast genesis and activation of the NF-kB pathway. The administration of anti-inflammatory cytokines such as TGF-β, IL-4, and IL-10 can inhibit immune responses and ensure effective tissue repair and regeneration. Liu et al. [66] constructed a PLG-based scaffold including TGF-β, resulting in a 40% reduction in the production of the cytokines IL-12, TNF-α, and MCP-1. The IL-1 and IL-33 families exhibit twain anti-inflammatory and pro-inflammatory effects, contingent upon the prevailing immunological milieu. IL-33 has been demonstrated to engage with cells including Th2, CD4+, macrophages, and Tregs, which convey the IL-33 receptor ST2. Liu et al. [75] constructed a PLG scaffold for the localized delivery of IL-33 into epididymal adipose tissue, resulting in an upsurge in the local up-regulation of graft-protective T cells, Foxp3 + population of CD4+ T cells, and a reduction in graft-destructive CD8+ T cells. SDF-1α can elicit an inflammatory response whilst facilitating tendon repair by attracting endogenous cells [66]. SDF-1 promotes osteochondral regeneration by enhancing cell homing, particularly BMSCs [76]. Kim et al. [66] developed a hypercritical lCO2–HFIP co-solvent solution encapsulating TGF-β3, which promoted cartilage production and inhibited differentiated chondrocyte hypertrophy. Anti-inflammatory medications have also been utilized. Flavonoids have anti-inflammatory and antioxidant effects and can modulate the equilibrium of osteoblasts and osteoclasts, rendering them effective for bone and cartilage regeneration applications. The research conducted by Kim, D. K. and his colleagues revealed that a 1 mg/ml curcumin/silk scaffold displayed the best cell survival and substantial extracellular matrix production capability [77]. Zhou et al. [78] constructed a scaffold utilizing both materials, mitigating the chondrocyte inflammatory response by the down-regulation of IL-1β. Macrophages have two polarization phenotypes (M1 and M2), which are crucial in the immunomodulation process. Amplifying the release of interferon-gamma (IFNg) facilitates the differentiation of the M1 phenotype, subsequently leading to the polarization of the M2 phenotype induced by IL-4 [66].

Drugs included into scaffolds could be uniformly blended and released in alignment with scaffold degradation, potentially resulting in unpredictable drug release. Consequently, drug delivery systems utilizing nanostructures, including emulsions, liposomes, and polymeric nanoparticles, function as carrier to transport target entities to specific locations with adjustable unleash rates in pharmaceutical administration. Furthermore, a composite carrier system can enhance the medication loading capacity. Dendrimers have a highly branching 3D structure. The hydrophobic interior architecture and interstitial holes inside branches facilitate the encapsulation of hydrophobic medicines such as paclitaxel, doxorubicin, camptothecin, and paclitaxel [79]. The polymeric materials' chemical bonds ought to be used to tether with proactive compounds. Moreover, medicines with inadequate characteristics can be altered by creating polymer–drug conjugates. Conventionally, medicines with low hydrophilicity are ineffective in transdermal applications; however, improving their solubility by the incorporation of water-soluble polymers might mitigate this challenge [80]. A liposome is a minuscule carrier with an aqueous core encased in phospholipids akin to bilayered frameworks. Liposomes can transport both hydrophilic and lipid-soluble pharmaceuticals owing to their biphasic characteristics. MSCs have immunomodulatory capabilities through interactions with inflammatory cells, in addition to their differentiation potential. MSCs encapsulated in PEG hydrogels have shown a beneficial impact in mitigating foreign body reactions. The hydrogels diminished mobilized primary murine macrophages by decreasing gene expression and pro-inflammatory cytokine release in vitro, whereas the existence of MSCs decreased fibrous capsule thickness in vivo [81].

7 In Vivo and In Vitro Investigations of Bone Scaffolds

Researchers are able to identify mechanical characteristics, osteoconductivity, and biocompatibility that are necessary for effective BTE through the use of in vivo and in vitro investigations, which are vital in the evaluation of bone scaffolds [82].

7.1 In Vivo Studies for Bone Scaffolds

In vivo studies assess the biological efficacy of bone scaffolds inside a live creature, often an animal model. These investigations examine osteoinductivity, biodegradability, osteoconductivity, biocompatibility, and mechanical durability within the intricate interconnections of a living system.

7.1.1 Animal Model Selection

Before clinical trials with humans, an optimal bone replacement must undergo in vivo evaluations to ensure its efficacy and safety. Consequently, establishing an appropriate animal model is a crucial phase in assessing the mechanical properties and biocompatibility of bone replacement biomaterials. This study examines the suitability of several varieties for evaluating biomaterials used as substitutes for bone defects in distinct anatomical locations, including the calvaria, femora, and ulna. We assessed the merits and drawbacks of each species for assessing certain flaws, investigated and contrasted the resemblances betwixt human clinical scenarios, and animal models, and highlighted the elements to contemplate whilst electing animals [83].

Various animal test models, including dogs, rabbits, goats, rats, and pigs, were established to replicate human in vivo environmental and physical circumstances for evaluating the efficacy and rapport of bone replacement biomaterials [84–87]. To replicate diverse orthopaedic scenarios, several defect locations have been investigated, including the calvaria, femora, and ulna. The pivotal size defect is cited as the minimal bone injury that fails to repair autonomously over an extended duration. For practical reasons, the absence of a mineralized region of $\geq 30\%$ at 52 weeks indicates that full bone regeneration is unattainable. The threshold for classifying a lesion as 'critical' remains ambiguous; nonetheless, it is characterized as a segmental bone shortage surpassing 2–2.5 times the damaged bone's diameter [83]. Multiple criteria must be evaluated when deciding on a certain animal species as a testing model. The selected animal model must have substantial physiological and pathophysiological similarities to humans. Secondly, it must be feasible to operate and monitor several research subjects post-surgery within a very short timeframe [88]. Beyond selection, variables involve purchase and care costs, animal availability, societal acceptability, tolerance to confinement, and housing feasibility [89]. As per the international standard, it is mandatory to evaluate the dimensions of the implant test samples, the implants' number per animal, the predicted time span of the test, and any interspecies variations in biological reactions [83].

7.1.2 Implantation Sites

Bone scaffold plantation sites are constructed microenvironments that facilitate cell proliferation, tissue regeneration, and vascularization in BTE constructions. These locations are essential for osteogenesis by providing an appropriate matrix for the attachment, proliferation, and differentiation of cells, including osteoblasts and MSCs, into bone tissue [90]. Biofunctionalization is the process of enhancing the biological properties of materials to improve their interaction with living tissues, especially when used in medical implants or tissue engineering. By incorporating biological molecules, bioactive elements, or functional groups onto a material's surface, biofunctionalization aims to encourage specific cellular responses, such as adhesion, proliferation, or differentiation. This process is particularly valuable in applications involving bone and tissue regeneration, where a scaffold or implant must not only support mechanical structure but also actively encourage healing and integration with the surrounding tissue. Key approaches for biofunctionalization include surface coating with biomolecules, incorporation of growth factors, chemical modification, use of nano- and microstructures, and immune response modulation [91, 92].

Cell seeding and implantation strategies in bone scaffolds are essential in tissue engineering, intending to fill scaffolds with suitable cells that would ultimately develop into functional bone tissue. Efficient seeding guarantees uniform cell distribution, preservation of viability, and commencement of bone production. Effective cell integration and adhesion inside the scaffold are crucial for the successful development of a bone graft or implant. There are several approaches that are used in cell seeding within bone scaffolds which are shown in Fig. 6; amongst them, 3D bioprinting is a sophisticated method in which cells and biomaterials are deposited layer by layer to form scaffold structures with predetermined designs and cellular arrangements. This method enables exact regulation of cell positioning inside the scaffold, fostering intricate tissue architecture that resembles genuine bone structure. 3D bioprinting is exceptionally customizable and can integrate various cell types, growth agents, and biomaterials, presenting considerable promise for individualized bone implants [48, 93].

Scaffold–host interaction and integration are critical for the success of bone scaffolds in TE [94]. For a bone scaffold to be effective, it must not only execute as an ephemeral structure for new bone growth but also actively engage with the host tissue to facilitate recovery, integration, and eventually the regeneration of functional bone. This interaction depends on several factors, including the scaffold's material properties, its bioactivity, local immune response, and the scaffold's ability to support new blood vessel formation (vascularization).

Biocompatibility is the scaffold's ability to exist in the body without triggering a negative response. A successful scaffold integrates seamlessly with surrounding tissue, avoiding chronic inflammation or rejection by the immune system. Osteoconductivity refers to the scaffold's property of supporting bone growth along its surface. The scaffold provides a framework for osteoblasts (bone-forming cells) and progenitor cells to attach, proliferate, and create new bone tissue. Materials like HaP and TCP are often used because they mimic the natural bone matrix and are conducive to bone cell activity. It is vital for bone growth that stem cells change into osteoblasts, which are guided by osteoinductive supports. Growth factors, like bone morphogenetic proteins (BMPs), can be put inside the support to give it this trait. Signalling molecules like BMPs can encourage stem cells and immature cells in the surrounding tissue to turn into bone cells. This helps the support integrate better and speeds up the mending process [95].

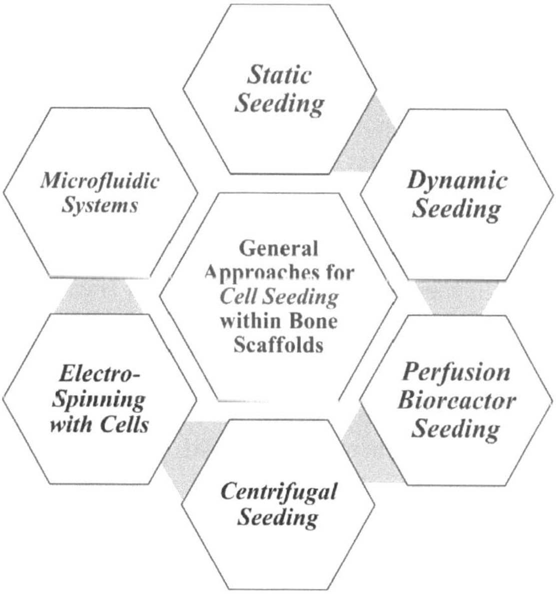

Fig. 6 Diagrammatic representation of several methods for cell seeding in bone scaffolds

Supporting the growth of new blood vessels is significant for a scaffold to integrate properly. These vessels bring air, nutrients, and trash away from the body. Vascularization is necessary in bigger bone holes where diffusion alone is not enough for cells to stay alive. Vascularization can be sped up by adding things like VEGF or making the scaffolding have holes that connect to each other and help blood vessels grow. In order to avoid collapse or displacement when subjected to pressure, bone scaffolds have to be mechanically stable and able to resist physiological pressures. The ideal situation would be for the elastic modulus (stiffness) of the scaffold to be the same as that of the adjacent bone in order to prevent stress shielding, which happens when the implant is too stiff and weakens the bone around it. Stability and support for bone ingrowth during resorption and subsequent restoration by native tissue are guaranteed by the mechanical compatibility of the scaffold [96]. A scaffold's degradation rate is a key factor in its integration. Ideally, the scaffold should gradually degrade as new bone forms, eventually being resorbed entirely and replaced by natural bone. Materials such as biodegradable polymers (e.g. PLGA) and bioresorbable ceramics are commonly used because their degradation products can be safely metabolized by the body. The degradation rate should align with the rate of new bone formation to maintain structural support without inhibiting natural bone remodelling [97]. After a scaffold is implanted, it interacts with the host immune system, which triggers a series of events that determine whether the material ought to be accepted or rejected. Some inflammation is normal and even helpful; it brings in healing cells. Integration and bone production may have been hindered by severe inflammation. It is feasible to cover scaffold materials with anti-inflammatory medicines to encourage a healing response that is favourable to tissue integration or to modify the scaffolding itself to reduce adverse immune responses.

Cell adhesion, proliferation, and differentiation are greatly influenced by the topography and surface chemistry of the scaffold. Improving cell–scaffold interactions allows for more efficient cell attachment and spreading; surface changes like adding proteins or making nanoscale structures may achieve this. For example, osteoblasts thrive on rough or porous surfaces because they enhance the area for cell attachment, which in turn promotes integration with host tissue. The scaffold's ongoing assistance throughout bone remodelling, the process by which new tissue replaces old tissue, proves crucial after initial integration, and the specifics for scaffold-hold integration are shown below in Fig. 7. For the integrated bone to preserve its health and structural integrity over time, the ideal scaffolding would let this natural process take place. If the scaffold is to facilitate early healing and interact harmoniously with the native bone over a longer time frame, then it must be able to support long-term remodelling [98].

7.2 In Vitro Studies for Bone Scaffolds

In vitro studies in BTE focus on developing biomaterials and methodologies to repair, regenerate, or replace damaged bone tissue. These studies typically involve

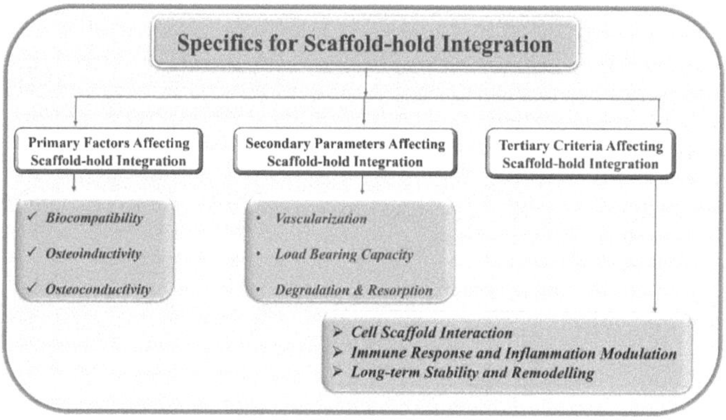

Fig. 7 Factors affecting scaffold-hold interaction and integration

testing cell–material interactions, osteogenesis, and the mechanical and biochemical properties of engineered constructs in controlled laboratory environments. The main aspects and components studies include cell types, bioreactors and dynamic cultures, growth factors and signalling molecules, and gene/protein expression analysis [99].

In vitro studies on bone scaffolds involve various cell types to mimic the complex cellular environment of bone tissue. Each cell type contributes uniquely to understanding scaffold performance in terms of osteogenesis, cell attachment, proliferation, and tissue remodelling. Various cell types in bone scaffolds and their functions are shown diagrammatically in Fig. 8. The capacity of bone scaffolds to assist with bone cell processes, including proliferation, differentiation, and matrix production, is assessed in vitro using gene and protein expression profiling [99]. Researchers can figure out how effectively cells respond to scaffolds by examining certain markers, which provide information on osteogenesis and the scaffold's ultimate efficacy in bone regeneration. Figure 9 lists the main markers utilized in gene and protein expression investigations for bone scaffolds, and analytical techniques are shown in Fig. 10.

Runt-related transcription factor 2 (Runx2) is a critical transcription factor for early osteogenic differentiation and is essential for promoting the expression of other osteogenic genes. High Runx2 levels indicate that MSCs or osteoprogenitor cells are beginning to differentiate into osteoblasts [100]. Alkaline phosphatase (ALP) is an enzyme produced in the early stages of bone formation and is involved in mineralization. When expressed at elevated levels, it serves as an indicator of early osteogenic activity, which can be quantified through assays or visualized with staining techniques [101]. Mature osteoblasts produce osteocalcin (OCN), a non-collagenous protein that is present in bone. OCN levels rise during matrix mineralization, signifying operational scaffold support for osteoblast development and functioning as a marker of later-stage bone formation [102]. Osteopontin (OPN) has a role in matrix mineralization and cell adhesion. Mature bone cells usually have a connection to their expression, implying scaffold support for the constant creation of bone matrix

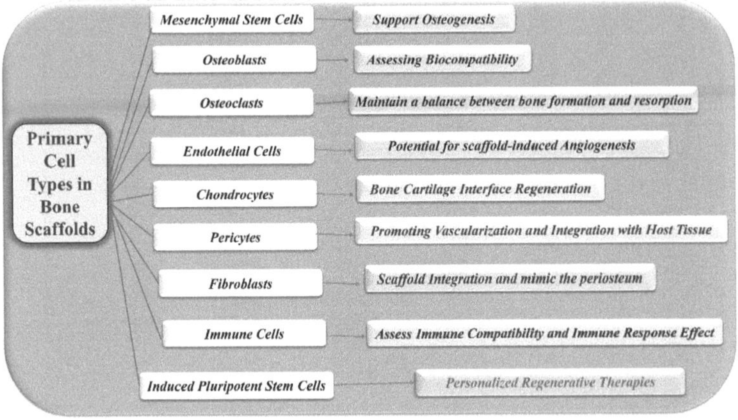

Fig. 8 Types of primary cells in bone scaffolds and their role

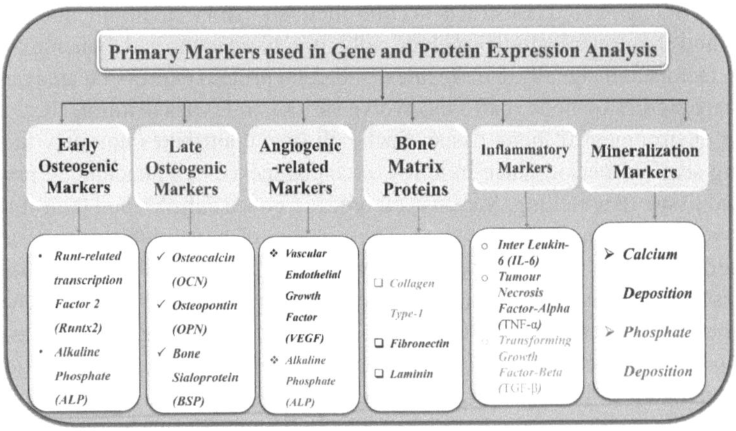

Fig. 9 Vital variables for the analysis of gene and protein expression

[103]. Another late-stage marker connected to the development of mineralized bone tissue is bone sialoprotein (BSP), which is a part of the extracellular matrix of bones. It promotes to the nucleation of HaP crystals, which are essential for the mineralization process of the scaffold [104].

The most abundant protein in bone, collagen type I, forms the scaffold of the bone extracellular matrix. High collagen expression confirms that cells on the scaffold are synthesizing an organic matrix compatible with bone structure. Although not exclusive to bone tissue, fibronectin and laminin are involved in cell adhesion and matrix assembly. Their expression can improve cell attachment and scaffold integration with the host tissue [105]. VEGF plays an essential function in vascularization by encouraging the development of blood vessels, which is needed for bone

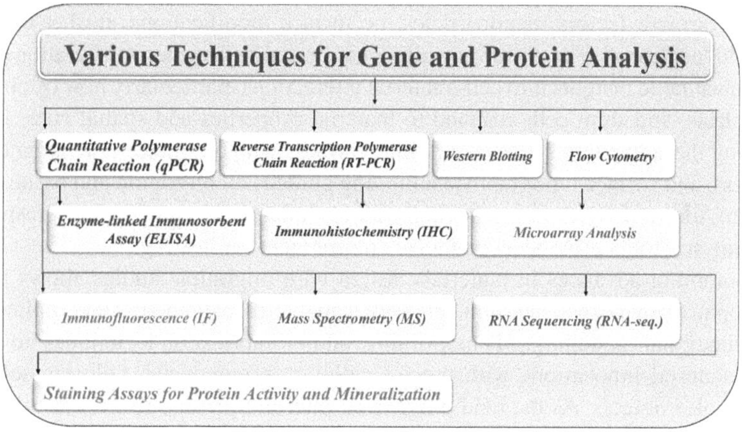

Fig. 10 Multiple techniques for gene and protein anatomization

survival and scaffold integration. According to its expression, the scaffold might encourage angiogenesis, which would be particularly advantageous for big bone transplants [106]. Along with VEGF, the markers angiopoietin-1 and angiopoietin-2 aid in the development and stabilization of new blood vessels. A scaffold environment that is suitable for vascularization is indicated by high levels of these indicators in co-culture systems (e.g. MSCs with endothelial cells). IL-6 and TNF-α are the markers that indicate an inflammatory response. Controlled inflammation is necessary for tissue healing, but high levels can indicate a scaffold's potential to provoke an immune reaction. Monitoring IL-6 and TNF-α helps gauge scaffold biocompatibility. TGF-β is involved in bone remodelling and immune response. Increased TGF-β can signal active bone remodelling or scaffold-induced modulation of the immune response [107]. The function of bone tissue is determined by mineralization, which is clearly shown by calcium deposits [108]. Alizarin red staining or calcium tests are employed to quantify calcium, which aids in evaluating the scaffold's ability to promote the development of a mineralized matrix. Identical to calcium, phosphate deposition must occur for mineralization and may be seen by von Kossa staining, which validates scaffold mineralization visually [109].

8 Summary

Materials design and migration studies have a transformative impact in expediting BTE by addressing the critical challenges of mimicking natural bone structure and function. The strategic design of biomaterials emphasizes biocompatibility, biodegradability, mechanical properties, and osteoinductive capabilities, ensuring that scaffolds support cellular growth and integration whilst degrading at a rate synchronized with new tissue formation. The bioactive components incorporation, such as growth factors, nanoparticles, or surface modifications, further enhances scaffold functionality by promoting osteogenesis and angiogenesis. Migration studies offer invaluable insights into cell–material interactions, particularly how osteoblasts, osteoclasts, and stem cells respond to material properties and spatial cues. Understanding these dynamics aids in creating scaffolds with optimized pore architecture, stiffness, and surface topography, facilitating guided cell migration and tissue regeneration. Additionally, these studies help address inflammation and immune responses, ensuring scaffolds provide a conducive environment for healing.

Combining advances in materials design with migration studies allows for the development of next-generation scaffolds capable of patient-specific applications. Computational modelling, 3D bioprinting, and biofabrication techniques now integrate material innovations with precise cellular arrangements, tailoring solutions for complex defects. As the field progresses, interdisciplinary efforts encompassing material science, biostatistics, biomechanics, bioinformatics, biophysics, neurophysics and computational neuroscience, bioenergetics, and engineering are essential for translating research into clinical applications. By bridging the gap between scaffold design and biological responses, materials design and migration studies collectively pave the way for safer, more effective, and personalized therapies in BTE, promising improved outcomes for patients with skeletal injuries or disorders.

References

1. Ikada Y, Soc JR (2006) Interface 3:589
2. Whitney GA, Jayaraman K, Dennis JE, Mansour JM (2017) J Tissue Eng Regen Med 11:412
3. Kanwar S, Vijayavenkataraman S (2021) Bioprinting 24:e00167
4. Cole JH, van der Meulen MCH (2011) Clin Orthop Relat Res 469:2139
5. Öhman C, Baleani M, Pani C, Taddei F, Alberghini M, Viceconti M, Manfrini M (2011) Bone 49:769
6. Wu S, Liu X, Yeung KWK, Liu C, Yang X (2014) Mater Sci Eng R Reports 80:1
7. Jiang D, Ning F, Wang Y (2021) J Mater Process Technol 289:116952
8. Elnaggar MA, El-Fawal HAN, Allam NK (2021) Mater Sci Eng C 119:111550
9. Eltom A, Zhong G, Muhammad A (2019) Adv Mater Sci Eng 2019:3429527
10. Yang G, Liu J, Li F, Pan Z, Ni X, Shen Y, Xu H, Huang Q (2014) Mater Sci Eng C 35:70
11. Alvarez K, Nakajima H (2009) Materials (Basel) 2:790
12. Van Hengel IAJ, Riool M, Fratila-Apachitei LE, Witte-Bouma J, Farrell E, Zadpoor AA, Zaat SAJ, Apachitei I (2017) Biomaterials 140:1
13. Kaur M, Singh K (2019) Mater Sci Eng C 102:844
14. Bornapour M, Celikin M, Pekguleryuz M (2015) Mater Sci Eng C 46:16
15. Laurencin C, Deng M (2014) Natural and synthetic biomedical polymers. Newnes
16. Ahsan SM, Thomas M, Reddy KK, Sooraparaju SG, Asthana A, Bhatnagar I (2018) Int J Biol Macromol 110:97
17. Jayaraman P, Gandhimathi C, Venugopal JR, Becker DL, Ramakrishna S, Srinivasan DK (2015) Adv Drug Deliv Rev 94:77
18. Caballé-Serrano J, Munar-Frau A, Ortiz-Puigpelat O, Soto-Penaloza D, Peñarrocha M, Hernández-Alfaro F (2018) J Clin Exp Dent 10:e477
19. Kadkhodapour J, Montazerian H, Darabi AC, Anaraki AP, Ahmadi SM, Zadpoor AA, Schmauder S (2015) J Mech Behav Biomed Mater 50:180

20. Lerebours C, Thomas CDL, Clement JG, Buenzli PR, Pivonka P (2015) Bone 72:109
21. Nune KC, Kumar A, Misra RDK, Li SJ, Hao YL, Yang R (2017) Coll Surf B Biointerfaces 150:78
22. Han C, Li Y, Wang Q, Wen S, Wei Q, Yan C, Hao L, Liu J, Shi Y (2018) J Mech Behav Biomed Mater 80:119
23. Osanov M, Guest JK (2016) Annu Rev Mater Res 46:211
24. Tang Y, Zhao YF (2016) Rapid Prototyp. J. 22:569
25. Panesar A, Abdi M, Hickman D, Ashcroft I (2018) Addit Manuf 19:81
26. Fantini M, Curto M (2018) Int J Interact Des Manuf 12:585
27. Du Y, Liang H, Xie D, Mao N, Zhao J, Tian Z, Wang C, Shen L (2020) Mater Chem Phys 239:121968
28. Fantini M, Curto M, De Crescenzio F (2016) Virtual Phys. Prototyp. 11:77
29. Gómez S, Vlad MD, López J, Fernández E (2016) Acta Biomater 42:341
30. Feng J, Fu J, Shang C, Lin Z, Li B (2018) Comput Methods Appl Mech Eng 336:333
31. Poltue T, Karuna C, Khrueaduangkham S, Seehanam S, Promoppatum P (2021) Int J Mech Sci 211:106762
32. Barba D, Alabort E, Reed RC (2019) Acta Biomater 97:637
33. Castro APG, Ruben RB, Gonçalves SB, Pinheiro J, Guedes JM, Fernandes PR (2019) Comput Methods Biomech Biomed Engin 22:567
34. Shen M, Li Y, Lu F, Gou Y, Zhong C, He S, Zhao C, Yang G, Zhang L, Yang X (2023) Bioact. Mater. 25:374
35. Liu F, Zhang DZ, Zhang P, Zhao M, Jafar S (2018) Materials (Basel). 11:374
36. Yoo D-J (2014) Int J Precis Eng Manuf 15:1657
37. Afshar M, Anaraki AP, Montazerian H, Kadkhodapour J (2016) J Mech Behav Biomed Mater 62:481
38. Wang L, Kang J, Sun C, Li D, Cao Y, Jin Z (2017) Mater Des 133:62
39. Feng Q, Tang Q, Liu Z, Liu Y, Setchi R (2018) Proc Inst Mech Eng Part B J Eng Manuf 232:1719
40. Yánez A, Herrera A, Martel O, Monopoli D, Afonso H (2016) Mater Sci Eng C 68:445
41. Zhang B, Pei X, Zhou C, Fan Y, Jiang Q, Ronca A, D'Amora U, Chen Y, Li H, Sun Y (2018) Mater Des 152:30
42. Li Y, Jahr H, Lietaert K, Pavanram P, Yilmaz A, Fockaert LI, Leeflang MA, Pouran B, Gonzalez-Garcia Y, Weinans H (2018) Acta Biomater 77:380
43. Quevedo González FJ, Nuño N (2016) Comput Methods Biomech Biomed Engin 19:845
44. Nune KC, Misra RDK, Li SJ, Hao YL, Yang R (2017) J Biomed Mater Res Part A 105:859
45. Zhang Q, Yang X, Li P, Huang G, Feng S, Shen C, Han B, Zhang X, Jin F, Xu F (2015) Prog Mater Sci 74:332
46. Orciani M, Fini M, Di Primio R, Mattioli-Belmonte M (2017) Front Bioeng Biotechnol 5:17
47. El-Rashidy AA, Roether JA, Harhaus L, Kneser U, Boccaccini AR (2017) Acta Biomater 62:1
48. Ho-Shui-Ling A, Bolander J, Rustom LE, Johnson AW, Luyten FP, Picart C (2018) Biomaterials 180:143
49. Thangaraju P, Varthya SB (2022) Medical device guidelines and regulations handbook. Springer, pp 163
50. Terauchi M, Tamura A, Arisaka Y, Masuda H, Yoda T, Yui N (2021) Pharmaceutics 13:136
51. Kim J, Mirando AC, Popel AS, Green JJ (2017) Adv Drug Deliv Rev 119:20
52. Lee C-S, Singh RK, Hwang HS, Lee N-H, Kurian AG, Lee J-H, Kim HS, Lee M, Kim H-W (2023) Prog Mater Sci 135:101087
53. Grimaudo MA, Concheiro A, Alvarez-Lorenzo C (2019) J Control Release 313:148
54. Babilotte J, Martin B, Guduric V, Bareille R, Agniel R, Roques S, Héroguez V, Dussauze M, Gaudon M, Le Nihouannen D (2021) Mater Sci Eng C 118:111334
55. Thi THH, Chai F, Leprêtre S, Blanchemain N, Martel B, Siepmann F, Hildebrand HF, Siepmann J, Flament MP (2010) Int J Pharm 400:74

56. Jacobsen PAL, Rafaelsen J, Nielsen JL, Juhl MV, Theilgaard N, Larsen KL (2013) Microporous mesoporous Mater 168:132
57. Chai F, Abdelkarim M, Laurent T, Tabary N, Degoutin S, Simon N, Peters F, Blanchemain N, Martel B, Hildebrand HF (2014) J Biomed Mater Res Part B Appl Biomater 102:1130
58. Abdelaziz AG, Nageh H, Abdalla MS, Abdo SM, Amer AA, Loutfy SA, Abdel Fattah NF, Alsalme A, Cornu D, Bechelany M (2024) Front Chem 12:1417407
59. Amiri E, Khakbiz M, Bakhshandeh B, Ranjbar N, Mohammadnejad J (2024) Mol Syst Des Eng 9:581
60. Tian C, Li K, Chu F, Wei Q, Xu S, Qiang L, Gou X (2024) RSC Adv 14:22420
61. Anderson JM, Rodriguez A, Chang DT (2008) In Seminars in immunology. Elsevier 20:86
62. Chung L, Maestas DR Jr, Housseau F, Elisseeff JH (2017) Adv Drug Deliv Rev 114:184
63. Chen Z, Klein T, Murray RZ, Crawford R, Chang J, Wu C, Xiao Y (2016) Mater Today 19:304
64. Dacoba TG, Olivera A, Torres D, Crecente-Campo J, Alonso MJ (2017) In Seminars in immunology. Elsevier 34:78
65. Nair A, Tang L (2017) In seminars in immunology. Elsevier 29:62
66. Yang D, Xiao J, Wang B, Li L, Kong X, Liao J (2019) Mater Sci Eng C 104:109927
67. Rodríguez-Cabello JC, de Torre IG, Ibañez-Fonseca A, Alonso M (2018) Adv Drug Deliv Rev 129:118
68. Fujihara Y, Takato T, Hoshi K (2010) Biomaterials 31:1227
69. Lin D, Chai Y, Ma Y, Duan B, Yuan Y, Liu C (2019) Biomaterials 196:122
70. Liu H, Cheng Y, Chen J, Chang F, Wang J, Ding J, Chen X (2018) Acta Biomater 73:103
71. Gümüşderelioğlu M, Tunçay EÖ, Kaynak G, Demirtaş TT, Aydın ST (2015) J Trace Elem Med Biol 31:120
72. Julier Z, Park AJ, Briquez PS, Martino MM (2017) Acta Biomater 53:13
73. Vasconcelos DM, Goncalves RM, Almeida CR, Pereira IO, Oliveira MI, Neves N, Silva AM, Ribeiro AC, Cunha C, Almeida AR (2016) Biomaterials 111:163
74. Li WA, Lu BY, Gu L, Choi Y, Kim J, Mooney DJ (2016) Biomaterials 83:249
75. Liu JMH, Zhang X, Joe S, Luo X, Shea LD (2018) J. Immunol. Regen. Med. 1:1
76. Chen P, Tao J, Zhu S, Cai Y, Mao Q, Yu D, Dai J, Ouyang H (2015) Biomaterials 39:114
77. Kim DK, Kim JI, Sim BR, Khang G (2017) Mater Sci Eng C 78:571
78. Zhou F, Zhang X, Cai D, Li J, Mu Q, Zhang W, Zhu S, Jiang Y, Shen W, Zhang S (2017) Acta Biomater 63:64
79. Guo Y, Zong S, Pu Y, Xu B, Zhang T, Wang B (2018) Molecules 23:1622
80. Wang N, Cheng X, Li N, Wang H, Chen H (2019) Adv Healthc Mater 8:1801002
81. Swartzlander MD, Blakney AK, Amer LD, Hankenson KD, Kyriakides TR, Bryant SJ (2015) Biomaterials 41:79
82. Mauney JR, Jaquiéry C, Volloch V, Heberer M, Martin I, Kaplan DL (2005) Biomaterials 26:3173
83. Li Y, Chen S-K, Li L, Qin L, Wang X-L, Lai Y-X (2015) J Orthop Transl 3:95
84. Zwingenberger S, Niederlohmann E, Vater C, Rammelt S, Matthys R, Bernhardt R, Valladares RD, Goodman SB, Stiehler M (2013) J Surg Res 181:e7
85. Zhu XS, Zhang ZM, Mao HQ, Geng DC, Zou J, Wang GL, Zhang ZG, Wang JH, Chen L, Yang HL (2011) J Mater Sci Mater Med 22:159
86. Koëter S, Tigchelaar SJ, Farla P, Driessen L, van Kampen A, Buma P (2009) J Biomed Mater Res Part B Appl Biomater 90:116
87. Kröpil P, Hakimi AR, Jungbluth P, Riegger C, Rubbert C, Miese F, Lanzman RS, Wild M, Schek A, Scherer A (2012) Acad Radiol 19:320
88. Noordin NNFNM, Jaafar M, Ahmad N, Sulaiman AR, Yahaya BH, Hamid ZAA (2022) Mater Today Proc 66:3031
89. Pearce AI, Richards RG, Milz S, Schneider E, Pearce SG (2007) Eur Cell Mater 13:1
90. Wei L, Wu S, Kuss M, Jiang X, Sun R, Reid P, Qin X, Duan B (2019) Bioact Mater 4:256
91. Stewart C, Akhavan B, Wise SG, Bilek MMM (2019) Prog Mater Sci 106:100588
92. Li L, Lu H, Zhao Y, Luo J, Yang L, Liu W, He Q (2019) Mater Sci Eng C 98:1241
93. Agarwal R, García AJ (2015) Adv Drug Deliv Rev 94:53

94. Koushik TM, Miller CM, Antunes E (2023) Adv Healthc Mater 12:2202766
95. Agrawal S, Srivastava R (2020) Racing for the surface: antimicrobial and interface tissue engineering, 355
96. Yin S, Zhang W, Zhang Z, Jiang X (2019) Adv Healthc Mater 8:1801433
97. Tajvar S, Hadjizadeh A, Samandari SS (2023) Int Biodeterior Biodegradation 180:105599
98. Kim J-A, Lim J, Naren R, Yun H, Park EK (2016) Acta Biomater 44:155
99. Rogowska-Tylman J, Locs J, Salma I, Woźniak B, Pilmane M, Zalite V, Wojnarowicz J, Kędzierska-Sar A, Chudoba T, Szlązak K (2019) Mater Sci Eng C 99:669
100. Zhao P-P, Hu H-R, Liu J-Y, Ke Q-F, Peng X-Y, Ding H, Guo Y-P (2019) Chem Eng J 359:1120
101. Vimalraj S (2020) Gene 754:144855
102. Manolagas SC (2020) PLoS Genet 16:e1008714
103. Si J, Wang C, Zhang D, Wang B, Hou W, Zhou Y (2020) Med Sci Monit Int Med J Exp Clin Res 26:e919159
104. Foster BL (2024) J Struct Biol:108132
105. Halper J (2021) Progress in heritable soft connective tissue diseases, 105
106. Hu K, Olsen BR (2016) Bone 91:30
107. He J, Chen G, Liu M, Xu Z, Chen H, Yang L, Lv Y (2020) Mater Sci Eng C 108:110411
108. Mi X, Gupte MJ, Zhang Z, Swanson WB, McCauley LK, Ma PX, Appl ACS (2020) Mater Interfaces 12:32503
109. Trybuś B, Zieliński A, Beutner R, Seramak T, Scharnweber D (2017) Acta Bioeng Biomech 19

Chapter 9
Bioimplants with 3D and 4D Additive Manufacturing Processes

Nisha Shankhwar and Arbind Prasad

1 Introduction

Over the last 20 years, 3D printing, or additive manufacturing, has reshaped many industries, especially healthcare. Traditional implants are mass-produced, which forces surgeons to adjust them to fit each patient. With 3D bioprinting (3DBioPr), we can now create custom implants tailored to an individual's anatomy. This personalized approach improves how well the implant works and how smoothly it integrates with the body. The demand for organ transplants has never been higher. Thousands of patients urgently need life-saving organs, but there aren't enough donors to meet the need. Long waiting lists are common, and sadly, many patients don't survive long enough to get their transplant [1]. This critical shortage underscores the need for innovative solutions, and 3DBioPr offers a game-changing way forward.

Using bioinks made from living cells, growth factors, and biomaterials, 3DBioPr can recreate complex tissues and organs, layer by layer, to closely mimic the structure and function of real organs. But despite its promise, this technology faces some tough challenges. One of the biggest hurdles is getting different types of cells positioned exactly where they need to be [2]. Since organs rely on a mix of cells working together, any misplacement can throw off their function. Figure 1 breaks down the key components of 3DBioPr: bioinks, bioprinters, scaffolds, and cells. It also touches on 4D bioprinting (4DBioPr) [3]. Unlike its predecessor, 4DBioPr uses materials that can change and adapt over time in response to things like temperature or mechanical

N. Shankhwar
Special Centre for Nanoscience, Jawaharlal Nehru University, New Delhi, India

A. Prasad (✉)
Mechanical Engineering Department, Katihar Engineering College (Under Department of Science, Technology and Technical Education, Government of Bihar), Katihar, Bihar, India
e-mail: arbind.iitg@gmail.com

A. Prasad and V. Katiyar (eds.), *Resorbable Polymers for Bioimplants and Fixation Devices*, Materials Horizons: From Nature to Nanomaterials,
https://doi.org/10.1007/978-981-96-9306-1_9

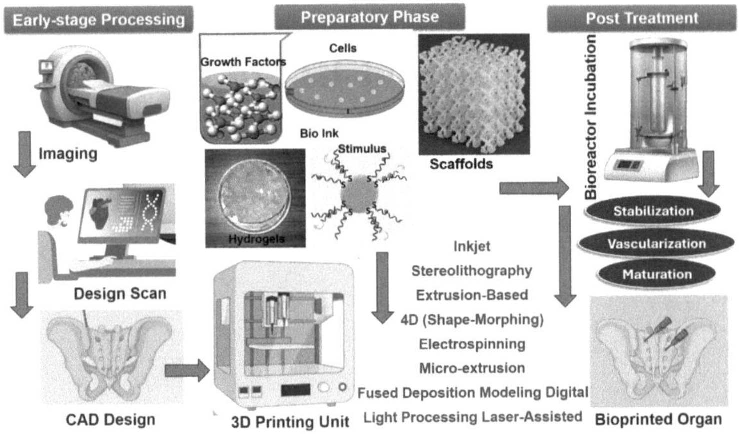

Fig. 1 Key elements of 3D bioprinting and the dynamic evolution to 4D bioprinting

stress. This innovation is pushing the boundaries of what printed structures can do, making them more dynamic and interactive.

Getting cells to behave as they should is crucial in bioprinting, and this often involves using growth factors and other signaling molecules. But that's just one part of the challenge. Once bioprinted tissues are implanted in the body, they need to maintain their structure and keep working long-term, which isn't easy. For 3DBioPr organs to actually replace traditional transplants, they must do more than just perform the right biological functions. They also need to be tough enough to handle the physical demands of the human body, like pressure and movement, without breaking down. Researchers are making significant progress, and bioengineered organs are edging closer to becoming a viable option. One of the key elements in building functional tissues is the extracellular matrix (ECM) [4]. This network of proteins and carbohydrates provides support to cells and helps them communicate. In bioprinting, recreating the ECM is critical to giving tissues both the strength and functionality of real organs. Advances in the field now include using decellularized ECM (dECM), which is tissue stripped of its cells but still retains its natural structure and biochemical features. This dECM can be infused with a patient's own cells, helping new tissues grow and ensuring they're compatible with the body. Researchers have already seen success using dECM from heart tissue to create bioinks, combining it with mesenchymal stem cells and cardiac progenitor cells [5].

As shown in Fig. 2, the bioprinting process integrates these living cells into a dECM-based bioink, laying the groundwork for creating fully functional, personalized tissues. These bioinks are used to print cardiac patches that support blood vessel formation and enhance cell-to-cell communication. Techniques like photo crosslinking are used to improve the mechanical properties of the printed tissue, ensuring it can endure the physical forces exerted by the heart and other organs.

3DBioPr is still in its early stages, but its potential to reshape organ transplantation and tissue regeneration is huge. Imagine being able to print organs like hearts, livers,

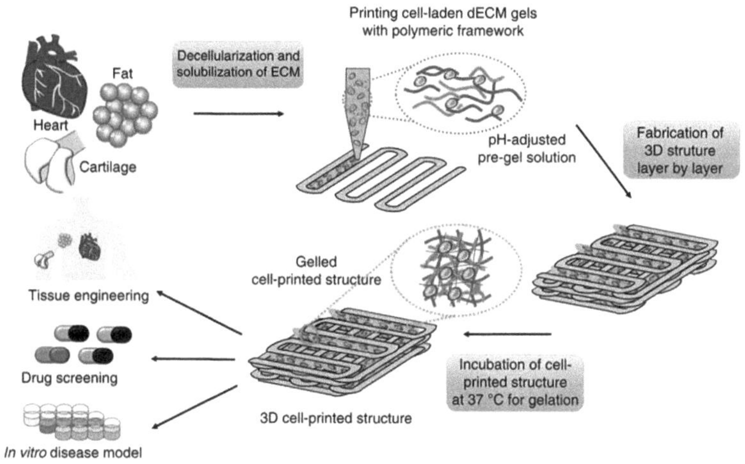

Fig. 2 Bioprinting process utilizing bioink composed of living cells embedded in decellularized extracellular matrix (dECM) [5]

or blood vessels, reducing the need for donor organs and making treatments more tailored to individual patients. As research and technology progress, bioprinting could help solve the global organ shortage and bring new hope to millions waiting for life-saving transplants [6]. Looking ahead, 4DBioPr takes things even further. It introduces the idea of implants that can adapt and change based on environmental factors like heat, moisture, or pH levels. 4DBioPr stents could expand when blood pressure shifts, or scaffolds could transform to better support tissue growth [2]. This next level of technology opens the door to medical devices that not only fit the patient perfectly but also adapt to the body's changing needs over time. Figure 3 shows how different stimuli interact with these smart materials, which are key to 4DBioPr and its role in advanced manufacturing techniques.

This chapter will explore how 3DBioPr/4DBioPr are changing the way we approach bioimplant design and functionality. These technologies are pushing the boundaries of what's possible in personalized healthcare, opening up new, more effective treatment options for patients. As these innovations continue to develop, they could revolutionize healthcare, addressing challenges that traditional implant designs have struggled to overcome.

1.1 Introduction to Bioimplants

Bioimplants are medical devices designed to replace, repair, or support damaged tissues in the body. They help restore movement, provide structural support, or assist in healing. These implants are made from materials like metals, ceramics, polymers, and composites, chosen for their strength, durability, and ability to work well with

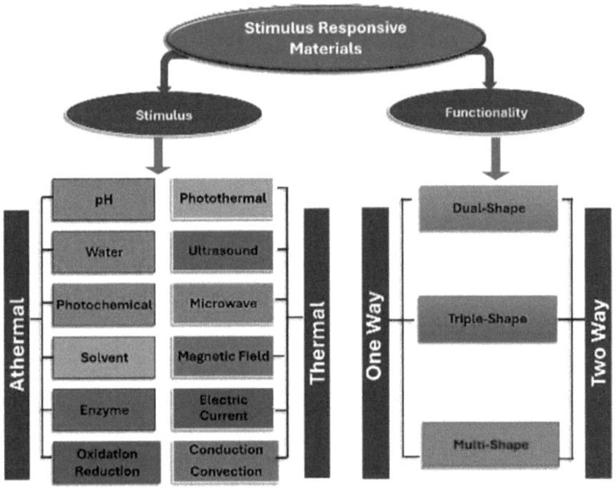

Fig. 3 Various stimuli and their corresponding functionalities in stimulus-responsive materials utilized in 4D bioprinting within AM processes

the body [7, 8]. Bioimplants are used in many areas of medicine. In orthopedics, they include things like hip replacements and spinal devices to help joints and bones work properly again. Dental implants are used to replace missing teeth or support the jaw [9, 10]. In cardiology, implants like stents and artificial heart valves ensure blood flows properly and keep the heart functioning. Regenerative medicine uses tissue scaffolds to encourage new tissue growth, while neural implants like cochlear devices help restore hearing or movement.

1.2 Key Factors for Bioimplant Success

The success of bioimplants depends on three critical factors: biocompatibility, structural integrity, and functionality. Biocompatibility ensures that the implant interacts with the body without causing adverse reactions, such as inflammation or rejection. Materials like titanium and certain polymers are widely used due to their compatibility and minimal risk of immune response. Structural integrity is essential for withstanding the mechanical stresses of the body, such as the repetitive loads placed on orthopedic or cardiovascular implants [11]. Finally, functionality is crucial for the implant to perform its intended role, whether replicating the natural movement of a joint or supporting tissue regeneration. The design, material selection, and manufacturing of bioimplants are all tailored to optimize these factors, ensuring the devices meet clinical performance standards.

Table 1 Comparison of traditional versus additive manufacturing for bioimplants

Aspect	Traditional manufacturing	Additive manufacturing (3D/4D printing)
Customization	Limited, standardized sizes and shapes	Fully customizable, patient-specific designs
Complexity of structure	Limited to simple, geometrically-simple shapes	Capable of creating complex, porous, and intricate structures
Material flexibility	Limited to a few materials	Wide range of materials, including biodegradable polymers and bioinks
Surface precision	Requires secondary processes for surface modification	High precision in surface properties for better cell adhesion and biointegration
Production time	Longer, requires molds and casting	Faster, from digital model to print
Cost	Higher for custom or complex designs	Cost-effective for small, complex batches

1.3 Challenges of Traditional Bioimplant Manufacturing

Traditional methods like casting, forging, and machining have been reliable for making bioimplants, but they come with limitations. These techniques usually produce implants in standard sizes, which might not fit perfectly for patients with unique anatomical needs. Customizing implants through these methods can be costly and take a lot of time, often leading to multiple surgeries [12, 13]. Another challenge is creating complex or porous designs, which are important for encouraging tissue growth and helping the implant bond well with the body. On top of that, traditional methods struggle with advanced materials like biodegradable polymers, which are hard to work with [14, 15]. Even achieving the right surface texture for better cell attachment and tissue integration often requires extra steps, adding more time and expense. Table 1 highlights the differences between traditional manufacturing and newer approaches like 3DBioPr/4DBioPr. These advanced methods offer more flexibility in design, a wider range of materials, and quicker production times, making it easier to create customized implants that better meet individual patient needs.

1.4 The Future of Bioimplants: Additive Manufacturing

3DBioPr/4DBioPr is changing the game for bioimplants. These advancements could transform how bioimplants are made, offering personalized solutions that improve recovery and long-term success [1]. By overcoming the limits of older methods, 3D and 4D bioprinting are set to push the field forward and deliver better results for patients. It offers several key advantages over traditional methods:

Custom Fit for Each Patient: Implants can be tailored to match a person's unique anatomy, leading to a better fit and improved performance.
Flexible Designs: Complex and porous structures that mimic natural tissues are easy to create, which helps the implant integrate with the body and speeds up healing.
Wider Range of Materials: Additive manufacturing allows for the use of advanced materials, like biodegradable polymers, giving more options for regenerative treatments.
Precise Surfaces: It enables fine control over surface textures, improving how cells attach and grow. This reduces the need for extra processing steps.

2 Design Considerations for 3D and 4D Printed Bioimplants

Designing bioimplants with 3DBioPr/4DBioPr technologies requires a meticulous approach to ensure that each implant is tailored to the individual patient's anatomy and optimized for mechanical and biological compatibility. The transformative power of 4DBioPr comes from its ability to change shape over time, driven by various external stimuli [2, 16]. The primary mechanisms behind 4DBioPr's dynamic behavior are summarized in Table 2, highlighting the diverse stimuli and their applications.

Several key considerations are crucial to achieving a high-performance bioimplant that integrates seamlessly with the body and can withstand long-term use. Below is an in-depth look at these design factors:

2.1 Customization and Patient-Specific Design Based on Imaging Data

One of the biggest breakthroughs of 3DBioPr/4DBioPr in bioimplants is the ability to design patient-specific devices. This level of customization starts with detailed anatomical data gathered from advanced imaging techniques like computed tomography (CT) and magnetic resonance imaging (MRI) [17]. These scans provide an exact map of the patient's unique bone, tissue, or organ structures, forming the foundation for implants that fit seamlessly. CT scans are ideal for capturing the fine details of bones, offering high-resolution images that are crucial for orthopedic implants. The dense tissues show up clearly, enabling precise measurements and modeling. On the other hand, MRI excels in imaging soft tissues such as cartilage, tendons, and blood vessels. This is particularly valuable for creating implants like vascular grafts or neural devices, where capturing subtle tissue variations is essential. After gathering imaging data, software like Mimics or 3D Slicer turns it into a 3D model of the patient's body. This model acts as a digital twin, letting engineers design an implant

Table 2 External stimuli in 4D printing, their mechanisms, features, and applications, enabling precise control of time-dependent shape transformations

Stimulus	Mechanism	Key features	Applications
Hydro-mechanics	Water absorption causes swelling in hydrophilic regions; reverts upon drying	Shape change driven by water; reversible upon drying	Soft robotics, self-healing materials
Thermo-mechanics	Shape change triggered by temperature shifts (with or without external load)	Utilizes two temperature points; reversible shape memory	Thermal sensors, adaptive structures
Hydro-thermal mechanics	Combined influence of water and heat on swelling/deswelling behavior	Enhanced control over complex shape transformations	Smart textiles, biomedical devices
pH-mechanics	Changes in environmental pH induce deformation	Material behavior tailored to specific pH levels	Drug delivery, pH-sensitive sensors
Thermo-photo-mechanics	Light and heat act together to induce shape changes	Dual stimuli for enhanced responsiveness	Light-activated actuators, soft robotics
Osmosis-mechanics	Osmotic pressure from solvent diffusion drives deformation	Solvent-specific shape shifting; controlled by concentration gradients	Fluid-driven actuators, filtration systems
Dissolution-mechanics	Selective dissolution in solvents reshapes the structure; can be reconfigured upon re-immersion	Dissolution of non-cross-linked regions enables reversible transformations	Reconfigurable devices, temporary implants
Cell origami	Cellular forces (actin polymerization, actomyosin interaction) fold structures into shapes	Biologically driven deformation; supports key biological processes	Tissue engineering, wound healing
Electro-responsive materials	Shape change under electric fields	Fast, precise control; responsive to electrical signals	Flexible electronics, sensors
Magneto-responsive materials	Shape change induced by magnetic fields	Enables remote actuation; non-contact manipulation	Targeted drug delivery, magnetic soft robots

that matches the natural shape of the area it will replace or support [10]. The result is an implant that fits precisely, often reducing the need for adjustments during surgery. This not only shortens surgery time but also boosts the implant's stability, lowering the chance of problems like loosening or dislocation. Custom implants also improve comfort and function. Since they are designed to match the patient's unique anatomy, they allow for more natural movement and alignment, especially in complex areas like joints or the jaw [10, 17]. This personalized approach leads to a better patient experience, quicker recovery, and better long-term results.

2.2 *Computational Modeling and Simulation in Bioimplant Design*

Finite Element Analysis (FEA): FEA is a tool used to predict how bioimplants will react under stress, deformations, and strain. By simulating real-world actions like walking or lifting, engineers can understand how the implant will perform under different conditions. This helps improve the implant's design, making it stronger, more durable, and flexible [18]. FEA is especially important for load-bearing implants, like hip or knee replacements, where maintaining stability is key to proper function.

Dynamic and Adaptive Simulation for 4D Bioimplants: For 4D bioimplants, simulations go further than just looking at static stress. They test how implants react to changes in temperature, pH, or pressure. This helps refine implants that need to adjust their shape, stiffness, or porosity over time [19]. These changes improve how the implant integrates with tissues and supports healing.

Biomechanical Modeling: Biomechanical modeling uses CT or MRI scans to see how an implant will work with the body's bones, muscles, and tissues. This is crucial for implants that carry weight, especially in joints, where alignment and load distribution matter [19]. The simulations help predict possible problems, such as implant wear or bone loss, and allow engineers to plan the best placement and design.

Fluid Dynamics and Vascular Modeling: For cardiovascular implants like stents or heart valves, computational fluid dynamics (CFD) models blood flow through the implant. This helps ensure smooth blood flow, reduce turbulence, and minimize the risk of clots [20]. By improving these designs, engineers make implants safer and more effective, leading to better long-term outcomes for patients.

2.3 *Patterning, Topography, and Texturization*

3DBioPr/4DBioPr offer an unparalleled ability to create intricate, porous structures that closely replicate the natural architecture of tissues like bone. These pores aren't just cosmetic—they play a crucial role in supporting vascularization, which is vital for thicker tissues. In complex organs like the brain and liver, where cells are intricately organized, bioprinting enables the precise placement of cells to mimic these arrangements. Even in scaffolds without cells, surface properties are key to promoting cell adhesion [21]. Nanoscale surface textures can enhance how cells attach, align, and eventually organize themselves post-printing.

Bioprinting stands out for its ability to arrange cell spheroids closely enough so they can naturally fuse together through surface tension. The design of nanostructures, especially those around 100 nm in size, plays a key role in how cells behave. Smaller features tend to enhance cell adhesion, making it easier for cells to stick to surfaces. Surface topographies, such as grooves and ridges, guide cells to align themselves along these features in a process known as contact guidance [22]. Modifying the surface with chemical and physical changes, like roughness or microchannels, can boost cell growth and help direct cell orientation, which supports better tissue formation. The rigidity of the surface is also important. Softer surfaces promote quicker cell migration and differentiation, while stiffer surfaces support stronger cell growth. Scaffold designs can also influence cellular responses, including inflammation. Materials like chitosan, with larger pores, can speed up healing by triggering beneficial pro-inflammatory signals. Nature-inspired surface structures, like nanopillars found on cicada wings, offer antibacterial properties [23]. These designs help prevent biofilm formation on implants, lowering the risk of infection and implant failure. Laser-induced forward transfer (LIFT) is an advanced bioprinting technique that uses a laser to precisely transfer cells or bioinks onto a substrate. This approach allows for the creation of complex 3D structures with high precision, improved cell viability, and excellent adaptability. Grid patterns and customizable gradients in cell concentration can mimic vascular networks, while advanced scaffolds ensure optimal distribution of nutrients, oxygen, and growth factors [24]. Textured 3D structures can also enhance mechanical properties, incorporating materials like anisotropic fibers or nanocrystals for added strength. Future innovations, such as auxetic metamaterials, may offer even greater control over scaffold flexibility, cell density, and nutrient flow, pushing the boundaries of tissue regeneration and functional healing.

3 Recent Breakthroughs in 3D and 4D Bioprinting for Bioimplants

3DBioPr/4DBioPr enables the creation of highly customized implants designed to meet the specific needs of individual patients. By printing materials layer by layer, researchers can design implants that not only function better but also integrate more naturally with the body [25]. New innovations focus on improving the strength and compatibility of these implants, while also incorporating living cells to support tissue regeneration [26]. As this technology evolves, 3DBioPr is poised to make surgeries more effective and reduce reliance on organ donations. Taking the next step, 4DBioPr introduces materials capable of changing their shape, properties, or behavior in response to external factors like temperature, light, or pH. This added flexibility allows 4DBioPr bioimplants to adapt over time, offering new possibilities for applications where dynamic adjustments are crucial to patient care [2, 3, 16]. These developments in 3DBioPr/4DBioPr represent a major leap forward, with the potential to significantly improve healthcare outcomes in the years to come.

3.1 Vascularized Tumor Models for Cancer Research

Recent progress in 3DBioPr/4DBioPr technologies has made it possible to create more accurate tumor models, which are important for cancer research and testing treatments. By printing clusters of cancer cells, scientists can better replicate the complex environment of a tumor compared to traditional 2D cell cultures [27]. These printed models not only resemble the structure and function of tumors but also include blood vessel-like networks, which are essential for the growth and spread of tumors. Researchers create these blood vessel-like structures by adding specific cells that help form tiny blood vessels, similar to those found in tumors [28]. These printed tumor clusters help researchers study how cancer cells behave, including how they interact with nearby tissues and blood vessels. This approach provides a more reliable way to test cancer treatments, as it shows how drugs move through and affect the tumor's blood supply. By adding growth factors like VEGF, which encourage the development of blood vessels, scientists can create models that better mimic the conditions of real tumors [29]. The use of 4DBioPr adds the ability to track changes over time, giving researchers insight into how tumors evolve, respond to treatments, and change as they grow. As this technology continues to improve, creating patient-specific tumor models will help make cancer treatments more personalized and accurate, offering better results for drug development and therapy testing.

3.2 Bone Tissue Engineering

Bone tissue engineering is an important area of research aimed at solving bone-related problems and creating implants that work well. Metals are commonly used in joint replacements, but they can corrode over time, which limits their effectiveness for long-term bone repair. To overcome this issue, scientists have developed special materials that contain living cells, which can grow in the body and help with healing [30]. These materials are made from natural substances and minerals, along with growth factors that support cell survival and regrowth. Researchers are working on improving how cells are placed and making the printing process faster, while still keeping the cells alive. New hybrid materials, such as biopolymers mixed with ceramics, support bone cells like osteoblasts and stem cells (MSCs and ADSCs), which contribute to growth-regulating cytokine secretion and ECM deposition. Bone's complex hierarchical structure—ranging from cortical bone to trabecular bone and from micro to nanoscale features—requires specialized scaffolds [31]. 3DBioPr allows for the production of implants with detailed surface features that improve how cells interact with the material. By using patterns at the micro- and nanoscale, these implants enhance cell attachment and protein absorption [32]. The surface designs, such as hexagonal and circular pores, play a crucial role in encouraging important cellular activities like movement, attachment, and tissue growth.

3.3 *Orthopedic Implants for Joint Replacement*

Traditional orthopedic implants often use rigid materials that can cause discomfort or lead to implant loosening over time. However, 4D-printed orthopedic implants are changing the game by using smart materials that adapt to changes in the surrounding bone structure or load-bearing conditions [33]. Knee replacements made from these materials become softer under low stress to minimize damage to surrounding tissues, while they stiffen when more load is applied, ensuring stability [34]. This adaptability improves long-term outcomes, enhances comfort, and reduces the risk of complications, offering a promising solution for joint replacement surgeries.

3.4 *Cell-Based Bioprinting in Dentistry and Reconstructive Surgery*

In dentistry, 3D bioprinting is opening up new possibilities for regenerating essential dental tissues, such as the periodontium, dentin-pulp complex, and even entire teeth. While still in its early stages, research is advancing with the use of materials like polycaprolactone (PCL), GelMA, and bioceramic blends. These materials are chosen because they closely mimic the natural properties of dental tissues, supporting cell growth and tissue regeneration. This approach could eventually lead to effective replacements for damaged or missing dental structures [35]. One promising study involved printing stem cells from the apical papilla (SCAP) of human third molars. These cells are known for their regenerative abilities. The research showed that printing SCAP cells didn't interfere with the shear-thinning property of the material, which is vital for the printing process. Shear-thinning helps the bioinks remain fluid while being printed and then solidifies after placement, ensuring precise tissue formation. This breakthrough in printing SCAP cells is a major step forward, indicating that with further advancements, 3D bioprinting could one day allow for the regeneration of not just dental tissues but entire teeth [36]. While challenges like ensuring proper blood supply and long-term tissue survival remain, the potential for 3D bioprinting to revolutionize dental care is clear.

3.5 Cartilage Tissue Engineering

Cartilage is soft, flexible, and viscoelastic, unlike rigid bone. It serves various functions depending on its location. Cartilage, being avascular and aneural, has limited self-healing abilities. There are three main types of cartilage: hyaline cartilage, which is smooth and found in joints, reducing friction and absorbing shock; elastic cartilage, which provides flexibility in structures like the ear; and fibrocartilage, which is tough and supports high-stress areas like intervertebral discs. Cartilage lacks a direct blood supply, relying on diffusion for nutrient delivery. It consists primarily of water (up to 80%), along with collagen, proteoglycans, and chondrocytes—specialized cells that maintain the ECM [37]. This unique composition gives cartilage its ability to resist compression and maintain flexibility. 3DBiopR provides a way to precisely replicate cartilage's native cell–matrix structure, enhancing regeneration. A major breakthrough involves printing cartilage progenitor cells into 3D spheroids and placing them on filament structures made with coaxially deposited alginate and its crosslinker [38]. Other techniques, such as combining electrospinning and inkjet printing, have produced mechanically stable cartilage with chondrocytes suspended in hydrogel [39]. UV crosslinking methods and the use of growth factors like TGF-b1 and FGF-2 have also proven effective in enhancing cartilage formation.

3.6 Lung Tissue Engineering

Lung tissue engineering is still in its early stages, with a few pioneering studies showing promise. Researchers have used 3DBioPr to create an air-blood barrier by printing endothelial and epithelial cells on a porous membrane, maintaining over 95% cell viability. Though clinical applications are still being explored, PLA is a common material used for creating patient-specific lung tissue models [40].

3.7 Tissue Engineering Scaffolds

Traditional scaffolds help guide tissue growth, but they often lack the ability to adapt to the changing conditions inside the body. This is where 4D-printed scaffolds offer an advantage. These scaffolds can respond to biological signals like changes in pH, temperature, or the presence of growth factors [16]. They can adjust their shape or stiffness in real-time, which helps improve tissue integration and healing [41]. This adaptability is especially valuable for bone and cartilage regeneration. Studies show how 4D scaffolds can release growth factors or adjust stiffness to promote better healing and tissue formation [42]. As research progresses, these 4D scaffolds promise to improve regenerative medicine by more closely mimicking the body's natural healing processes.

3.8 Wound Healing Implants

Wound dressings designed with 4D printing offer more than simple protection; they adapt to the wound's changing condition. These advanced dressings can adjust their shape, expand to cover larger areas, or release healing agents such as antibiotics and growth factors when infection or inflammation is detected, accelerating the healing process. 3D bioprinting and 4D-printed smart dressings are poised to transform wound care, speeding up recovery and lowering infection risks [43]. Crafted from biocompatible shape-memory polymers, these dressings are initially compressed and placed onto the wound. As they warm to body temperature, they expand to their original form, applying gentle pressure to the wound and encouraging clot formation, which supports faster healing. This technology is particularly effective for chronic wounds like diabetic ulcers and burns. Integrating shape-memory polymers with 3D bioprinting enables the creation of materials that offer temporary support and scaffolding for tissue regeneration, mimicking the mechanical properties of natural tissues to foster cell growth and tissue repair [41, 44].

3.9 Liver Tissue Engineering

Human pluripotent stem cells (hPSCs) show great potential for creating functional liver tissue due to their ability to replicate indefinitely and differentiate into various cell types, including hepatocytes. Recent studies have successfully bioprinted hepatocyte-like cells in alginate matrices, maintaining a morphology similar to native liver tissue [45]. These models can be used to improve surgical planning and potentially reduce risks for liver-related procedures, though the time and costs involved remain challenges for broader adoption [46]. Igami et al. investigated using a 3D-printed liver model to assist in hepatectomy for small tumors that are difficult to detect with intraoperative ultrasound. This approach made the procedure more practical and streamlined [47]. The liver model was made from rigid acrylic resin, with water-soluble resin used for support (Fig. 4).

After printing, the surface was coated with a support material (Fig. 4a), which was then removed using an ultrasonic cleaner. The hepatic veins were cleared of any residual mold, the surface was smoothed, painted with urethane resin, and air-dried. Indigo Carmine dye was injected to color the veins (Fig. 4b), while the portal veins stayed whitish due to residual mold charge (Fig. 4c). Before surgery, the 3D-printed liver was vacuum-sealed in a sterilized nylon bag (Fig. 4d), ensuring that no fragments would be lost in the abdominal cavity if the model broke during the procedure (Fig. 4e). This innovative tool brings a new level of precision and safety to complex surgeries (Fig. 4).

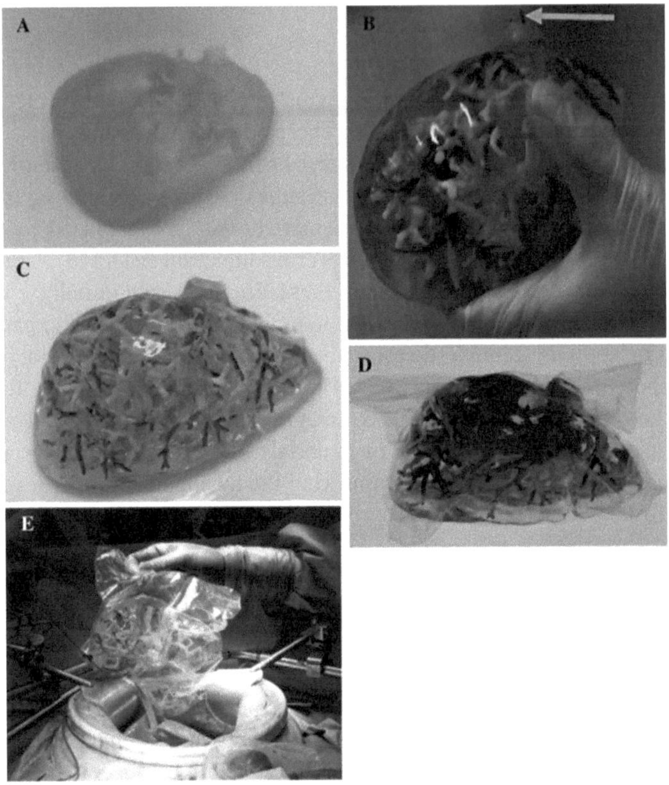

Fig. 4 3D-printed liver (70% scale) was coated with acrylic resin for support, with dyed hepatic (blue) and portal veins (whitish). It was vacuum-sealed for surgical use (yellow arrow: angiographic catheter) [47]

3.10 Cardiac Tissue and Heart Valve Engineering

Repairing cardiac tissue after injury is no easy feat, especially since mammalian cardiomyocytes stop regenerating after birth. Once damaged, the heart must rely on the remaining muscle cells for recovery. However, recent advances offer hope. One promising strategy involves pre-vascularized patches made from stem cells like human c-kit + cardiac progenitor cells (hCPCs) or mesenchymal stem cells (hMSCs), combined with heart-derived dECM [48]. These bioinks, enriched with vascular endothelial growth factor (VEGF), create an environment that supports cell survival and aids tissue repair. Vitamin B2 helps with the hardening of materials under UV light, while heat-based methods provide the necessary strength for repairing the heart. Bioinks made from decellularized tissue have been shown to promote muscle cell development and significantly increase the activity of heart-related genes—up to 20 times more than traditional collagen-based bioinks [49]. Bioinks made from alginate, which resemble the natural structure of tissues, also improve cell attachment and show great potential for use in medical treatments. Studies in living organisms confirm that bioprinted patches with heart stem cells, mixed with a gel-like material, help repair damaged heart tissue after a heart attack. However, there are still challenges, such as limited growth of heart cells and difficulty in mimicking the natural alignment of heart muscle fibers [48]. Creating blood vessels in the tissue is also essential to ensure proper nutrient and oxygen flow. Induced stem cells are another promising option because they can turn into heart muscle cells and form small heart structures for testing. Adding gold particles to a gel material has improved communication between cells, helping them contract together, which is important for heart tissue function [50]. PVA and polycaprolactone mixed with carbon nanotubes are also showing promise in supporting heart tissue repair [51]. In heart valve research, progress is slower. While artificial valves work in the short term, they face issues like hardening and poor remodeling over time. Research is focused on regenerating heart valves, especially the aortic and pulmonary valves, using materials that help cells spread and reshape. New developments in gel-like and elastic materials are bringing us closer to creating personalized, long-lasting heart valves for patients.

3.11 Nervous System

Nerve damage can lead to serious health issues, and while autologous grafting provides a potential solution, it comes with risks such as infection and the need for additional surgeries. Biological grafts that mimic natural tissue offer a promising alternative for nerve repair. The human nervous system includes the central nervous system (CNS) and the peripheral nervous system (PNS), both of which are connected by neurons. Neural stem cells (NSCs) are capable of regeneration and can move to injury sites, making them suitable for repairing the CNS. Water-based bioinks mixed with NSCs help deliver the cells to the site of injury. Biodegradable polyurethane combined with PCL and PLLA creates a gel that supports NSC survival, maintaining over 50% cell viability after 24 h [52]. Systems that release growth factors further enhance NSC migration and growth, supporting better tissue regeneration. Collagen and fibrin-based bioinks help maintain NSC viability, while conductive polysaccharide bioinks like carboxymethylchitosan and graphene aid NSC differentiation into neurons [53]. In nerve grafting, Schwann cells (SC) assist with neuron regeneration but struggle with adhesion. Combining SC with bone marrow stem cells (BMSC), which adhere more readily, overcomes this issue. Scaffold-free grafts printed with BMSC and SC support axon regrowth, confirmed by electrophysiological testing. Research is also exploring the replication of brain structures using a layer-by-layer approach. Using gellan gum, which prevents cell aggregation, researchers have printed a simplified brain-like architecture for in vitro applications.

3.12 Pancreas Tissue

In type I diabetes, the immune system destroys insulin-producing β cells, often requiring islet transplantation. However, transplantation success is limited by complications such as inflammation and reduced vascularization. Studies demonstrated success of bioprinting large cell aggregates of human islets within an alginate-gelatin matrix. While this approach holds promise, high viscosity can impair nutrient diffusion and affect islet function [54]. Smaller cell aggregates may improve outcomes by reducing diffusion limitations. Using pancreas-derived dECM to replicate the native microenvironment could restore islet viability and insulin production. Co-culturing islets with endothelial cells (HUVECs) has been shown to enhance vascularization and insulin secretion [54]. Induced pluripotent stem cells (iPSCs) provide an alternative, offering an endless supply of insulin-producing cells, addressing donor organ shortages. Directed differentiation of iPSCs into pancreatic cells within a dECM environment is promising, but shear forces during printing may affect cell viability [55]. Alternative methods, like differentiating bone marrow-derived mesenchymal stem cells (bMSCs) into insulin-producing aggregates, show potential for pancreatic tissue regeneration.

3.13 Skin Tissue

Human skin serves as a complex protective barrier, regulating temperature and sensation. Effective skin regeneration is crucial after injury, but traditional grafts have limitations. Bioprinting is helping replicate the native skin structure using key cells like keratinocytes (KCs) and fibroblasts (FBs). One method uses a robotic bioprinting system to create a collagen-based skin construct, where KCs form the epidermis and FBs create the dermis [56]. This approach ensures high cell viability and even distribution. While collagen, alginate, and fibrin are commonly used bioinks, they fail to fully replicate the complexity of the ECM. Porcine skin-derived dECM bioinks improve mechanical properties and support cell proliferation, enhancing construct quality. Inkjet printing techniques, using low-viscosity bioinks, help prevent shear stress and clogging [57]. Laser-assisted bioprinting, with energy between 30 to 60 μJ, offers a promising method to maintain cell viability while printing skin constructs.

3.14 Neuroprosthetic Implants

Neuroprosthetics aim to restore lost functions by interfacing with the nervous system. Traditional devices are rigid and unable to adapt to changes in nerve tissue. 4D-printed neuroprosthetics, created from materials that respond to stimuli, can change shape or stiffness when exposed to factors like temperature or pH [3]. This ability helps the implants better integrate with nerve tissue. These neuroprosthetics have shown potential in improving nerve regeneration, particularly in spinal cord injuries and peripheral nerve damage [58]. The implants adjust to the surrounding tissue, supporting nerve growth and enhancing neural signaling. This technology opens up new possibilities for treating nerve injuries more effectively.

3.15 Self-expanding Stents

Stents are essential in cardiovascular procedures, helping to open blocked or narrowed arteries. Traditional stents, while effective, are rigid and don't adapt once implanted. 4D-printed self-expanding stents, made from shape-memory alloys like nickel-titanium (NiTi), provide a solution to the challenge of stent placement [59]. These stents are compressed for easy insertion and expand once exposed to body temperature. This ability ensures they fit properly, reducing the risk of restenosis (re-narrowing of the artery) and offering a more effective, minimally invasive treatment option. By adapting to body temperature, these stents expand to fit the artery perfectly, reducing the need for post-operative interventions [60].

3.16 Drug Delivery Systems

Conventional drug delivery implants release medication at a constant rate, which does not adapt to changes in the patient's physiological condition. In contrast, 4D-printed drug delivery systems offer a more responsive solution by adjusting the release of drugs based on environmental factors like temperature, pH, or the presence of specific enzymes. These systems utilize materials like temperature-sensitive hydrogels, which undergo structural changes in response to heat, allowing precise control over the drug release rate [61]. Hydrogel might swell or shrink as the temperature changes, controlling the rate at which the drug is delivered. This technology is especially beneficial for chronic conditions, like diabetes or cancer treatments, where drug release needs to be aligned with disease progression. In cancer treatment, for instance, the drug can be released more effectively during periods of rapid tumor growth, ensuring that the therapeutic dose is delivered precisely when it's most needed [62]. This method can minimize the risk of side effects, such as toxicity, that often arise from constant drug release.

3.17 Bioprinting Blood Vessels

One of the major challenges in organ bioprinting is creating blood vessels that facilitate nutrient transport, waste removal, and cell viability. Hollow, tubular structures mimicking blood vessels have been successfully printed using coaxial nozzles, with alginate [63]. The core of the tube is filled with a crosslinking agent, while the outer layer contains cell-laden bioink. However, the cell viability in these printed vessels is highly dependent on factors like hydrogel viscosity, nozzle diameter, and dispensing pressure [64]. High alginate concentrations can cause shear stress, reducing cell viability, but post-printing incubation can help improve cell survival. Using a triaxial nozzle to layer endothelial cells with smooth muscle cells mimics natural blood vessels, enhancing vascularization [65]. This process, combined with drug-loaded stem cells in vascularized hydrogel matrices, has shown significant promise in improving the viability and functionality of printed blood vessels.

3.18 Bionic Organs

Bionic organs integrate biological tissues with functional electronics to create devices that interact with the human body. However, this presents significant challenges: the mechanical properties of electronics and biological tissues differ drastically, fabrication processes are distinct, and electronic wafers are flat, while organs have complex 3D shapes [66]. To overcome these hurdles, advancements in 3DBioPr and functional nanomaterials are proving invaluable. Chondrocyte-laden alginate hydrogel, printed with silver nanoparticles, was used to create a human ear with an inductive coil antenna. When connected to cochlea-shaped electrodes on silicone, this device improved auditory function, showcasing how bioprinting can help bridge the gap between biology and electronics, creating more functional, complex bionic organs [67]. Table 3 illustrates the key technological innovations in 3DBioPrand 4DBioPr, showcasing the materials and techniques employed in creating a range of engineered tissues and organs.

4 Concluding Remarks and Future Directions

3DBioPr/4DBioPr are changing the way we approach healthcare, especially when it comes to implants and creating organs. Traditional implants often don't fit perfectly and require adjustments during surgery. With 3D printing, we can now create implants that are tailored to each patient's unique anatomy, improving both the fit and function [12, 35, 40]. 3DBioPr holds great potential for improving medical treatments by reducing the need for multiple surgeries and speeding up recovery. By printing tissues and even organs using living cells and specialized materials, we are closer to solving the global shortage of organs for transplant. Engineered tissues, like heart patches or blood vessels, can replicate the structure of natural organs and help restore their function [68]. Using tissue from donors to create scaffolds for these cells helps us build structures that could one day replace the need for organ transplants. There are still challenges, however. Placing cells accurately, keeping printed tissues intact,

Table 3 Summary of bioprinting techniques, materials, and developed tissues/organs

Advancement area	Materials	Key features	Challenges	Recent innovations	Developed organs/ Tissues
Bioinks for tissue engineering	Alginate, GelMA, collagen, chitosan	Biocompatibility, supports cell growth, customizable for different tissues	Maintaining cell viability, improving mechanical properties	Development of hybrid bioinks, enhanced cell-laden bioinks	Cartilage, skin, bone, muscle
Polymer-based implants	PCL, PLA, PHA, TPU	Biodegradable, customizable, and provides structural support	Limited material strength, post-processing requirements	Innovations in polymer composite materials for stronger bioimplants	Bone, cartilage, vascular scaffolds
Ceramics and bioceramics for bone implants	Tricalcium phosphate, hydroxyapatite, calcium phosphate	Mimics bone structure, supports osteogenesis, biodegradable	Brittleness, limited vascularization, material degradation	Hybrid ceramic-composite materials for enhanced mechanical properties	Bone, bone grafts, dental implants
Natural polymers for soft tissue regeneration	Collagen, hyaluronic acid, fibrinogen	Mimics extracellular matrix, promotes cell adhesion and growth	Mechanical limitations, complex tissue structures	Use of decellularized tissues, bioinks mimicking natural ECM	Cartilage, tendons, skin, nerve tissues
Smart materials for adaptive implants	Shape-memory alloys (NiTi), Smart hydrogels (PVA, PEG)	Respond to external stimuli (e.g., pH, temperature)	Complexity in material processing and stability under physiological conditions	Self-expanding stents, dynamic scaffolds, and adaptive implants	Blood vessels, heart valves, stents
Stimuli-responsive hydrogel composites	PVA, PEG, GelMA	Controlled drug release, self-healing, and ability to change shape	Controlled release over time, ensuring long-term stability	Development of hydrogels responsive to pH, temperature, and ions	Soft tissues, wound healing, cartilage

(continued)

Table 3 (continued)

Advancement area	Materials	Key features	Challenges	Recent innovations	Developed organs/ Tissues
Vascularization in tissue engineering	Smart hydrogels, PCL, bioceramic composite	Mimics natural tissue response, allows better nutrient exchange, adapts to environmental conditions	Ensuring long-term functionality and integration in the body	Vascularization strategies using self-assembling and dynamic scaffolds	Vascular tissues, organs with vascular networks
Drug delivery and therapeutic implants	Hydrogel composites (e.g., PVA, PEG), bioceramics	Sustained release, biocompatibility, drug loading potential	Achieving sustained release, ensuring stability and safety	Development of 4D-printed drug delivery systems responsive to stimuli	Bone, soft tissue, cancer treatment, organs for localized therapies
Tissue-mimicking implants	Biocomposites, bioinks (cell-laden)	Mimics native tissue architecture, cell proliferation	Cell integration, tissue maturation and complexity	Integration of multiple cell types for complex tissue structures	Skin, cartilage, bone, organs for transplant preparation
Bioactive implants for regeneration	Shape-memory polymers, bioceramic composite	Responds to mechanical or environmental cues, accelerates healing	Complexity in bioactive material development and stability	Bioactive polymers that respond to tissue regeneration needs	Cartilage, bone, tendon, vascular tissues
Customizable implants for personalized medicine	Hybrid bioinks (natural + synthetic), Advanced printing techniques	Tailored for individual patients, customizable shape and function	Need for precise control over materials and printing techniques	Progress in personalized 3D printing for bone, cartilage, and dental implants	Bone, cartilage, dental implants, custom prosthetics

and ensuring implants work over the long term are hurdles that need to be addressed. Advances in materials are making printed tissues and scaffolds stronger and more compatible with the body. New techniques, like modifying surface textures, help improve the effectiveness of these implants. Looking ahead, 4DBioPr has the potential to take this technology even further. Unlike traditional implants, 4D-printed devices can change shape or adapt to the body's needs. For example, stents can expand when blood pressure increases, or scaffolds can shift to better support tissue growth [69]. This adaptability could reshape how we think about implants and regenerative medicine. Computational modeling tools are also playing a crucial role in improving these innovations. By simulating how implants behave under different conditions or interact with biological tissues, engineers can design devices that not only fit more precisely but are also stronger and more durable [70]. As 3DBioPr/4DBioPr technologies continue to evolve, regulatory and ethical considerations will play an essential role. Agencies like the FDA need to ensure that bioprinted implants meet strict safety and performance standards before being used in patients [71]. The complexity of working with living cells and bioinks introduces concerns about long-term safety, potential side effects, and the risk of the body rejecting or reacting to the implants. Ethical issues also arise around the use of human cells, the creation of lab-grown tissues, and the printing of entire organs. Addressing these concerns through informed consent, privacy protections, and clear guidelines will be key to ensuring the responsible development of these technologies [72]. Balancing innovation with careful oversight will help make bioprinted implants and therapies both safe and effective for patients.

The future of healthcare is being shaped by the marriage of 3DBioPr/4DBioPr. These technologies are enabling personalized, adaptable, and functional implants that promise to improve patient care on an unprecedented scale. While there are challenges ahead, the breakthroughs happening now are laying the foundation for a new era in medicine—one where custom, biologically-integrated implants could transform the way we treat everything from joint replacements to organ transplants.

Acknowledgements The author extends sincere gratitude for the generous financial support provided by the Department of Biotechnology (DBT), Ministry of Science and Technology, Government of India, under project No. DBT/2022-23/July/N/2240, as part of the DBT-RA Program. Heartfelt gratitude goes to Prof. A. Srinivasan for his expert guidance, and constant encouragement. Special thanks to Mohit Kumar for his unwavering support and valuable insights.

References

1. Amaya-Rivas JL et al (2024) Future trends of additive manufacturing in medical applications: an overview. Heliyon 10(5):e26641. https://doi.org/10.1016/j.heliyon.2024.e26641
2. Patil AN, Sarje SH (2021) Additive manufacturing with shape changing/memory materials: a review on 4D printing technology. Mater Today Proc 44:1744–1749. https://doi.org/10.1016/j.matpr.2020.11.907

3. Kim J, Gouripriya DA, Debnath P, Saha P (2024) Smart multi-responsive biomaterials and their applications for 4d bioprinting. Biomimetics 9(8):484. https://doi.org/10.3390/biomimetics9080484
4. Hoshiba T et al (2016) Decellularized extracellular matrix as an in vitro model to study the comprehensive roles of the ECM in stem cell differentiation. Stem Cells Int 2016(1). https://doi.org/10.1155/2016/6397820
5. Pati F et al (2014) Printing three-dimensional tissue analogues with decellularized extracellular matrix bioink. Nat Commun 5(1):3935. https://doi.org/10.1038/ncomms4935
6. Mirshafiei M, Rashedi H, Yazdian F, Rahdar A, Baino F (2024) Advancements in tissue and organ 3D bioprinting: current techniques, applications, and future perspectives. Mater Des 240:112853. https://doi.org/10.1016/j.matdes.2024.112853
7. Do A, Khorsand B, Geary SM, Salem AK (2015) 3D printing of scaffolds for tissue regeneration applications. Adv Healthc Mater 4(12):1742–1762. https://doi.org/10.1002/adhm.201500168
8. Prashar G, Vasudev H, Bhuddhi D (2023) Additive manufacturing: expanding 3D printing horizon in industry 4.0. Int J Interact Des Manuf (IJIDeM) 17(5):2221–2235. https://doi.org/10.1007/s12008-022-00956-4
9. Rouf S et al (2022) Functionally graded additive manufacturing for orthopedic applications. J Orthop 33:70–80. https://doi.org/10.1016/j.jor.2022.06.013
10. Haleem A, Javaid M (2018) Role of CT and MRI in the design and development of orthopaedic model using additive manufacturing. J Clin Orthop Trauma 9(3):213–217. https://doi.org/10.1016/j.jcot.2018.07.002
11. Pesode P, Barve S (2022) Additive manufacturing of metallic biomaterials and its biocompatibility. Mater Today Proc. https://doi.org/10.1016/j.matpr.2022.11.248
12. Tshephe TS, Akinwamide SO, Olevsky E, Olubambi PA (2022) Additive manufacturing of titanium-based alloys—a review of methods, properties, challenges, and prospects. Heliyon 8(3):e09041. https://doi.org/10.1016/j.heliyon.2022.e09041
13. Pettersson ABV et al (2024) Core legal challenges for medical 3d printing in the EU. Healthcare 12(11):1114. https://doi.org/10.3390/healthcare12111114
14. Adugna YW, Akessa AD, Lemu HG (2021) Overview study on challenges of additive manufacturing for a healthcare application. IOP Conf Ser Mater Sci Eng 1201(1):012041. https://doi.org/10.1088/1757-899X/1201/1/012041
15. Ngo TD, Kashani A, Imbalzano G, Nguyen KTQ, Hui D (2018) Additive manufacturing (3D printing): a review of materials, methods, applications and challenges. Compos B Eng 143:172–196. https://doi.org/10.1016/j.compositesb.2018.02.012
16. Lai J et al (2024) 4D bioprinting of programmed dynamic tissues. Bioact Mater 37:348–377. https://doi.org/10.1016/j.bioactmat.2024.03.033
17. Martín-Noguerol T, Paulano-Godino F, Riascos RF, Calabia-del-Campo J, Márquez-Rivas J, Luna A (2019) Hybrid computed tomography and magnetic resonance imaging 3D printed models for neurosurgery planning. Ann Transl Med 7(22):684–684. https://doi.org/10.21037/atm.2019.10.109
18. Filipovic N (2021) Computational modeling and simulation examples in bioengineering. Wiley-IEEE Press
19. Imran R, Al Rashid A, Koç M (2022) Review on computational modeling for the property, process, product and performance (PPPP) characteristics of additively manufactured porous magnesium implants. Bioprinting 28:e00236. https://doi.org/10.1016/j.bprint.2022.e00236
20. Acuna A et al (2018) Computational fluid dynamics of vascular disease in animal models. J Biomech Eng 140(8). https://doi.org/10.1115/1.4039678
21. Tripathi S, Mandal SS, Bauri S, Maiti P (2023) 3D bioprinting and its innovative approach for biomedical applications. MedComm (Beijing) 4(1). https://doi.org/10.1002/mco2.194
22. Blersch DM, Kardel K, Carrano AL, Kaur M (2017) Customized 3D-printed surface topography governs species attachment preferences in a fresh water periphyton community. Algal Res 21:52–57. https://doi.org/10.1016/j.algal.2016.10.027
23. Lee M-S et al (2023) Nature-inspired surface structures design for antimicrobial applications. Int J Mol Sci 24(2):1348. https://doi.org/10.3390/ijms24021348

24. Chang J, Sun X (2023) Laser-induced forward transfer based laser bioprinting in biomedical applications. Front Bioeng Biotechnol 11. https://doi.org/10.3389/fbioe.2023.1255782
25. Aufa AN et al (2024) Current trends in additive manufacturing of selective laser melting for biomedical implant applications. J Market Res 31:213–243. https://doi.org/10.1016/j.jmrt.2024.06.041
26. Mobarak MH et al (2023) Recent advances of additive manufacturing in implant fabrication—a review. Appl Surf Sci Adv 18:100462. https://doi.org/10.1016/j.apsadv.2023.100462
27. Chen EP, Toksoy Z, Davis BA, Geibel JP (2021) 3D bioprinting of vascularized tissues for in vitro and in vivo applications. Front Bioeng Biotechnol 9. https://doi.org/10.3389/fbioe.2021.664188
28. Zhang Y et al (2021) Recent advances in 3D bioprinting of vascularized tissues. Mater Des 199:109398. https://doi.org/10.1016/j.matdes.2020.109398
29. Shukla AK, Yoon S, Oh S-O, Lee D, Ahn M, Kim BS (2024) Advancement in cancer vasculogenesis modeling through 3d bioprinting technology. Biomimetics 9(5):306. https://doi.org/10.3390/biomimetics9050306
30. Prem Ananth K, Jayram ND (2024) A comprehensive review of 3D printing techniques for biomaterial-based scaffold fabrication in bone tissue engineering. Ann 3D Printed Med 13:100141.https://doi.org/10.1016/j.stlm.2023.100141
31. Yazdanpanah Z, Johnston JD, Cooper DML, Chen X (2022) 3D bioprinted scaffolds for bone tissue engineering: state-of-the-art and emerging technologies. Front Bioeng Biotechnol 10. https://doi.org/10.3389/fbioe.2022.824156
32. Zhang B et al (2021) 3D printed bone tissue regenerative PLA/HA scaffolds with comprehensive performance optimizations. Mater Des 201:109490. https://doi.org/10.1016/j.matdes.2021.109490
33. Wang CJ, Hazlehurst KB (2019) Orthopedic implant design and analysis: potential of 3D/4D bioprinting. In: 3D and 4D printing in biomedical applications. Wiley, pp 423–442. https://doi.org/10.1002/9783527813704.ch16
34. Ramavath D, Yeole SN, Kode JP, Pothula N, Devana SR (2023) Development of patient-specific 3D printed implants for total knee arthroplasty. Explor Med:1033–1047. https://doi.org/10.37349/emed.2023.00193
35. Jessop ZM, Al-Sabah A, Gardiner MD, Combellack E, Hawkins K, Whitaker IS (2017) 3D bioprinting for reconstructive surgery: principles, applications and challenges. J Plast Reconstr Aesthet Surg 70(9):1155–1170. https://doi.org/10.1016/j.bjps.2017.06.001
36. Yang P, Ju Y, Hu Y, Xie X, Fang B, Lei L (2023) Emerging 3D bioprinting applications in plastic surgery. Biomater Res 27(1). https://doi.org/10.1186/s40824-022-00338-7
37. Zhou J, Li Q, Tian Z, Yao Q, Zhang M (2023) Recent advances in 3D bioprinted cartilage-mimicking constructs for applications in tissue engineering. Mater Today Bio 23:100870. https://doi.org/10.1016/j.mtbio.2023.100870
38. Qasim M, Chae DS, Lee NY (2019) Advancements and frontiers in nano-based 3D and 4D scaffolds for bone and cartilage tissue engineering. Int J Nanomedicine 14:4333–4351. https://doi.org/10.2147/IJN.S209431
39. Shankhwar N, Kumar M, Mandal BB, Srinivasan A (2016) Novel polyvinyl alcohol-bioglass 45S5 based composite nanofibrous membranes as bone scaffolds. Mater Sci Eng, C 69:1167–1174. https://doi.org/10.1016/j.msec.2016.08.018
40. Mahfouzi SH, Safiabadi Tali SH, Amoabediny G (2021) 3D bioprinting for lung and tracheal tissue engineering: criteria, advances, challenges, and future directions. Bioprinting 21:e00124. https://doi.org/10.1016/j.bprint.2020.e00124
41. Shankhwar N, Unal A, Singh S (2025) Nanomedicine in patient-specific clinical trend: shape memory polymers for emerging biomedical applications and their future prospects. In: Nanomedicine in translational research. Elsevier, pp 309–338. https://doi.org/10.1016/B978-0-443-22257-3.00014-3
42. Arif ZU, Khalid MY, Ahmed W, Arshad H (2022) A review on four-dimensional (4D) bioprinting in pursuit of advanced tissue engineering applications. Bioprinting 27:e00203. https://doi.org/10.1016/j.bprint.2022.e00203

43. Naghib SM, Hosseini SN, Beigi A (2024) 3D/4D printing of chitosan-based materials for wound healing with chitosan-based materials, which provide a fresh method for creating customized scaffolds and wound dressings applications. Carbohyd Polym Technol Appl:100594. https://doi.org/10.1016/j.carpta.2024.100594
44. Shankhwar N, Kumar M, Mandal BB, Robi PS, Srinivasan A (2016) Electrospun polyvinyl alcohol-polyvinyl pyrrolidone nanofibrous membranes for interactive wound dressing application. J Biomater Sci Polym Ed 27(3):247–262. https://doi.org/10.1080/09205063.2015.1120474
45. Wang Q et al (2024) Innovations in 3D bioprinting and biomaterials for liver tissue engineering: paving the way for tissue-engineered liver. iLIVER 3(1):100080. https://doi.org/10.1016/j.iliver.2024.100080
46. Sun L, Wang Y, Zhang S, Yang H, Mao Y (2023) 3D bioprinted liver tissue and disease models: current advances and future perspectives. Biomater Adv 152:213499. https://doi.org/10.1016/j.bioadv.2023.213499
47. Igami T et al (2014) Application of a three-dimensional print of a liver in hepatectomy for small tumors invisible by intraoperative ultrasonography: preliminary experience. World J Surg 38(12):3163–3166. https://doi.org/10.1007/s00268-014-2740-7
48. Prakash A, Sridhar SB, Wali AF, Talath S, Shareef J, Malviya R (2024) 3D and 4D printed materials for cardiac transplantation: advances in biogenerative engineering. Bioprinting 43:e00362. https://doi.org/10.1016/j.bprint.2024.e00362
49. Liu N et al (2021) Advances in 3D bioprinting technology for cardiac tissue engineering and regeneration. Bioact Mater 6(5):1388–1401. https://doi.org/10.1016/j.bioactmat.2020.10.021
50. Xu Q et al (2024) Hydrogel-based cardiac repair and regeneration function in the treatment of myocardial infarction. Mater Today Bio 25:100978. https://doi.org/10.1016/j.mtbio.2024.100978
51. Ho CMB et al (2017) 3D printed polycaprolactone carbon nanotube composite scaffolds for cardiac tissue engineering. Macromol Biosci 17(4). https://doi.org/10.1002/mabi.201600250
52. Zennifer A, Thangadurai M, Sundaramurthi D, Sethuraman S (2023) Additive manufacturing of peripheral nerve conduits—fabrication methods, design considerations and clinical challenges. SLAS Technol 28(3):102–126. https://doi.org/10.1016/j.slast.2023.03.006
53. Qiu B et al (2020) Bioprinting neural systems to model central nervous system diseases. Adv Funct Mater 30(44). https://doi.org/10.1002/adfm.201910250
54. Parvaneh S, Kemény L, Ghaffarinia A, Yarani R, Veréb Z (2023) Three-dimensional bioprinting of functional β-islet-like constructs. Int J Bioprint 9(2):665. https://doi.org/10.18063/ijb.v9i2.665
55. Ionescu-Tirgoviste C et al (2015) A 3D map of the islet routes throughout the healthy human pancreas. Sci Rep 5(1):14634. https://doi.org/10.1038/srep14634
56. Lee V et al (2014) Design and fabrication of human skin by three-dimensional bioprinting. Tissue Eng Part C Methods 20(6):473–484. https://doi.org/10.1089/ten.tec.2013.0335
57. Zhang H et al (2023) Strategies for improving the 3D printability of decellularized extracellular matrix bioink. Theranostics 13(8):2562–2587. https://doi.org/10.7150/thno.81785
58. Wan Z, Zhang P, Liu Y, Lv L, Zhou Y (2020) Four-dimensional bioprinting: current developments and applications in bone tissue engineering. Acta Biomater 101:26–42. https://doi.org/10.1016/j.actbio.2019.10.038
59. Hatami H, Almahmeed W, Kesharwani P, Sahebkar A (2024) Exploring the potential of 3D and 4D printing in advancing stent manufacturing for cardiovascular diseases. Eur Polym J 212:113035. https://doi.org/10.1016/j.eurpolymj.2024.113035
60. Ge Q, Sakhaei AH, Lee H, Dunn CK, Fang NX, Dunn ML (2016) Multimaterial 4d printing with tailorable shape memory polymers. Sci Rep 6(1):31110. https://doi.org/10.1038/srep31110
61. Noroozi R et al (2023) 3D and 4D bioprinting technologies: a game changer for the biomedical sector? Ann Biomed Eng 51(8):1683–1712. https://doi.org/10.1007/s10439-023-03243-9
62. Solanki R, Bhatia D (2024) Stimulus-responsive hydrogels for targeted cancer therapy. Gels 10(7):440. https://doi.org/10.3390/gels10070440

63. Ng HY, Lee K-XA, Kuo C-N, Shen Y-F (2024) Bioprinting of artificial blood vessels. Int J Bioprint 4(2):140. https://doi.org/10.18063/ijb.v4i2.140
64. Puppi D, Chiellini F (2020) Biodegradable polymers for biomedical additive manufacturing. Appl Mater Today 20:100700. https://doi.org/10.1016/j.apmt.2020.100700
65. Hou Y-C et al (2023) Three-dimensional bioprinting of artificial blood vessel: process, bioinks, and challenges. Int J Bioprint 9(4):740. https://doi.org/10.18063/ijb.740
66. Wang X (2019) Bioartificial organ manufacturing technologies. Cell Transplant 28(1):5–17. https://doi.org/10.1177/0963689718809918
67. Dwivedi R, Yadav PK, Pandey R, Mehrotra D (2022) Auricular reconstruction via 3D bioprinting strategies: an update. J Oral Biol Craniofac Res 12(5):580–588. https://doi.org/10.1016/j.jobcr.2022.07.014
68. Taylor S, Mueller E, Jones LR, Makela AV, Ashammakhi N (2024) Translational aspects of 3D and 4D printing and bioprinting. Adv Healthc Mater 13(27). https://doi.org/10.1002/adhm.202400463
69. Ashammakhi N et al (2018) Advances and future perspectives in 4d bioprinting. Biotechnol J 13(12). https://doi.org/10.1002/biot.201800148
70. Arif ZU, Khalid MY, Zolfagharian A, Bodaghi M (2022) 4D bioprinting of smart polymers for biomedical applications: recent progress, challenges, and future perspectives. React Funct Polym 179:105374. https://doi.org/10.1016/j.reactfunctpolym.2022.105374
71. The Ministry of Health, Labour and Welfare (MHLW) and the P. Japan: the Ministry of Health. Regulations and approval/certification of medical devices [Online]. Available: https://www.pmda.go.jp/english/review-services/reviews/0004.html
72. Rafi K, Zhonghong Liu A, Di Prima M, Bates P, Seifi M (2022) Regulatory and standards development in medical additive manufacturing. MRS Bull 47(1):98–105. https://doi.org/10.1557/s43577-022-00266-w

Chapter 10
Resorbable Polymer-Based Additive Manufacturing Techniques for Bioimplants

Revathi Devulapalli, Subhasree Panda, and S. K. Khadheer Pasha

1 Introduction

Resorbable polymers (RPs) are synthetic materials that break down and are ready to be absorbed into the body over time. RPs are frequently used in preparing medical implants and tissue engineering. They are also called as bioresorbable polymers due to their nature and compatibility with the human body. RPs provide temporary structural support and allow tissues to regenerate naturally. Tailoring the implant to specific patient needs and enhancing its functionality with complex geometries are possible with additive manufacturing (AM) or three-dimensional (3D) printing. The additive manufactured resorbable bioimplants result in faster healing with reduced pain and do not necessitate the process of a second surgery for implant removal. The resorbable bioimplants help in tissue regeneration, gradually breaking down by the body when they are no longer needed and subsequently eliminated by the body's metabolism. RPs break down through a process called hydrolysis. RPs can be combined with ceramics, metals, and other polymers to substantially increase their structural and physicochemical properties [1]. Moreover, several biomaterials can be used in the 3D printing of resorbable patient-specific implants (PSIs) such as magnesium (Mg), zinc (Zn), molybdenum (Mo), cobalt (Co), calcium phosphate (CaP), poly(lactic acid) (PLA), glycolic acid, titanium alloy, stainless steels, silicate, etc. Technological advancements led to tremendous digitalization in PSIs and amended clinical results with polymer composite resorbable bioimplants, where conventional implants do not work in treating complex pathologies [2]. RPs are extensively used to fabricate different biomedical devices like implants, sutures, and scaffolds for bone repair devices, drug carriers, tissue engineering, etc. Nanofiber-based RPs have

R. Devulapalli · S. Panda · S. K. Khadheer Pasha (✉)
Functional Nanomaterials and Polymer Nanocomposite Laboratory, Department of Physics, VIT-AP University, Amaravati, Guntur, Andhra Pradesh, India
e-mail: khadheerpasha.sk@vitap.ac.in

A. Prasad and V. Katiyar (eds.), *Resorbable Polymers for Bioimplants and Fixation Devices*, Materials Horizons: From Nature to Nanomaterials,
https://doi.org/10.1007/978-981-96-9306-1_10

promising surface-to-volume ratios and frequently possess great pore interconnectivity and porous structures [3]. AM plays a vital role in designing patient-specific complex structure bioresorbable polymer (bio-RPs) implants for appropriate targeted sites to mimic the structure of natural bone or tissue. As an integrated process, bio-RPs are developed by combining selected materials, structural design, and manufacturing strategies. Laser cutting, micro-injection molding, braiding technique, and knitting are some of the AM techniques used in the preparation of RPs for bioimplant materials/devices [4]. However, biocompatibility is an important criterion for bioimplant materials to avoid undesirable effects like cytotoxicity or immunogenicity. Chitosan, poly(lactic acid) (PLA), collagen, poly (glycolic acid) (PGA), poly(ethylene glycol) (PEG), etc., are the most frequently used RPs. By mixing RPs with other substances including ceramics, metals, and resorbable or non-resorbable polymers to create composites, their qualities can be improved [1].

2 Resorbable Polymers (RPs) for Bioimplants

Polymers are preferred for their flexibility. They can be formed into a variety of shapes according to their applications, including solids, fibers, films, and fabrics [5]. Materials used as biomedical implants must have properties like biocompatibility and nontoxicity. The biostable materials are meant to remain in the body and continue to work as intended for the duration of the patient's life. The most commonly used resorbable materials are from nature or nature-derived materials in the form of ceramics, metals, synthetic polymers, and composites. Mostly metals and their alloys are utilized as implants for dental and orthopedic operations because of their toughness, ductility, greater strength, and other mechanical features. However, metallic implants may cause complications (e.g. cause irritation due to erosion and corrosion in the biotic environment, aseptic loosening of implants, stress shielding may increase the risk of a second fracture of the operated bone) leading to a secondary operation to remove the implanted device. Ceramics and glasses are widely used in biomedical implants, particularly in bone replacement. However, they are not appropriate for load-bearing purposes because of their brittle nature. In comparison to these metals, glasses, and ceramics, polymers are widely and successfully used in biomedical field as, replacement, support, augmentation, scaffolds, fixation, and bio-mimicking [6]. Biomaterials are classified as biostable, bioabsorbable, or bioactive based on how they behave in the biological environment. The transition from biostable prostheses to bio-resorbable implants during the last three decades has been one of the most significant advances in bio-materials research. Among the polymeric materials derived, synthetically degradable polymers and aliphatic polyesters like, PLA, Polyanhydride, PGA, Polydioxanone, Polyurethane, Polycaprolactone (PCL), etc., are commonly used RPs and natural polymers which are readily available in nature such as various polysaccharides, collagen, microbial polyesters have been extensively used as RPs. A class of polymers is prepared by mixing the natural and synthetic polymers in order to tailor sustainable and suitable properties for bio-RPs

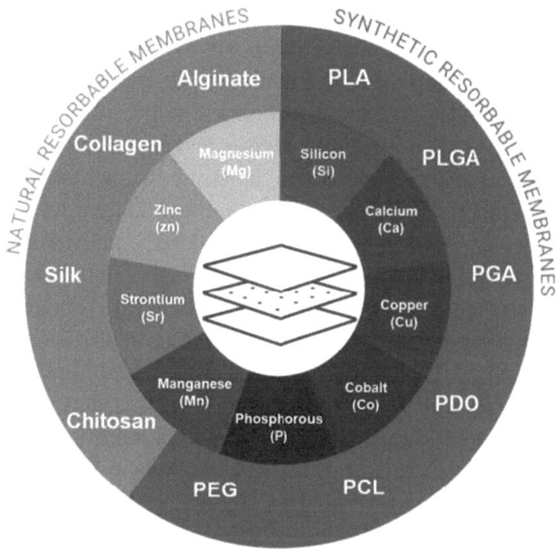

Fig. 1 Classification of RPs and their composites for biomedical implants. Adapted from [7] Copyright © 2023 The Authors. Published by American Chemical Society

for instance, Collagen-PLA, Hyaluronic acid-chitosan, Gelatin-PCL. Various properties can be improved by incorporating various fillers into polymers to enhance their surface to volume ratio and modify their mechanical and structural behavior reasonable to the bio-environment [6]. Figure 1 depicts the classification of RPs and composite-RPs for diverse biomedical applications.

2.1 Synthetic RPs

Historically, synthetic polymers have been acting as the material of choice for implants [8]. The class of biodegradable, biocompatible synthetic polymers used in the field of medicine and biomedical engineering are called synthetic RPs. These polymers are meant to dissolve and absorb into the human body gradually, thus eliminating the need for surgical removal. However, synthetic RPs are versatile and the degradation rate of these polymers can be precisely controlled allowing for tailored release of drugs and support tissue regeneration. Synthetic RPs are often made through chemical structures optimized to maximize the physical characteristics of biomedical materials, as well as purity and well-defined compositions that are superior to those achieved with natural polymers. Nylon and Polyvinyl alcohol (PVA) are synthetic polymers manufactured in 1930s, an era of synthetic medical polymers. There are various proven chemical categories of biodegradable synthetic polymers that are exceptionally biocompatible and can be utilized to change resorption, biodegradation, and mechanical strength rates. PLA, PGA, and their copolymers are widely used bio-RPs that decay in the body through simple hydrolysis of the ester backbone, resulting in non-toxic chemicals. The breakdown products are either excreted through kidneys or removed as CO_2 and water via well-known excretion mechanisms. These are mostly used as surgical sutures, drug encapsulation and resorbable implants [7]. PLA is the furthermost commonly used RP in medical applications today ranging from tissue engineering to drug delivery systems, temporary to long-term implantable devices and expanding its wings to new fields. From 1970, resorbable suture materials came into use, which are synthesized via melt spinning of synthetic RPs like, PGA popularly known as Dexon™ [9] and polyglycolide-L-lactide or (polyglactin 910) popularly known as Vicryl™ [10]. Many thermoplastic polymers have been used as medical implants such as, polydioxanone (PDO), poly(ether–ether–ketone) (PEEK), PCL, polyethylene terephthalate (PET), polyhydroxy butyrate (PHB), etc., PCL is known for its resorbable nature in both in vivo and in vitro by hydrolysis [11]. PEEK-based implants are used in AM via fused filament fabrication (FFF) technique or fused deposition modeling (FDM) to prepare customized implants which cannot be achieved in traditional methods to fabricate complex structured biomedical implants, especially orthopedic implants due to its valuable properties such as rigidity, lightweight nature, tissue-compatibility, and longer-lifespan [12]. Poly (methyl methacrylate) (PMMA) is utilized in ophthalmic implants, silicone polymers in breast implantation, ultra-high molecular weight polyethylene in hip joints, and so on. Silicone implants are known for their flexibility, durability, and stability, leading to enduring success and higher rate of patient satisfaction. Silicones are ideal for ophthalmic implants, particularly lenses, owing to their high optical transparency and flexibility in the ophthalmic environment. These features improve visual outcomes for people undergoing cataract correction surgery or operations to correct refractive error [13]. For brain implants, researchers looked into the usage of poly (3,4-ethylenedioxythiophene) (PEDOT) known as a conjugated polymer. They discovered that PEDOT has good conductivity towards electricity, biocompatibility and can be applied as a coating to enhance the functionality and durability of neuronal implants [14]. Other synthetic RPs, such as polypropylene carbonate (PPC) and polybutylene adipate terephthalate (PBAT), are less common, but their use has increased in recent years [15, 16]. PVA and PLGA shuttles are used as bio-RP neural probes [17]. Table 1 depicts the use of synthetic RPs in biomedical application and their resorption rates.

2.2 Natural RPs

The biological materials or biopolymers are a category of macromolecules, acquired from biotic sources like plants, animals, and microorganisms. These biological polymers are also called natural polymers. Most of the natural polymers available in nature are biodegradable. Collagen, fibrin, elastin, hyaluronic acid, gelatin, alginate, chitosan, cellulose, and polyhydroxyalkanoate, are some of the examples of natural RPs. The characteristics of these polymers make them attractive materials for a number of medical and pharmaceutical applications.

Table 1 Synthetic RPs and their resorption rate

Material name	Functioning	Factors affecting resorption	Resorption rate	References
Poly(lactic acid) (PLA)	Growth development in cells	Molecular weight, crystallinity, processing conditions	Slow to moderate	[18]
Poly(ethylene glycol) (PEG)	Coating on implants	Molecular weight	Low	[19]
Poly(glycolic acid) (PGA)	Anterior cruciate ligament, tissue defects	Molecular weight, crystallinity, processing conditions	Fast	[20]
Poly(vinyl alcohol) (PVA)	Flexible neural probes	Degree of hydrolysis, molecular weight, enzymes	Slow	[17]
Polyurethane (PU)	Grafts, cardiovascular tissue regeneration	Hydrophilicity, ester, carbonate bonds	Varies with composition and structure	[21]
Polycaprolactone (PCL)	Cell adhesion Drug delivery	Molecular weight, crystallinity	Slow	[22]
Poly(hydroxybutyrate) (PHB)	Periodontal tissue	Molecular weight, crystallinity	Moderate	[23]
Poly(lactic-co-glycolic acid) (PLGA)	Drug delivery	Molecular weight, crystallinity, processing conditions	Tunable	[24]

Chitosan: It is the deacetylated version of chitin. Chitin is found in the exoskeleton of crustacea, like shrimp, crab, lobsters, and some fungi. It is a linear polysaccharide, originated from N-acetylglucosamine (GlcNAc) molecules. It is used in a wide variety of biomedical and non-biomedical arenas due to its excellent hydrophobicity and the presence of free amine groups to make them soluble in acids. Wound healing, waste water treatment, and drug delivery are promising applications along with sensing and energy storage applications [25, 26]. Usually, natural RPs act as drug release carrier due to their porous network. Chitosan nanoparticles are used to develop transparent scaffolds for functionalizing corneal endothelial cells using solvent casting method [27].

Cellulose: An organic compound, a polysaccharide abundantly available in nature. The main constituent of plant cell walls, extensively available in wood, cotton, linseed, bacteria, algae, sea weed, xylan, etc., is a complex carbohydrate molecule containing a long chain of glucose molecules linked together with beta-1,4 glycosidic bonds. These long chains of glucose units' hydrogen bond to each other, forming strong intermolecular forces that give cellulose its strength and rigidity. Cellulose has promising applications in textile productions, food additives, wound dressings, medical implants, biofuel productions, dye absorbents, sensing, and composite materials [28, 29].

Gelatin: Gelatin-based scaffolds are extensively used in drug delivery and tissue engineering. Yao Ding et al. synthesized hyaluronic acid and gelatin nanocomposite hydrogel coatings on titanium-based implants to alleviate biofilm infection and exaggerated allergic reactions [30]. Gelatin-based hydrogels made from collagen hydrolysis are widely employed in pharmaceutical and medical applications because they are biocompatible and biodegradable. Because of their capacity to promote cell adhesion and proliferation, gelatin-based hydrogels are employed in medicine administration and tissue engineering. Due to their excellent fluid absorption capabilities, these hydrogels can also be used as wound dressings. UV stereo-lithography as well as two-photon polymerization can be used to construct structures with highly photosensitive gelatin-based hydrogels [31].

Collagen: Numerous functions of collagen are involved in the formation of blood vessel walls, cartilage, and bone. Additionally, this protein may slow down the aging process of human tissue. It is well recognized that collagen is biocompatible, non-cytotoxic, and biodegradable. During the phases of hemostasis, inflammation, proliferation, and remodeling, it serves as a foundation or mechanical support scaffold for the development of contemporary as well as progressive tissue. Additionally, collagen has an extensive range of applications in various tissue engineering, artificial blood vessels, skin and bone replacements, etc. Collagen finds extensive application in medical implants. In tissue engineering, collagen-based scaffolds are widely acknowledged as an excellent substrate for promoting cell adhesion, growth, and tissue repair. These scaffolds are essential for restoring and growing tissues that have been damaged or impacted by disease [32]. Figure 2 shows the examples of natural and synthetic RPs.

2.3 *Hybrid RPs*

Hybrid resorbable polymers (HRPs) are materials which are produced as the result of the combination of two or more types of molecules, at least one of them is polymer. They can be blends or composites and possess improved physical and structural properties as well as biocompatibility when compared to natural or synthetic RPs alone. HRPs are also used in preparing scaffolds for tissue engineering applications, such as nerve tissue constructs for enhanced mechanical stability compared to natural RPs and better biodegradability compared to synthetic RPs. A notable example of a hybrid polymer is the fibrous nerve conduit made of gelatin and PCL. Gelatin stabilizes PCL fibers for efficient cell attachment, whereas fibrous PCL offers a better framework for cell migration. Neuronal development in the hybrid conduit showed enhanced neuronal differentiation and motility. Polyhydroxyalkanoate (PHA) hybrid

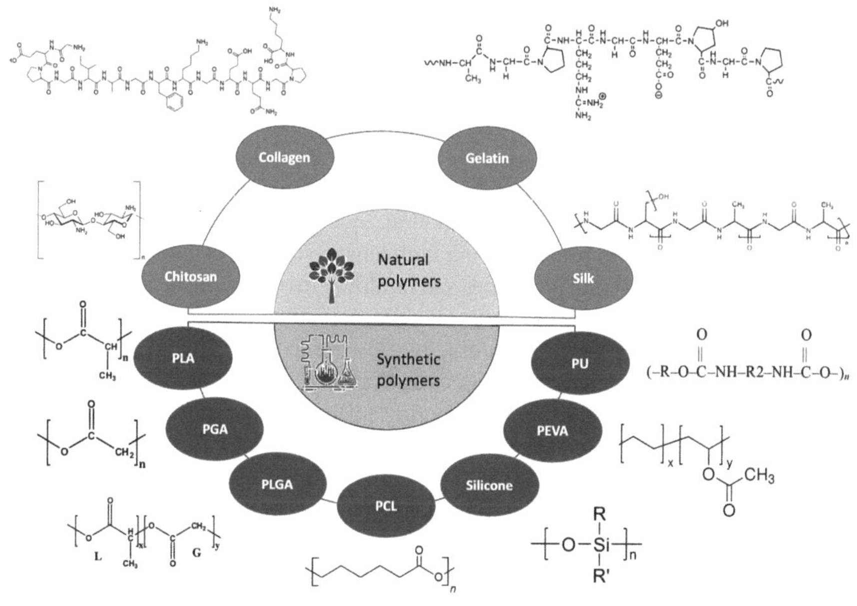

Fig. 2 Examples of natural and synthetic RPs. Reproduced with permission from [33], © 2024 The Korean Society of Industrial and Engineering Chemistry. Published by Elsevier B.V. All rights reserved

biomaterials with better physical, chemical, mechanical, and piezoelectric properties along with controlled decomposition have attracted a lot of attention in neural tissue engineering [34]. The bio-inspired polymers are a category of polymers consisting of man-made or non-naturally occurring proteins, made through biosynthesis. They can be manufactured under controlled environments and have control over properties to overcome the consequences in naturally originated polymers like polysaccharides and collagen which suffer from source-to-source variable properties and are easily prone to bacterial and viral contaminations. Due to the possibility of producing materials that mimic the functionality and structure of body's own structural polymers [8]. Other specialized hybrid polymers include zwitterionic polymers. Polymers like these have chains of molecules with homogeneous distribution of anionic and cationic functional groups. The above features result in significant hydrophilic and anti fouling qualities, allowing them to effectively reject inappropriate protein adsorption, hinder bacterial adhesion, and restrict biofilm formation. Consequently, zwitterionic polymers have considerable potential for application in a wide range of biomedical and ecological fields, such as in vivo drug delivery systems, blood-contacting sensors, and coatings that stop fouling for biomedical devices. [35]. The extraordinary properties of PCL and its incorporation with proteins, polysaccharides, or synthetic polymers have been exploited in various ways, including PCL/chitosan [27], PCL/PVA/silk fibroin [36], gelatin/hydroxyapatite composite [37], PCL/chitosan/polypyrrole [38], PCL/chitosan/gelatin, [39] etc.

2.4 Resorbable Polymer Composites (RPCs)

The addition of materials into host polymer to improve its properties and achieve desired characteristics in materials can be defined as polymer composites. They may be polymer–metal composites, polymer–ceramic composites, or polymer–polymer composites. The addition of nano-fillers to bio-RPs enhances the properties and is promising for AM. Jagan Mohan Dodda et al. synthesized PLA/PLGA- and PCL/PGLA-based RPCs/blend films showing high tensile strength extending from 9.5 to 11.8 MPa. PCL's exceptional properties and its ability to blend with proteins, polysaccharides, or synthetic RPs have been used in a variety of combinations, such as PCL/CS, PCL/Starch, PCL/PVA/Silk fibroin, etc. Owing to their film forming properties, many recent reports suggested them as the main key ingredient to synthesize biotic films, particularly in tissue engineering [40]. Metal-on-plastic compositions are the most frequently used configuration of artificial joints. Elin Akerlund et al. used PLA/PLC and hydroxyapatite biodegradable composites to advance biocompatible and bio-RP-based amalgamated filaments via extrusion for FFF printing. Here, PLA and PLC act as supportive polymer matrices while hydroxyapatite enhances the biological activity. They also possess greater mechanical capabilities in comparison to human trabecular bone, even after deterioration with a mass loss of around 30% in blended polymers and 60% in nanocomposite [41]. Biodegradable nanocomposites have recently piqued researchers' interest, and numerous types of biodegradable polymer matrices have been produced with enhanced qualities such as controlled release, sustained release, delayed release, and so on [42].

3 Additive Manufacturing Based Bioimplants

Layer-by-layer construction technique to prepare 3D shapes and products using computer-aided designs (CADs) is scientifically called AM or 3D printing. Metals, ceramics, polymers, and composites can be additively manufactured according to their purpose. Prototypes for the development of new products were built using solid freeform manufacturing processes. Conventional manufacturing approaches include substantial mold-making and casting stages. Some of the AM techniques used to produce RP-based implants include binder jetting, material jetting, directed energy deposition, material extrusion, powder bed fusion, sheet lamination, vat photopolymerization, 3D bioprinting, and electrospinning (Fig. 3) [43]. AM uses a variety of polymers, including elastomers, thermoplastics, thermosets, functional polymers, polymer blends, composites, hydrogels, and biological systems. Polymer design, additives along with processing parameters are reviewed in terms of boosting build speed whilst improving accuracy, performance, appearance, stability, mechanical properties, and porosity [43, 44]. However, AM is changing the landscape of the

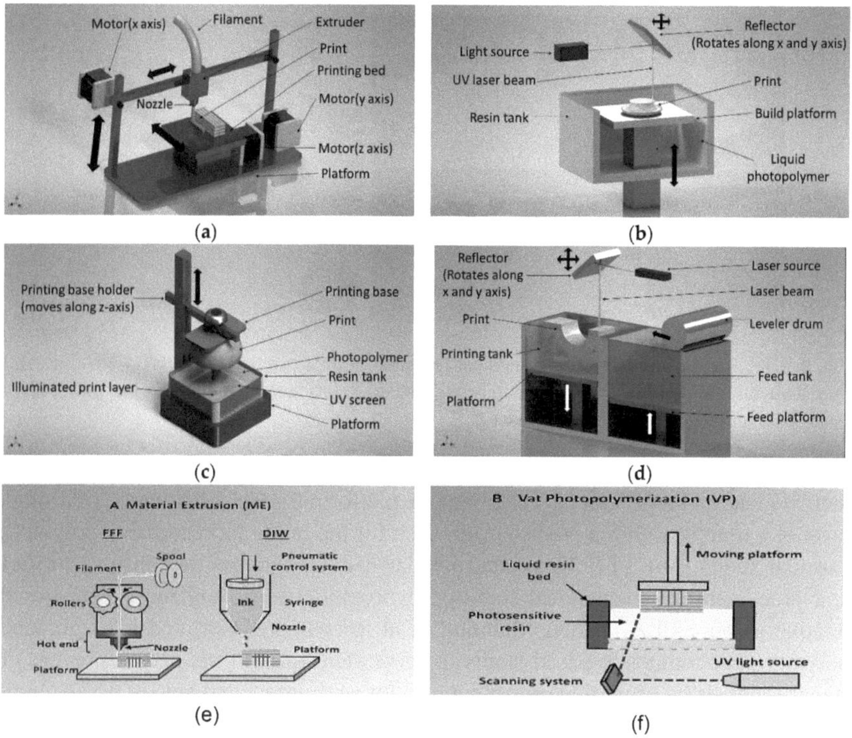

Fig. 3 AM techniques **a** The Fused filament fabrication (FFF) printers. **b** The stereolithography (SLA). **c** The digital light processing (DLP). (d) The selective laser sintering (SLS). Adapted from [45] © 2020 by the authors. Licensee MDPI, Basel, Switzerland. This article is an open access article distributed under the terms and conditions of the Creative Commons Attribution (CC BY) license. **e** Representing material extrusion (ME), a common process in the FFF-method, which involves melting the solid filament, extruding it through the nozzle, and then depositing it layer by layer, and Direct ink writing (DIW) utilizes viscoelastic ink, extruded through a tiny nozzle at regulated flow rates. **f** Vat photopolymerization (VP). Reproduced with permission from [44], Copyright: © 2021 Elsevier Inc

healthcare sector by enabling patient-specific devices with highly complex structures. Some of the AM techniques used for the fabrication of biomedical implants from RPs are discussed in the subsequent section.

3.1 *Fused Filament Fabrication (FFF)*

Fused filament fabrication (FFF) generally called Fused deposition modeling (FDM) is an AM technique that means building a 3D object by depositing melted plastic filament layer by layer, using a reel of plastic filament mechanically pressed with rollers and then sending through an extremely hot liquefier with a nozzle at one end. The

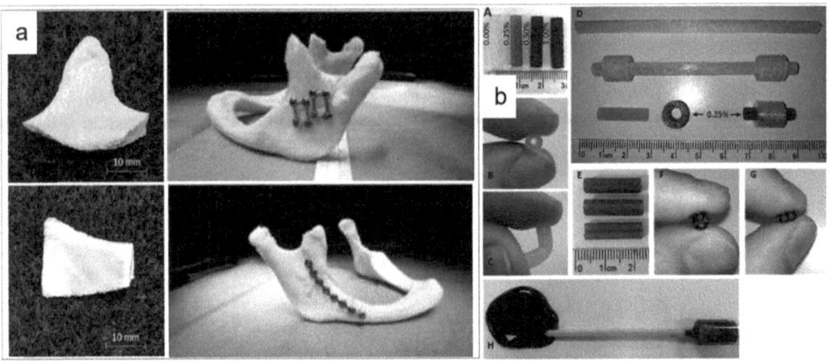

Fig. 4 Elastomeric FFF 3D printed medical implants: **a** dual printed TPU-PETG bio-functional maxillofacial implant, **b** anti-infective dialysis catheter. Reproduced with permission from [49], © 2023 The Authors. Published by Elsevier B.V

melted material comes out of it and onto the platform for the construction of a single layer at a time, creating a precise print, used for modeling bioresorbable polymeric vascular stents [46]. FFF 3D printed tubular nanocomposites demonstrated excellent processability, mechanical stability, hydrophilicity and enhanced cell growth relative to pure PLA. Zaheri Mahmud et al. prepared 3D printed biodegradable polymer nanocomposite-based stents to serve stent thrombosis and to prevent in-stent restenosis by providing proper support for vasculature and help in tissue regeneration using carbon quantum dots (CQDs) reinforced PLA [47]. Thermoplastic polyester-based materials and polyethers are excellent candidates for FFF 3D printing considering their tunable properties. In contrast to natural polymers, they maintain homogeneous viscose flow and integrity of the structure. For accurate and repeatable printing, the right cooling rate in relation to melting temperature, maximum purity, molecular weight, and degree of crystallinity are essential components [48]. Figure 4 depicts the biomedical implants fabricated by the FFF 3D printing method [49].

3.2 Electrospinning

It is a voltage-driven fabrication process guided by a special electrohydrodynamic phenomena. Small fibers are generated from a polymer solution. A high voltage is administered between a nozzle and a collector plate, then the electrical field causes the polymer solution to deform into a cone shape, known as a Taylor cone and the jet stretches due to electrostatic force and solvent evaporation. The jet solidifies and the result is collected on the grounded collector as solid fibers. This technique is used to prepare metallic, nano, hollow fibers and nested structures [50]. Wang et al. fabricated PCL/PLGA blend polymers ureteral stents using double-needle electrospinning [51]. Kucinska Lipika et al. fabricated PU-based biocompatible complex fibers by electrospinning procedure for cardiovascular tissue engineering [21].

3.3 *Direct Ink Writing (DIW)*

This is also an extrusion-based AM method that uses either piston, air pressure, or screw extruder to drive ink through the nozzle. After depositing the ink into build substrate, it undergoes photo-polymerization or thermal curing, compared to FFF printing, DIW can result in soft materials with hardness lesser than 85A [52, 53]. Rubina et al. fabricated nanocellulose and chitosan-based biocompatible hydrogel implants for mesh matrices via DIW. Mesh exhibited excellent mechanical properties demanding soft tissues [54]. Chao et al. fabricated tunable porous Fe bone implants using DIW and vacuum sublimation of Manganese. The ensuing growth of new bone tissue and adequate distribution of loads between the implant and normal bones paves the way for the medical application of Fe bone implants [55].

3.4 *VAT-Photopolymerization (VPP)*

A type of 3D printing method that polymerize light-sensitive polymer resin by using light with a certain wavelength and enough energy. The photo initiator present in resin reacts to light, leading to crosslinking and solidification on exposure region. Degradable RPs attained through polycondensation such as poly glycerol sebacate (PGS) have great potential in VPP [56]. It works with a variety of photopolymer resins, offering different material behavior like flexibility, strength, and heat resistance. In photo polymerization process, a photo sensitive polymer and a light-curable resin are combined in a vat and subjected to visible or ultraviolet (UV) light. The treatment with light initiates and manifests the polymerization reaction, which culminates in the production of polymer chains or crosslinks, producing a solid resin [57]. Scientists produced an UV-curable PMMA appropriate for VPP printing, and the microneedles were constructed with a hollow side structure to improve drug loading efficiency. Durability, medication loading efficiency, and skin penetration capability were all tested thoroughly [58]. Figure 5 shows the implants and scaffolds fabricated by VPP using various RPs and studied [59].

3.5 *Digital Light Processing (DLP)*

DLP AM is a promising technology for fabricating RP implants. Owing to its high resolution and accuracy, DLP permits precise control over the printing process, consequential in highly detailed and accurate implants. This is crucial for implants that need to fit specific anatomical structures. Popular AM methods are used to fabricate complex 3D shapes, enabling the design of implants that mimic natural tissues and promote healing. Rapid prototyping of implants, orthopedic, dental, drug delivery implants and 3D scaffolds can be fabricated via DLP. Scientists fabricated bio-RP 3D

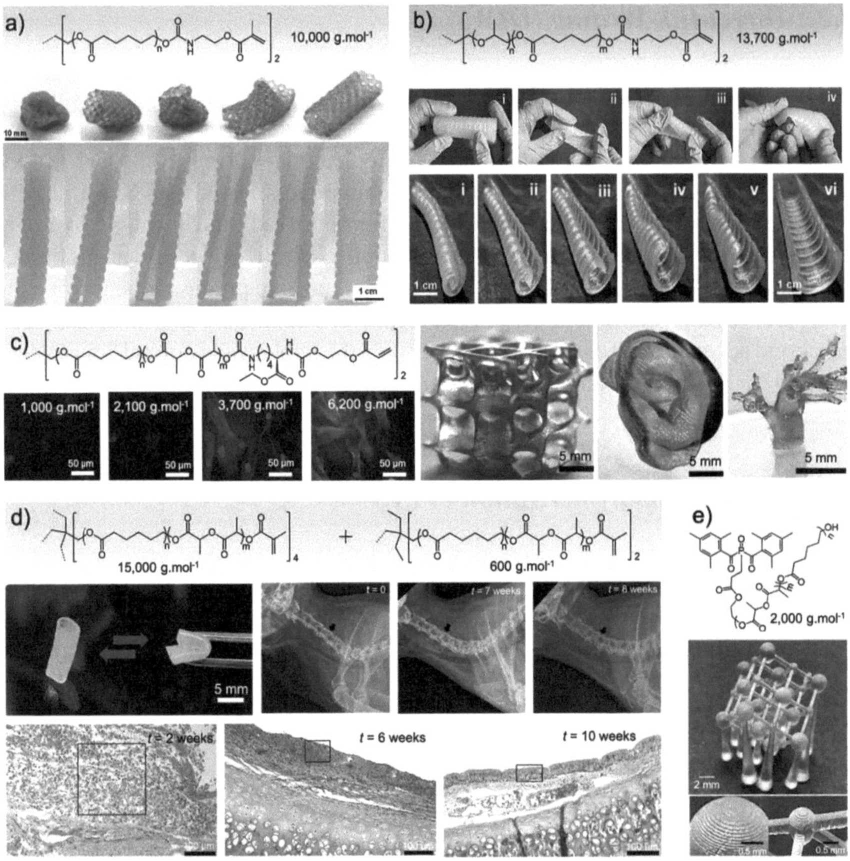

Fig. 5 Additive manufactured biomedical implants and scaffolds via heat-assisted VPP. **a**, **b** shape-memory stent prototypes, **c** scaffolds in gyroid, ear, and vascular model **d** customized airway stents and the healing mechanism after insertion in rabbit's trachea, and **e** biodegradable scaffolds in the form of NaCl crystal 3D lattice. Adapted from [59], Copyright © 1999–2024 John Wiley & Sons, Inc or related companies. All rights reserved, including rights for text and data mining and training of artificial intelligence technologies or similar technologies

printed vascular stents with nitric oxide (NO) donor to overcome metallic stents. The development of photo-cross-linkable polyester-based materials, such as methacrylated poly(dodecanediol citrate) (mPDC), for applications involving light-assisted AM methods, created opportunities for the production of bioresorbable vascular stents (BVS). Understanding the pro-endothelial and anti-thrombotic qualities of NO has resulted in the evolution of NO-releasing BVS (Fig. 6) [60].

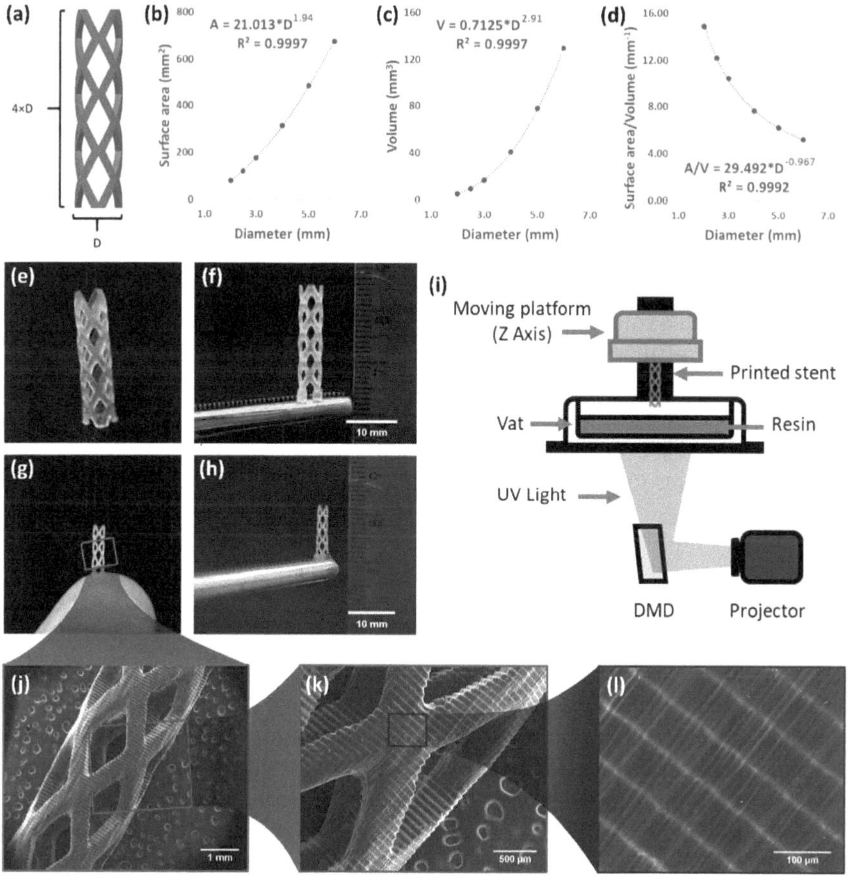

Fig. 6 **a** CAD model of additive manufactured mPDC stent, with accompanying **b** surface area vs diameter, **c** occupied volume vs diameter, and **d** surface/volume ratio vs diameter curves, as well as fitted curve equations and correlation coefficients (R^2), **e**, **f** photos of additive manufactured stents with diameters of 6 mm and **g**, **h** 3 mm, **i** illustration of the DLP 3D printer, **j**–**l** SEM of additive manufactured mPDC stent at various amplifications, from left–right, ranging millimeter to micrometer length scales. Reproduced with permission from [60], © 2021 Elsevier B.V. All rights reserved

3.6 Stereolithography

This is a solid free form procedure (SFF) proposed in 1980s. Although there are several AM techniques available today, this is the most prevailing and accomplished among all the SFF methods. Stereolithography is highly adaptable in terms of structure design and building scales: sub-micron-sized structures and decimeter-sized objects are constructed. These advancements in the biomedical sector have resulted in the creation of scaffolds, customized parts, mold-assisted implant production prototypes for individual patients, and assistance with challenging surgeries [61, 62].

3.7 Selective Laser Sintering (SLS)

One of the well-known AM techniques that involves sintering ultra-fine layers of powdered materials dispersed across the substrate of a printing bed via laser as a power source. SLS allows you to fabricate 3D forms layer on layer using a CAD model. In SLS printing, a very thin coating of powder is applied on a platform inside the manufacturing chamber. Laser beams are used to heat and selectively melt the powder, strengthening specific regions of the model. By screening and heating the fine powder beneath or at its point of melting, its fragments mechanically bond to generate a solid component [63]. The mechanical characteristics of porous implants made of titanium with the calcium phosphate biomimetic coating were evaluated using computed tomography (CT), a non-destructive testing method. AM technologies, such as SLS, direct laser metal sintering (DLMS), and electron beam melting (EBM), have proven potential in manufacturing implants for teeth and porous titanium constructs [64].

4 Biomedical Applications

RPs are utilized in practically every sector of medicine, from routine outpatient operations such as birthmark excision, intradermal sutures, ophthalmological procedures, microsurgeries, and complicated techniques. Surgical sutures are now the most commonly used typical absorbable therapeutic devices. The foremost deployed device was the ordinary catgut, but these days, surgical closures are made up of synthetic polymers, like PLGA, ε-caprolactone, glycolide, poly(p-dioxanone), or poly-4-hydroxybutyrate, which are captivated into body metabolisim through hydrolysis. The thickness of sutures varies from (0.01 to > 1.10 mm) and varies in types such as, monofilament, multi-filament depending on biopolymertype, resorption time ranges to maximum 180 days. Work relief can be offered by a wide range of suture constituents and suturing approaches, based on observations and medical treatments performed over many decades [65]. Synthetic RPs like PU, PEG, PEGMA, PCl, etc. are used in AM cardiovascular, which helps in tissue generation without effecting the blood stream and show enzymatic and hydrolytic degradation mechanisms. The mass reduction was 13% in 56 days for the hydrolytic breakdown and 95% in 30 days for enzymatic degradation. Electrospun Pus' tensile strength rose from 2.82 ± 0.16 MPa to 7.07 ± 0.44 MPa. Suture holding strength ranged from 2.48 ± 0.33 to 8.38 ± 0.35 N, and burst pressure strength was 72 ± 2 kPa to 172 ± 2 kPa. In general, increasing the solution concentration improved the physicomechanical properties of the electrospun PU tubular scaffold [50, 66]. Figure 7 and Table 2 show the applications of RP based biomedical implants fabricated by different AM methods.

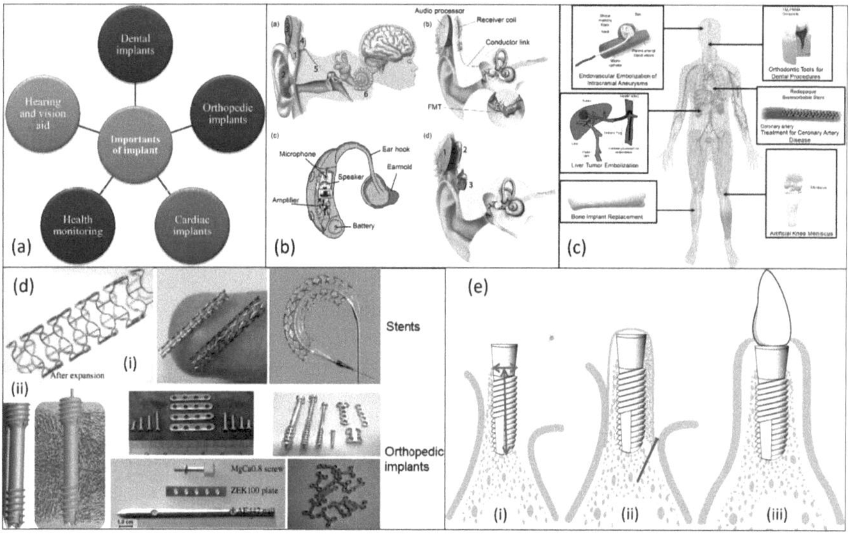

Fig. 7 **a** Applications of Implants. Reproduced with permission from [67], © 2023 The Author(s). Published by Elsevier B.V. **b** Schematic of cochlear implant, middle ear implant, hearing aid, and bone-anchored hearing aid. Adapted from [68], © 2021 by the authors. Licensee MDPI, Basel, Switzerland. This article is an open access article distributed under the terms and conditions of the Creative Commons Attribution (CC BY) license. **c** Examples of polymer blends and composites used in biomedical applications. Reproduced with permission from [69], © 2021 Elsevier Ltd. All rights reserved. **d** (i) Stents and (ii) orthopedic implants made of Mg alloys. Reproduced with permission from [70], © 2021 Chongqing University. Publishing services provided by Elsevier B.V. on behalf of KeAi Communications Co. Ltd.) (i) A sufficient volume of bone, both in height and width, is necessary for an implant procedure to be successful. (ii) Bone grafts and barrier membranes are used as bone replacements to accelerate bone formation. (iii) Final prosthesis is fabricated following the new bone formation. Reproduced with permission from [71], Copyright © 2013 Japan Prosthodontic Society. Published by Elsevier Ireland Ltd

5 Degradation Mechanisms

The time that a biomaterial requires to perform the desired function is called lifetime, which lies in between a few hours, days, several weeks to months, and years in some cases. Non-resorbable polymer bio-implants stay in the body throughout life with no need for surgical removal and non-degradable in nature to hold the tissues and give support. In contrast, the RP based polymers degrade in the body with a specific degradation rate. There are three types of degradation mechanisms: hydrolysis, heat degradation, and enzymatic degradation. The factors like, porosity, degree of crystallization, molecular weight, body temperature, exposure to enzymes and physiological pH effects the stability and durability of RP-based bioimplants [33]. Degradation in bio-RPs has inherent qualities that effect their biocompatibility, comprising material composition, molecular mass, dissolution, hydrophilicity/hydrophobicity, implant geometry, structure, absorption of water, breakdown and erosion mechanisms [78].

Table 2 Fabrication methods of biomedical devices from RPs and their biomedical applications

Application	Functionality	Lifetime/ degradation time	Manufacturing technique	Polymers	References
Bone regeneration	Anchoring system	NA	Hot pressing	UHMWPE	[72, 73]
		4 weeks	Stereolithography	Acrylate	
Scaffolds	Therapeutic implant	NA	3D printing Electrospinning	PCL/Chitosan/ 5-FU/Au NP PU	[21, 74]
Drug delivery	Drug-elution		Electrospinning		[75] [22]
		2 weeks		PU	
		1 week		PCL/Silicon	
Screws/ plates	Orthopedic use Dentistry	3–5 years	3D printing	PLA, PGA	[18, 71]
Mesh/stent	Ureteral stents Ureteral stents	2–5 weeks	FFF Double-needle electrospinning	PU	[51, 76]
		10 weeks		PCL/PLGA	
		10 weeks		Chitosan/ Epoxy	
		NA		Fe_3O_4/PLAUs	
Occlusion device	Cardiovascular tissue Prosthetic eye	NA	Electrospinning Selective laser sintering (SLS)	PU/Silicon	[21, 69, 77]
		1–2 months		PU	
		10 weeks		PLGA-PEGDA	
		NA		NA	

PLA is a biopolymer with great biocompatibility, owing to its self-degradation mechanism. Once the polymer comes in contact with the biological medium, it starts breaking down, generally by hydrolysis, creating lactic acid or carbon dioxide (CO_2) and water. These compounds are either processed within cells or eliminate waste through urine and breath. Infectious bacteria and foreign-body inflammation promote the breakdown of PLA by secreting enzymes that destroy the polymer matrix [79]. Wang et al. fabricated novel ureteral stents using PCL/PLGA with gradient composition consisting 3 segments namely, proximal, intermediate and distal which denotes the varying compositions of PCL to provide gradual degradation function from distal to proximal segment. These stents experimentally degraded completely in 10 weeks [51]. Some PGA fragments can be expelled directly by the kidneys themselves, and degradation by enzymes has been connected to the disintegration of polyester. Breakdown into acidic by-products can produce acidity of the area surrounding an implant, which can contribute to irritation. However, this has been observed to occur seldom and mainly in poorly vascularized tissues, such as cartilage, where there is little fluid flow to dilute the breakdown products [80]. Photodegradable biomaterials are commonly employed in tissue engineering, providing fine control over material characteristics and capturing extracellular matrix variability in synthetic scaffolds [81].

6 Advantages and Disadvantages

The most common RP implants used for anterior cruciate ligament reconstruction, shoulder surgery, fracture treatment and meniscal repair are interference screws, pins, plates, meniscal repair implants, suture anchors, and basic fracture fixation implants. However, many clinicians continue to rely on metallic fixation devices, considering its high mechanical strength and the issues associated with some of the present resorbable implant materials [82]. FFF is a versatile, cost-effective AM technique suitable for wide range of bio-RPs but it exhibits limited resolution and lower potential for layer-by-layer variations. Stereolithography has potential to design complex geometries. Whereas, DLP is able to create both high resolution and complex geometries. Degradable polymers attained from polycondensation (such as poly(glycerol sebacate)) also exhibit great potential in VPP [59]. The physiochemical features of primary components, their degradation behavior, and blending capabilities, which have a direct effect on the mechanical characteristics of the resulting films, are a key aspect in designing these biofilms [40]. While PLGA provides benefits for implantation, it's important to recognize its limitations. The variable deterioration rate makes it difficult to develop an accurate and predictable deterioration profile. PLGA implants can perform poor mechanical durability, particularly when using compositions with high breakdown rates. This is especially important for load-bearing or structurally reinforced implants. PLGA breakdowns can create an acidic milieu, affecting tissue responsiveness and medication stability [83, 84]. SLS may produce parts with lower surface finish and requires post-processing to remove excess powder [85]. AM techniques necessitate real-time monitoring to ensure production efficiency. Standardization of 3D printing is preeminent pressing issues facing the industry right now. The next step is to establish corresponding 3D printing equipment certification criteria, along with internal examination and closed-loop management of the entire manufacturing process, in order to achieve the perfect prediction model and effect. Preparing scaffolds with complicated characteristics is tough. Defective bone tissue contains both cortical and cancellous bone, causing stiffness gradients. As a result, the mechanical characteristics of integrated bone tissue scaffolds varies significantly, and their architectures gets more complicated, necessitating improved AM procedures [86]. The downsides of PVA-based scaffolds include decreased biodegradability after cross linking and reduced cell adhesion due to the polymer scaffold's hydrophilic moieties (i.e., hydroxyl groups). Blending PVA with other synthetic or natural materials may provide a remedy to these drawbacks [87]. One disadvantage of using additively produced medication delivery systems is the need to prioritize patient safety. Personalizing drug dosage can be advantageous, but it may also pose a risk to patients. Sterilization methods for additively produced medication delivery devices remain unknown [88].

7 Summary

Increased population and health issues require fast and sustainable technologies to overcome the health problems. Owing to the number of AM techniques available today to 3D print patient-specific complex geometric structures without hassle compared to subtraction manufacturing. With all efforts of the scientific community, several number of resorbable, non-toxic and patient-specific implants are possible to fabricate using AM (FFF, electrospinning, DIW, VPP, DLP, stereolithography, SLS, etc.). From neural implants to orthopedic implants, from scaffolds to stents RPs implants are replacing the metallic and ceramic implants. Polymers are known to their flexibility and mechanical strength, tailored metal on polymer and ceramic on polymer-based RPCs are grabbing researchers' attention to explore much in the field of biomedical resorbable implants. With the help of CAD and rapid proto typing, AM has become best option to achieve high-resolution, stable, flexible, and flawless implants. The rate of resorption and the degradation mechanisms of implants depends on the functionality and reaction with local pH, temperature. poly(glycerol) sebacate are highly adaptable and can be developed for specific biomedical and pharmacological applications, including drug delivery, bone regeneration, tissue engineering, cardiovascular diseases, and cartilage replacement. These resources have gained popularity in orthopedics, where temporary implants can outperform permanent ones. Bioabsorbable implants have improved with time, however they are often made from synthetic materials, which can cause chronic inflammation and toxicity. Natural materials can successfully address these issues. Current tissue engineering techniques utilize 3D hybrid scaffolds (half synthetic or partially natural) that well interact with human cells. This technique has possible benefits such as lower donor site morbidity, shorter operational time, easier healing, and mimicking the in vivo micro environment to support typical bone development. AM techniques' customization in hospitals benefits both patients and medical personnel by creating safer and more fitting medical devices.

References

1. Mulchandani N, Prasad A, Katiyar V (2019) Resorbable polymers in bone repair and regeneration, in materials for biomedical engineering. Elsevier, pp 87–125
2. Maintz M, Tourbier C, de Wild M, Cattin PC, Beyer M, Seiler D, Honigmann P, Sharma N, Thieringer FM (2024) Patient-specific implants made of 3D printed bioresorbable polymers at the point-of-care: material, technology, and scope of surgical application. 3D Print Med 10:13
3. Wu T, Ding M, Shi C, Qiao Y, Wang P, Qiao R, Wang X, Zhong J (2020) Resorbable polymer electrospun nanofibers: history, shapes and application for tissue engineering. Chin Chem Lett 31:617
4. Yasmin F, Vafadar A, Tolouei-Rad M (2024) Application of additive manufacturing in the development of polymeric bioresorbable cardiovascular stents: a review. Adv Mater Technol
5. Kang C-W, Fang F-Z (2018) State of the art of bioimplants manufacturing: part I. Adv Manuf 6:20

6. Pertici G (2017) Introduction to bioresorbable polymers for biomedical applications. In: Bioresorbable polymers for biomedical applications. Elsevier, pp 3–29
7. Abtahi S, Chen X, Shahabi S, Nasiri N (2023) Resorbable membranes for guided bone regeneration: critical features, potentials, and limitations. ACS Materials Au 3:394
8. Venkatraman S, Boey F, Lao LL (2008) Implanted cardiovascular polymers: natural, synthetic and bio-inspired. Prog Polym Sci 33:853
9. Frazza EJ, Schmitt EE (1971) A new absorbable suture. J Biomed Mater Res 5:43
10. Conn J, Oyasu R, Welsh M, Beal JM (1974) Vicryl (polyglactin 910) synthetic absorbable sutures. Am J Surgery 128:19
11. Youssef A, Hollister SJ, Dalton PD (2017) Additive manufacturing of polymer melts for implantable medical devices and scaffolds. Biofabrication 9:012002
12. Sankar S, Paulraj J, Chakraborti P (2023) Fused filament fabricated PEEK based polymer composites for orthopaedic implants: a review. Int J Mater Res 114:980
13. Subhash NE, Nair S, Srinivas SP, Theruveethi N, Bhandary SV, Guru B (2024) Development of a biodegradable polymer-based implant to release dual drugs for post-operative management of cataract surgery. Drug Deliv Transl Res
14. Kim D, Abidian M, Martin DC (2004) Conducting polymers grown in hydrogel scaffolds coated on neural prosthetic devices. J Biomed Mater Res A 71A:577
15. Bora D, Dutta H, Saha B, Kumar Reddy YA, Patel R, Verma SK, Sellamuthu PS, Sadiku R, Jayaramudu J (2023) A review on the modification of polypropylene carbonate (PPC) using different types of blends/composites and its advanced uses. Mater Today Commun 37:107304
16. Li X, Meng L, Zhang Y, Qin Z, Meng L, Li C, Liu M (2022) Research and application of polypropylene carbonate composite materials: a review. Polymers (Basel) 14:2159
17. Pas J et al (2018) A bilayered PVA/PLGA-bioresorbable shuttle to improve the implantation of flexible neural probes. J Neural Eng 15:065001
18. DeStefano V, Khan S, Tabada A (2020) Applications of PLA in modern medicine. Eng Regen 1:76
19. Asadollahi M, Gerashi E, Zohrevand M, Zarei M, Sayedain SS, Alizadeh R, Labbaf S, Atari M (2022) Improving mechanical properties and biocompatibility of 3D printed PLA by the addition of PEG and titanium particles, using a novel incorporation method. Bioprinting 27:e00228
20. Sharma U et al (2018) The development of bioresorbable composite polymeric implants with high mechanical strength. Nat Mater 17:96
21. Kucinska-Lipka J, Gubanska I, Janik H, Sienkiewicz M (2015) Fabrication of polyurethane and polyurethane based composite fibres by the electrospinning technique for soft tissue engineering of cardiovascular system. Mater Sci Eng, C 46:166
22. Irani YD, Tian Y, Wang M, Klebe S, McInnes SJ, Voelcker NH, Coffer JL, Williams KA (2015) A novel pressed porous silicon-polycaprolactone composite as a dual-purpose implant for the delivery of cells and drugs to the eye. Exp Eye Res 139.123
23. Phuegyod S, Pramual S, Wattanavichean N, Assawajaruwan S, Amornsakchai T, Sukho P, Svasti J, Surarit R, Niamsiri N (2023) Microbial poly(hydroxybutyrate-co-hydroxyvalerate) scaffold for periodontal tissue engineering. Polymers (Basel) 15:855
24. Yang F, Stahnke R, Lawal K, Mahnen C, Duffy P, Xu S, Durig T (2024) Development of poly (lactic-co-glycolic acid) (PLGA) based implants using hot melt extrusion (HME) for sustained release of drugs: the impacts of PLGA's material characteristics. Int J Pharm 663:124556
25. Bhattarai N, Gunn J, Zhang M (2010) Chitosan-based hydrogels for controlled, localized drug delivery. Adv Drug Deliv Rev 62:83
26. Jang J, Kang K, Raeis-Hosseini N, Ismukhanova A, Jeong H, Jung C, Kim B, Lee J-Y, Park I, Rho J (2020) Self-powered humidity sensor using chitosan-based plasmonic metal–hydrogel–metal filters. Adv Opt Mater 8:1901932
27. Tayebi T, Baradaran-Rafii A, Hajifathali A, Rahimpour A, Zali H, Shaabani A, Niknejad H (2021) Biofabrication of chitosan/chitosan nanoparticles/polycaprolactone transparent membrane for corneal endothelial tissue engineering. Sci Rep 11:7060

28. Omidian H, Akhzarmehr A, Chowdhury SD (2024) Advancements in cellulose-based superabsorbent hydrogels: sustainable solutions across industries. Gels 10:174
29. Mohan D, Teong ZK, Bakir AN, Sajab MS, Kaco H (2020) Extending cellulose-based polymers application in additive manufacturing technology: a review of recent approaches. Polymers (Basel) 12:1876
30. Ding Y, Ma R, Liu G, Li X, Xu K, Liu P, Cai K (2023) Fabrication of a new hyaluronic acid/gelatin nanocomposite hydrogel coating on titanium-based implants for treating biofilm infection and excessive inflammatory response. ACS Appl Mater Interfaces 15:13783
31. Jaipan P, Nguyen A, Narayan RJ (2017) Gelatin-based hydrogels for biomedical applications. MRS Commun 7:416
32. Furtado M, Chen L, Chen Z, Chen A, Cui W (2022) Development of fish collagen in tissue regeneration and drug delivery. Eng Regener 3:217
33. Ershad-Langroudi A, Babazadeh N, Alizadegan F, Mehdi Mousaei S, Moradi G (2024) Polymers for implantable devices. J Indus Eng Chem 137:61
34. Akhtar A, Farzam Rad V, Moradi A-R, Yar M, Bazzar M (2023) Emerging polymeric biomaterials and manufacturing-based tissue engineering approaches for neuro regeneration-a critical review on recent effective approaches. Smart Mater Med 4:337
35. Erathodiyil N, Chan H-M, Wu H, Ying JY (2020) Zwitterionic polymers and hydrogels for antibiofouling applications in implantable devices. Mater Today 38:84
36. Kalidas S, Sumathi S (2024) Electrospun gelatin-polyvinyl alcohol-silk fibre/hydroxyapatite scaffolds and their physicochemical and biological studies. Colloids Surf A Physicochem Eng Asp 684:133030
37. Dizaj SM, Rezaei Y, Namaki F, Sharifi S, Abdolahinia ED (2024) Effect of curcumin-containing nanofibrous gelatin-hydroxyapatite scaffold on proliferation and early osteogenic differentiation of dental pulp stem cells. Pharm Nanotechnol 12:262
38. Wan Y, Gao J, Zhang J, Peng W, Qiu G (2010) Biodegradability of conducting chitosan-g-polycaprolactone/polypyrrole conduits. Polym Degrad Stab 95:1994
39. Ezati M, Safavipour H, Houshmand B, Faghihi S (2018) Development of a PCL/gelatin/chitosan/β-TCP electrospun composite for guided bone regeneration. Prog Biomater 7:225
40. Dodda JM, Azar MG, Bělský P, Šlouf M, Gajdošová V, Kasi PB, Anerillas LO, Kovářík T (2023) Bioresorbable films of polycaprolactone blended with poly(lactic acid) or poly(lactic-co-glycolic acid). Int J Biol Macromol 248:126654
41. Åkerlund E, Diez-Escudero A, Grzeszczak A, Persson C (2022) The effect of PCL addition on 3d-printable PLA/HA composite filaments for the treatment of bone defects. Polymers (Basel) 14:3305
42. Armentano I, Puglia D, Luzi F, Arciola C, Morena F, Martino S, Torre L (2018) Nanocomposites based on biodegradable polymers. Materials 11:795
43. Ligon SC, Liska R, Stampfl J, Gurr M, Mülhaupt R (2017) Polymers for 3d printing and customized additive manufacturing. Chem Rev 117:10212
44. Park S, Shou W, Makatura L, Matusik W, Fu K (Kelvin) (2022) 3D printing of polymer composites: Materials, processes, and applications. Matter 5:43
45. Nath SD, Nilufar S (2020) An overview of additive manufacturing of polymers and associated composites. Polymers (Basel) 12:2719
46. Hua W, Shi W, Mitchell K, Raymond L, Coulter R, Zhao D, Jin Y (2022) 3D printing of biodegradable polymer vascular stents: a review. Chin J Mech Eng: Addit Manuf Front 1:100020
47. Mahmud Z, Nasrin A, Hassan M, Gomes VG (2022) 3D-printed polymer nanocomposites with carbon quantum dots for enhanced properties and in situ monitoring of cardiovascular stents. Polym Adv Technol 33:980
48. Mohseni M, Hutmacher D, Castro N (2018) Independent evaluation of medical-grade bioresorbable filaments for fused deposition modelling/fused filament fabrication of tissue engineered constructs. Polymers (Basel) 10:40
49. Duran MM, Moro G, Zhang Y, Islam A (2023) 3D printing of silicone and polyurethane elastomers for medical device application: a review. Adv Industr Manuf Eng 7:100125

50. Das SK, Chakraborty S, Naskar S, Rajabalaya R (2021) Techniques and methods used for the fabrication of bionanocomposites. In: Bionanocomposites in tissue engineering and regenerative medicine. Elsevier, pp 17–43
51. Wang X, Shan H, Wang J, Hou Y, Ding J, Chen Q, Guan J, Wang C, Chen X (2015) Characterization of nanostructured ureteral stent with gradient degradation in a porcine model. Int J Nanomedicine 3055
52. Zhou L, Fu J, He Y (2020) A review of 3d printing technologies for soft polymer materials. Adv Funct Mater 30
53. Miron VM, Lämmermann S, Çakmak U, Major Z (2021) Material Characterization of 3D-printed silicone elastomers. Procedia Struct Integrity 34:65
54. Ajdary R, Reyes G, Kuula J, Raussi-Lehto E, Mikkola TS, Kankuri E, Rojas OJ (2022) Direct ink writing of biocompatible nanocellulose and chitosan hydrogels for implant mesh matrices. ACS Polymers Au 2:97
55. Xu C, Yu S, Wu W, Liu Q, Ren L (2022) Direct ink writing of Fe bone implants with independently adjustable structural porosity and mechanical properties. Addit Manuf 51:102589
56. Wang P, Berry DB, Song Z, Kiratitanaporn W, Schimelman J, Moran A, He F, Xi B, Cai S, Chen S (2020) 3D printing of a biocompatible double network elastomer with digital control of mechanical properties. Adv Funct Mater 30
57. Pagac M, Hajnys J, Ma Q-P, Jancar L, Jansa J, Stefek P, Mesicek J (2021) A review of vat photopolymerization technology: materials, applications, challenges, and future trends of 3d printing. Polymers (Basel) 13:598
58. Che Ab Rahman A, Lee B-J, Lim S (2024) Optimizing polymethyl methacrylate (PMMA)-based stretchable microneedle arrays by vat photopolymerization for efficient drug loading. Addit Manuf 94:104472
59. Bao Y, Paunović N, Leroux J (2022) Challenges and opportunities in 3d printing of biodegradable medical devices by emerging photopolymerization techniques. Adv Funct Mater 32
60. de Oliveira MF, da Silva LCE, de Oliveira MG (2021) 3D printed bioresorbable nitric oxide-releasing vascular stents. Bioprinting 22:e00137
61. D'Urso PS, Effeney DJ, Earwaker WJ, Barker TM, Redmond MJ, Thompson RG, Tomlinson FH (2000) Custom cranioplasty using stereolithography and acrylic. Br J Plast Surg 53:200
62. Melchels FPW, Feijen J, Grijpma DW (2009) A poly(d, l-lactide) resin for the preparation of tissue engineering scaffolds by stereolithography. Biomaterials 30:3801
63. Song Y, Ghafari Y, Asefnejad A, Toghraie D (2024) An overview of selective laser sintering 3D printing technology for biomedical and sports device applications: processes, materials, and applications. Opt Laser Technol 171:110459
64. Moreno Madrid AP, Vrech SM, Sanchez MA, Rodriguez AP (2019) Advances in additive manufacturing for bone tissue engineering scaffolds. Mater Sci Eng: C 100:631
65. Li Y, Meng Q, Chen S, Ling P, Kuss MA, Duan B, Wu S (2023) Advances, challenges, and prospects for surgical suture materials. Acta Biomater 168:78
66. Han J, Cao R, Chen B, Ye L, Zhang A, Zhang J, Feng Z (2011) Electrospinning and biocompatibility evaluation of biodegradable polyurethanes based on L-lysine diisocyanate and L-lysine chain extender. J Biomed Mater Res A 96A:705
67. Mobarak MH, Islam MA, Hossain N, Al Mahmud MZ, Rayhan MT, Nishi NJ, Chowdhury MA (2023) Recent advances of additive manufacturing in implant fabrication—a review. Appl Surf Sci Adv 18:100462
68. Latif R, Noor MM, Yunas J, Hamzah AA (2021) Mechanical energy sensing and harvesting in micromachined polymer-based piezoelectric transducers for fully implanted hearing systems: a review. Polymers (Basel) 13:2276
69. Toh HW, Toong DWY, Ng JCK, Ow V, Lu S, Tan LP, Wong PEH, Venkatraman S, Huang Y, Ang HY (2021) Polymer blends and polymer composites for cardiovascular implants. Eur Polym J 146:110249

70. Tsakiris V, Tardei C, Clicinschi FM (2021) Biodegradable Mg alloys for orthopedic implants—a review. J Mag Alloys 9:1884
71. Rakhmatia YD, Ayukawa Y, Furuhashi A, Koyano K (2013) Current barrier membranes: titanium mesh and other membranes for guided bone regeneration in dental applications. J Prosthodont Res 57:3
72. Zou X, Huang L, Chen K, Jiang M, Zhang S, Wang M, Hua X, Shan A (2021) Surface structuring via additive manufacturing to improve the performance of metal and polymer joints. Metals (Basel) 11:567
73. Sheydaeian E, Ibhadode OO, Hu E, Pilliar R, Kandel R, Toyserkani E (2021) Additive manufacture of porous ceramic proximal interphalangeal (PIP) joint implant: design and process optimization. Int J Adv Manuf Technol 115:2825
74. Di Luca M, Hoskins C, Corduas F, Onchuru R, Oluwasanmi A, Mariotti D, Conti B, Lamprou DA (2022) 3D printed biodegradable multifunctional implants for effective breast cancer treatment. Int J Pharm 629:122363
75. Wienen D, Gries T, Cooper SL, Heath DE (2023) An overview of polyurethane biomaterials and their use in drug delivery. J Control Release 363:376
76. Jin L, Yao L, Yuan F, Dai G, Xue B (2021) Evaluation of a novel biodegradable ureteral stent produced from polyurethane and magnesium alloys. J Biomed Mater Res B Appl Biomater 109:665
77. Miechowicz S, Wojnarowska W, Majkut S, Trybulec J, Pijanka D, Piecuch T, Sochacki M, Kudasik T (2021) Method of designing and manufacturing craniofacial soft tissue prostheses using additive manufacturing: a case study. Biocybern Biomed Eng 41:854
78. Pires PC, Mascarenhas-Melo F, Pedrosa K, Lopes D, Lopes J, Macário-Soares A, Peixoto D, Giram PS, Veiga F, Paiva-Santos AC (2023) Polymer-based biomaterials for pharmaceutical and biomedical applications: a focus on topical drug administration. Eur Polym J 187:111868
79. da Silva D, Kaduri M, Poley M, Adir O, Krinsky N, Shainsky-Roitman J, Schroeder A (2018) Biocompatibility, biodegradation and excretion of polylactic acid (PLA) in medical implants and theranostic systems. Chem Eng J 340:9
80. Low YJ, Andriyana A, Ang BC, Zainal Abidin NI (2020) Bioresorbable and degradable behaviors of PGA: current state and future prospects. Polym Eng Sci 60:2657
81. Brown TE, Anseth KS (2017) Spatiotemporal hydrogel biomaterials for regenerative medicine. Chem Soc Rev 46:6532
82. Pina S, Ferreira JMF (2012) Bioresorbable plates and screws for clinical applications: a review. J Healthc Eng 3:243
83. Elmowafy EM, Tiboni M, Soliman ME (2019) Biocompatibility, biodegradation and biomedical applications of poly(lactic acid)/poly(lactic-co-glycolic acid) micro and nanoparticles. J Pharm Investig 49:347
84. Lu Y, Cheng D, Niu B, Wang X, Wu X, Wang A (2023) Properties of poly (lactic-co-glycolic acid) and progress of poly (lactic-co-glycolic acid)-based biodegradable materials in biomedical research. Pharmaceuticals 16:454
85. Wahab Hashmi A, Singh Mali H, Meena A (2023) Improving the surface characteristics of additively manufactured parts: a review. Mater Today Proc 81:723
86. Wu Y, Lu Y, Zhao M, Bosiakov S, Li L (2022) A critical review of additive manufacturing techniques and associated biomaterials used in bone tissue engineering. Polymers (Basel) 14:2117
87. Vach Agocsova S, Culenova M, Birova I, Omanikova L, Moncmanova B, Danisovic L, Ziaran S, Bakos D, Alexy P (2023) Resorbable biomaterials used for 3d scaffolds in tissue engineering: a review. Materials 16:4267
88. Panda S, Deshmukh K, Karthik TVK, Pasha SK (2023) Additive manufacturing techniques for drug delivery applications. In: Nanotechnology-based additive manufacturing. Wiley, pp 495–525

Chapter 11
Additive Manufacturing of Artificial Organs Using Resorbable Polymeric Composites

Arbind Prasad, Ranjit Barua, Bidyanand Mahto, and Sudipto Datta

1 Introduction

Stereolithography, more referred to as 3D printing or additive manufacturing (AM), has an array of sectors and is influential in the biomedical engineering area. In the past four decades, a variety of AM technologies have evolved as versatile methodologies for effective structure manufacturing by providing effective control of the shape, size, and characteristics of materials used [1]. For instance, the use of AM to fabricate artificial organs has received a lot of attention since it promises to resolve the increasing donor organ deficit along with enhancing the regenerative medicine field [2, 3]. The major hurdle in tissue engineering and organ transplantation involves finding appropriate materials that can facilitate cell attachment, possess mechanical and biological characteristics similar to natural tissue, and finally be metabolized in the body. Numerous drawbacks are associated with the standard techniques of organ transplantation, such as organ shortage, rejection by the body's immune system, and long waiting periods. In this regard, the use of 3D printing in fabricating artificial organs using resorbable polymeric composites has proven to be the best solution

A. Prasad (✉)
Mechanical Engineering Department, Katihar Engineering College (Under Department of Science, Technology and Technical Education, Government of Bihar), Katihar, Bihar, India
e-mail: arbind.iitg@gmail.com

R. Barua
Centre for Healthcare Science and Technology, Indian Institute of Engineering Science and Technology, Howrah, West Bengal, India

B. Mahto
Government Engineering College, Vaishali (Under Department of Science, Technology and Technical Education, Government of Bihar), Vaishali, Bihar, India

S. Datta
Indian Institute of Science, Bangalore, India

A. Prasad and V. Katiyar (eds.), *Resorbable Polymers for Bioimplants and Fixation Devices*, Materials Horizons: From Nature to Nanomaterials,
https://doi.org/10.1007/978-981-96-9306-1_11

[4]. These kinds of materials have numerous benefits to biocompatibility, biodegradability, and mechanical properties that can be adjusted according to the need, which makes these materials suitable for the formation of scaffolding structures that can be utilized to affect tissue and organ regeneration.

The incorporation of resorbable polymeric composites into additive manufacturing techniques makes it possible to create elaborate structures of the scaffolds that would closely resemble the organs of the body in terms of their structures and functionalities [5]. Compared to conventional manufacturing processes that involve the use of molds or structures formed in advance, AM gives the ability to place the materials on a layer-by-layer basis resulting in creation of very complicated structural forms that are extremely hard to produce using other technologies [6, 7]. This high level of customization is especially useful in the design of patient-specific scaffolds as it will improve the fit and performance of organ replacements needed. Polymers, including PLA, PCL, and PLA/GA copolymers are typical resorbable polymers often used in tissue engineering because of their property of biodegradation within the body [8]. These materials act as a kind of temporary framework to allow cells to grow and tissues to regenerate and dissipate as the actual tissue repair and recovery takes hold [9]. For instance, PLA is used commonly because of its good mechanical characteristics and processability, coupled with high crystallinity and slow degradation rates of PCL. PLGA is a PLA/PCL copolymer that can be easily tailored towards a desired degradation and mechanical behavior to fit certain TE applications [10, 11].

The opportunities for using 3D-printed artificial organs based on SMP composites are almost unlimited, and it is possible to give only a few examples: the creation of new bones, cartilage, and skin substitutes; the development of artificial livers, hearts, and kidneys [12]. It is important to note that the uses that follow do not only include organ transplantation alone but also include other related areas such as drug testing, disease modeling, and even individualized medicine [13–15]. The possibilities to produce vital and bio-degradable organs through 3D bio-printing offer new opportunities for drugs and medical devices for preclinical testing concerning human tissues. The major challenge that must be overcome is attaining the necessary mechanical properties, resolution, and structural design for artificial organs. Also, providing vascular networks for oxygen and nutrient delivery in the tissue and controlling the degradation rate of the scaffold to match cell/tissue growth is another growing frontier. Overcoming these challenges will be critical to achieving the long-term goals of additive manufacturing in creating artificial organs [16].

This chapter will seek to ask questions such as: What is additive manufacturing, and how is it related to resorbable polymeric composites and artificial organs? The types of resorbable materials that have been used and the benefits and disadvantages of these materials will also be looked at together with the various methods of 3D printing used in the modeling of artificial organs. In addition, we will briefly review the existing literature on this topic and the issues that have to be addressed as well as prospects for constructing functional, biodegradable, and personalized artificial organs. Figure 1 shows schematics of FDM printers. Finally, this chapter will focus on the role of additive manufacturing in the development of bio-artificial organs and on the subject of regenerative medicine in general.

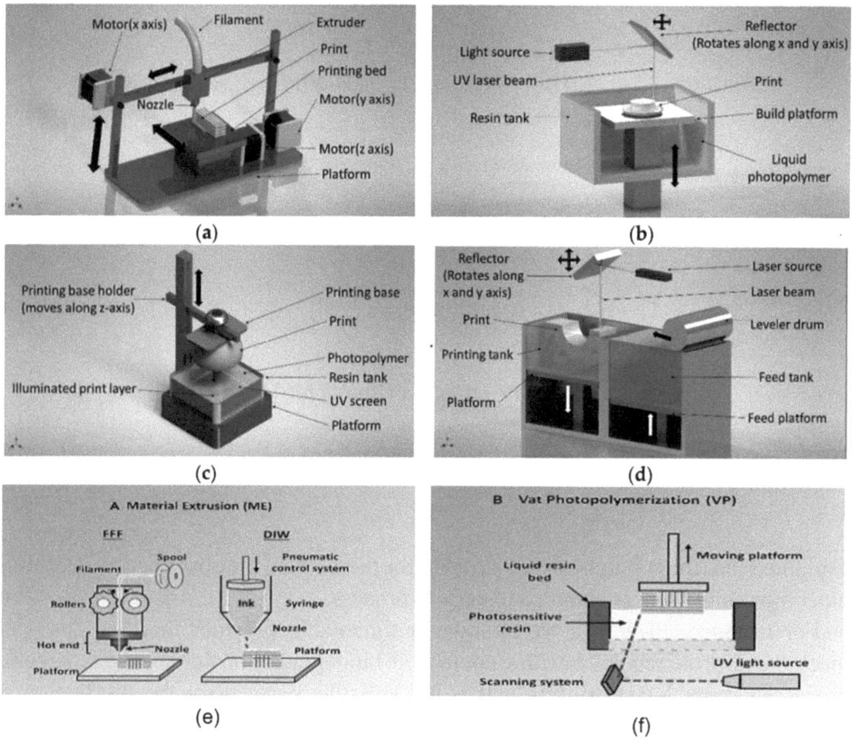

Fig. 1 **a** The materials extrusion principle is employed by Fused Deposition Modelling (FDM) printers to create desired pieces. **b** To achieve high-accuracy printing, the stereolithography (SLA) process cures liquid UV-curable polymer using a UV laser beam. **c** Compared to SLAs, digital light processing (DLP) printers use less time and use UV screens to create extremely accurate parts. **d** A laser beam and fine powder polymers are fused in selective laser sintering (SLS). On the other hand, the printed components have gritty or uneven surfaces. **e** During the polyjet printing process, several nozzles spray tiny droplets of polymers onto the printing bed, which are instantly cured by UV light. It can print quickly using multi-material deposition. **f** To create the finished product, sheets of polymers are carefully cut and layered in a process called sheet lamination (LOM). The LOM technique is recommended when quick printing and large-format printing are needed [17]

2 Resorbable Polymeric Composites for Additive Manufacturing

Biodegradable polymeric composites are advanced composition materials that contain polymers and reinforcements or bioactive fillers. Such composites are intended to disintegrate in the body progressively, permitting the gradual growth of body tissues in the areas in which the composites have been placed [4, 18, 19]. The common resorbable polymers in tissue engineering are polylactic acid (PLA), polycaprolactone (PCL), and poly (lactic-co-glycolic acid) (PLGA). Chung et al. [3]

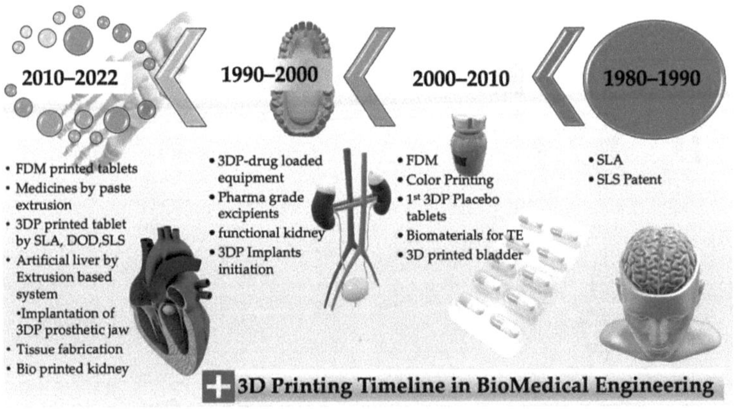

Fig. 2 This timeline illustrates the development of 3D printing and its key breakthroughs from the 1980s to the present [21]

explained that these polymers are suitable for biomedical purposes because they are biocompatible, biodegradable, and easy to process.

For instance, PLA is a well-known sedimentable polymer that has desirable mechanical proficiencies besides controlled biodegradation. Its ability to become processed through 3D printing makes it a suitable material for the production of structures like scaffolds and other body organs. PCL, yet other biodegradable polymers is more flexible and less hydrolytically unstable compared to PLA and, hence, useful in areas where longer degradation time is desirable [12]. PLGA is thus a copolymer of PLA and PCL, which brings about the flexibility of both types, modifiable mechanical characteristics, and degradation rates [20]. Figure 2 shows the timelines for biomedical Engineering.

3 Additive Manufacturing Techniques for Artificial Organs

AM, more famous as the third Industrial Revolution technology, 3D printing is a technology building products layer after layer. Based on the level of detail and degree of specificity it is always beneficial to use AM technologies for the fabrication of artificial organs. Other AM processes such as FDM, SLA, and SLS have also been used to manufacture scaffolding and tissue constructive to develop organs.

FDM is one of the most common forms of AM technology that uses an additive approach developed by building layers of thermoplastic material by drawing from a heated nozzle. PLA and PCL are widely used polymers in FDM because they are easy to extrude and offer a suitable print quality [4, 18, 19]. On the other hand, SLA employs ultraviolet light to solidify photopolymer resins thereby enabling high-precision structures to be built despite the levels of complexity in their shape. It has been used in designing complex scaffolds in tissue engineering applications for instance by [22]. Selective Laser Sintering involves the use of a laser to fuse powdered materials that can be resorbable polymeric composites to create solid parts. This technique has been applied in generating both large and small-scale scaffolding for artificial organs [16].

Concurrent advancement in AM has revolutionized the design and manufacture of MEDICAL DEVICES and IMPLANTS, in the development of artificial organs in particular. That AM allows the creation of multilevel structures out of bioresorbable materials with exceptional accuracy, making it an indispensable tool for tissue engineering and medicine [23]. As applied to organ constructs, AM enables design control of the scaffold structure and its mechanical characteristics as well as cell attachment mediating bioactive motor that can be included in the scaffold. Several direct manufacturing processes have been used to print scaffold structures for organs and tissues, each with its benefits and drawbacks [13–15, 24]. Such methods are Fused Deposition Modeling (FDM), Stereolithography (SLA), Selective Laser Sintering (SLS), Inkjet Printing, and Bioprinting [25] 26. A comprehension of the relevant principles, benefits, and drawbacks of each of these methods is indispensable for the further development of artificial organ production.

3.1 Fused Deposition Modeling (FDM)

Fused Deposition Modeling (FDM) is one of the most popular open-source technologies used in manufacturing of tissue scaffolds and implants [27]. FDM uses thermoplastic filaments, with biodegradable polymers such as PLA, PCL, or their composites where materials are heated and extruded through a nozzle and successively layered to form a structure [28]. This technique is easy to use commonly in 3D printing since it does not require highly skilled personnel and its cost is not expensive.

Principles and Process: FDM functions by melting a filament to a state where it becomes malleable, and then can be extruded through a printer head. The material is then extruded and deposited onto the build platform and undergoes rapid cooling to form a layer of the part. One on top of another builds the required 3D object by sticking to the existing layer beneath it. At each layer, the print process of the designed structure is repeated to form the final structure.

Advantages: FDM is flexible for use with many types of material, and it is possible to use resorbable polymers such as PLA, PCL, and PLGA. The method is particularly appropriate for fabricating scaffolds that need to be mechanically stronger and helpful in generating large cellular constructs such as bone or cartilage substitutes [29], 30. Also, FDM enables the determination of the location of the deposited materials; therefore, it can be effectively used to fabricate a scaffold with internal structures that are similar to the natural tissue. Its utility of forming composites with bioactive agents that foster the growth of tissues is a strong feature of tissue engineering. Figure 3 shows the sources and methods for creating an albumin scaffold.

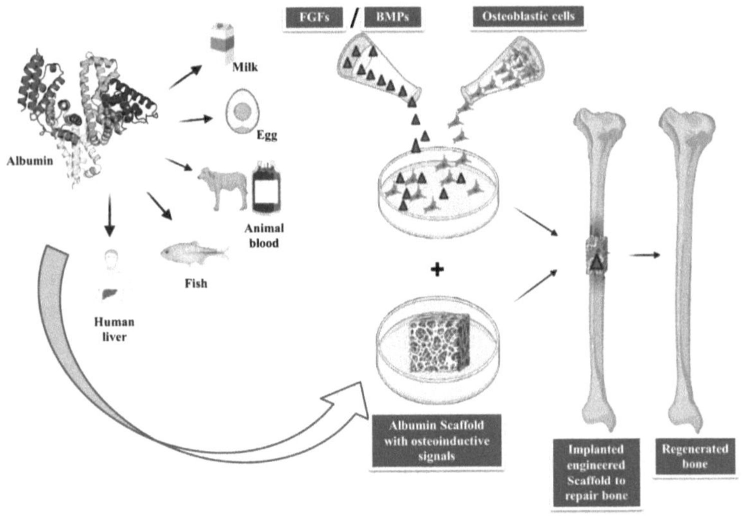

Fig. 3 Sources and methods for creating an albumin scaffold that promotes bone growth [31]

Limitations: The major drawback of FDM is that this method does not have a high enough resolution to replicate the detailed microstructure needed for some types of tissues [32]. However, the layer-by-layer deposition may cause anisotropy, meaning that mechanical properties depend on the angles of the layer formations. This may weaken the material reduce the physical properties of the material and result in diseased regions or regions of soft tissues, like in blood vessels.

3.2 *Stereolithography (SLA)*

Stereolithography (SLA) is a print technology of superior accuracy that uses a UV laser to harden liquid photopolymer resins and form layers of solid material [30]. SLA is regarded as having a high resolution and thus capable of representing fine geometries suitable for developing structures that are required in forming artificial organs.

Principles and Process: In SLA, a laser beam is projected on the surface of a liquid resin, which is, in general, a photocurable polymer. The laser cures the resin, making it harden through sequential layers. This building platform is then reduced in a stepwise manner to facilitate the curing of the following layer of the resin. It goes on until the final structure is printed out to completion [33].

Advantages: SLA delivers high resolution and accuracy in the construction of microstructures with compact and matt surface finish. This makes it suitable for uses where the geometry of the structure is complex and the resolution highly detailed, as in microvascular structures for example. SLA also permits the use of resorbable photopolymer materials, which may be programmed to dissolve over some time, and at the same time promote tissue formation.

Limitations: Enjoying a wide range of applications, SLA's main drawbacks are the kind of material to be printed and the size of a built object. Photopolymer resins, in general, are characterized by non-ideal mechanical properties for load-bearing applications. Therefore, SLA is less suitable for scaffolds where mechanical stress and load are required [34]. Furthermore, SLA systems may take more time than other AM methods, which is an important disadvantage while working with large and complex organs.

3.3 Selective Laser Sintering (SLS)

Selective Laser Sintering (SLS) is an AM process of using a laser to selectively melt and bond layers of powder particles together [35]. The HSS method is widely applied for the creation of functional elements and scaffolds in the biomedical industry.

Principles and Process: SLS operates by expelling a thin layer of powdered material—for example, resorbable polymers or composites—on a build platform. A laser is then focused on the powder, and only that section of the powder bed where the particles are to be fused into a solid layer is heated [36]. When one layer is done, the platform is pulled down, and another layer of the powder is poured in. This process going on for the final part is fabricated to its required complete level of fabrication.

Advantages: SLS can achieve very fine detail of the internals as well as surfaces of a part and intricate structures such as pores and hollows that are useful in the creation of tissue scaffolding. This technique can be used to prepare scaffolds with dual properties of mechanical strength and porosity that are essential to facilitate cell penetration and new tissue formation [37]. In addition, the use of resorbable polymeric powders, particularly PCL and PLA, makes the composite highly appropriate for biomedical purposes.

Limitations: The first key limitation of SLS is the need to clean up the printed part of the powder remaining on the crafted part [38]. On the other hand, SLS lets the creation of something intricate and detailed at a high resolution while maybe slightly lower than materials like resins used in SLA. In addition, SLS systems are generally somewhat more costly and sophisticated, ruling out some uses.

3.4 Inkjet Printing

Inkjet Printing is yet another novel AM technique used to develop artificial organs and tissue scaffolds. This technique involves the use of a droplet-on-substrate technique to gradually construct multilayers of material such as resorbable polymers and bioactive substances [39].

Principles and Process: Inkjet works by spraying material droplets with a nozzle towards a build platform; the material has to harden to form the layer. The droplets are placed in a way that the material in a microsystem is arranged in a specific pattern [40]. In some cases, it is also possible to incorporate several materials at once, for example, polymer composite containing cells or bioactive molecules.

Advantages: Compared to other forms of printing, inkjet printing is very adaptable to various types of materials; it can accommodate bioinks that include living cells for constructing various tissues [41]. It is especially effective in a case where different materials have to be printed at the same time: this is useful when building composites that are filled with cells, growth factors, and resorbable polymers [39]. Another advantage of the technique is its high resolution and rather fast printing compared to other technologies.

Limitations: The main limitation of inkjet printing is the viscoelastic property of the materials that are being used. This significantly restricts the range of materials that can be printed because highly viscous fluids do not pass through the inkjet nozzle well. Furthermore, the resolution may not be as high as required for large-scale fabrication of scaffolds or organs, and non-uniformity in the distribution of cells within printed structures can compromise tissue quality.

3.5 Bioprinting

Bioprinting can be considered a subset of additive manufacturing in which biomaterials, cells, growth factors, and extracellular matrix constituents are printed complementary to traditional 3D printing to build tissue mimetic structures or even organs [42].

Principles and Process: Bioprinting is the process of sequentially layering biological materials—including living cells—within a substrate like bioink or in conjunction with other such printable materials as resorbable polymers [43]. The bioinks are then printed using some or most of the known AM techniques, such as extrusion, inkjet, and laser-assisted techniques, among others [13–15, 44]. The cells in the bioink are not killed during the printing process and are included in the scaffold, which provides structural support as well as encourages cell adhesion, division, and specialization.

Advantages: The capabilities of building organs, which can look and work like natural organs, give bioprinting a particular advantage unlike any other technique. This technique can be used to generate a high level of structural organization of tissue construct for the inclusion of vascular networks for supporting the larger tissue construct or organ [45]. Figure 4 shows MSC-assisted bone regeneration procedure.

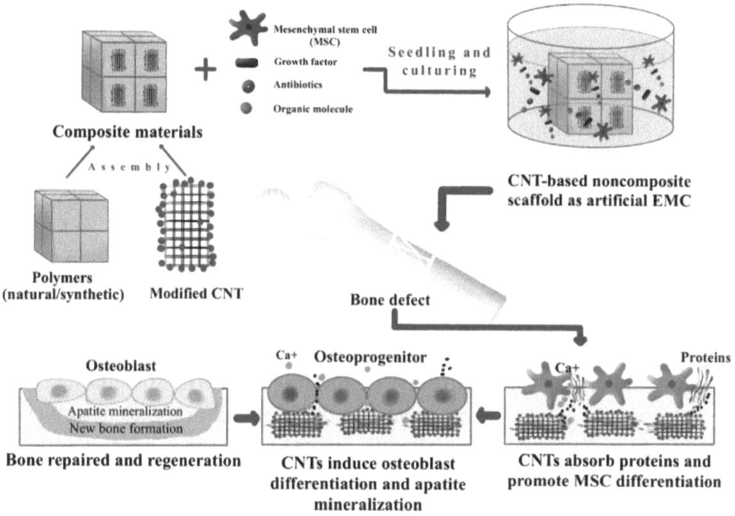

Fig. 4 The MSC-assisted bone regeneration procedure [46]

In addition, bioprinting affords better control over the placement of cells, where tissue-specific structures can be created individually for a particular patient.

Limitations: Still, one of the major issues that are inherent to bioprinting technology is the necessity of keeping the cells' viability intact all through the bio-printing process, as some of the cells could be negatively affected with the shear forces or temperature changes inherent in the printing process. Furthermore, the question of vascularization, which is critical when working with large tissues or organs, remains a challenge; it is still difficult to create functioning blood vessels within the printed tissues.

Different techniques of AM have different pros and cons, especially in the application of artificial organs using resorbable polymeric composites [47]. Stereolithography has the highest quality and enables highly detailed design work, the main reason for this is Fused Deposition Modeling which is used most often because it is inexpensive and suited for very complex designs. Inkjet printing offers the potential to make direct composite scaffolds of cells and bioactive materials; thereby, bioprinting is an advanced technique of tissue engineering and is far ahead of other methods in producing living, functional tissues [48]. These technologies and materials are still advancing along with the progress in different techniques of bioprinting and vascularization, which will be the key points in the fabrication of artificial organs.

4 Properties and Performance of Resorbable Polymeric Composites in Tissue Engineering

The mechanical performance of fabricated artificial organs using resorbable polymeric composites is determined by the following aspects: mechanical strength, degradation behavior, and bioactivity. A typical requirement in scaffold design is that the mechanical properties of the scaffold must relate to the biomechanical properties of the target tissue, such as stiffness, elasticity, and tensile strength of bone, cartilage, or muscle tissues [49]. The degradation rate of the polymer is another important factor that must also be considered. The scaffold has to be degradable in a manner to facilitate the new tissue to form in the place of the scaffold. If the rate favored for degradation is high, then the scaffold structure might lose its structural competence before tissue formation. On the other hand, if the degradation rate is slow, it implies that the scaffold will remain within the body and elicit the wrong effects. Another factor about activity must also be mentioned, this being bioactivity. Traditional polymeric matrices used in resorbable composites can be reinforced with bioactive additives like hydroxyapatite, growth factors, and collagen to facilitate cell adhesion and proliferation as well as proper differentiation. Such bioactive components can easily enhance the functional and tissue response of the implanted artificial organ [50].

The problems of tissue engineering and potential applications such as the creation of artificial organs strongly depend on the materials for scaffold production. The class of resorbable polymeric composites has been highly regarded for this application because they dissolve within the body while healing is taking place. These composites consist of biocompatible polymers with the positive impact of fillers on the properties of composites or other materials, which make them suitable for developing scaffolds that are not only intended to support the affected tissue mechanically but also to release signals that stimulate cell growth. Mechanical and physical properties, thermal stability, degradation behavior, and bioactive component incorporation of resorbable polymeric composites used in tissue engineering are discussed here.

(a) **Biocompatibility and Biodegradability**

Biocompatibility is the primary concern of any material employed in tissue engineering because the material must not elicit a negative immune response when placed inside the body. The polymeric composites are resorbable by nature hence, they do not elicit serious inflammation or immune response, which is an essential requirement for their integration into biomedical devices, implants, or scaffolds. Polymeric composites that resorb in the body offer promising behavior that has tremendous application in the case of organ prostheses [51]. It gradually dissolves in the body through hydrolysis or enzymatic digestion, after which it disappears from the body. This degradation process is important because the scaffold has to offer structural support at first, then gradually break apart, and the body replaces the scaffold with new functional tissue. PLA and PCL are two common types of resorbable polymers that are preferred when they dissolve at a certain rate in the body because resorbable polymers can be designed according to the requirements of tissue engineering.

(b) **Mechanical Properties**

Mechanical properties are key in tissue engineering because the scaffold has to possess similar stiffness, strength, and elasticity to the tissue it is designed for. These mechanical properties make sure that the loaded scaffold can bear physiological load, stimulate the formation of tissue, and not fail prematurely. The resorbable polymeric composites may be Tailor-made in terms of material strength and stiffness to suit a particular application. For instance, bone scaffolds need materials that are compressively strong and stiff, in contrast, the scaffolds for soft tissues like skin or muscles need to be tough but significantly less stiff. Additional properties of the composite can be coaxed by the use of reinforcing agents like HA calcium phosphates or carbon-based fillers in polymer matrix for increased mechanical characteristics of the composite for load-bearing applications [52]. Furthermore, the mechanical performance can be tuned concerning the weight, glass transition temperature, and solubility parameter of the selected polymer matrix, crosslink density, as well as the choice and concentration of filler material. In the case of soft tissue replacements, especially cartilage or skin substitutes, the scaffold material should be capable of assuming a highly elastic property to be able to accommodate for relative deformations under the various mechanical forces exerted on such tissues [53]. For instance, to complement: PCL-based polymers possess a low modulus and high flexibility to conform to elasticity or deformability. The incorporation of specific bioactive agents or fillers can also alter the elasticity of the resultant composite, thus developing scaffolds with mechanical properties of the natural soft tissues. While tensile strength is also crucial for tissues that undergo tension, for example, skin or muscles. The tensile properties of the resorbable composites can be modified to attain the tensile properties of the target tissue by modifying the molecular weight and the level of crystallinity of the polymers or by the addition of fillers such as carbon nanotubes or nanofibers.

(c) **Degradation Rate and Control**

The rate of degradation is one of the most important factors that affect the performance of resorbable polymeric composites for tissue engineering applications. This kind of degradation may be disadvantageous since the scaffold should provide an appropriate surface for the tissue to grow on. Instead, it fails prematurely. On the other hand, if the degradation rate is slow, the scaffold stays in the body for a long time, which may lead to inflammation or fibrosis. Resorbable polymers are known to degrade mainly by hydrolysis, through which the water molecules cleave the Polymer chains into smaller pieces, and/or enzymatic action, where specific enzymes break down the material. PLA and PCL exhibit different degradation rates. PCL is generally considered to degrade at a slower rate than PLA because while both polymers undergo hydrolysis degradation, the crystallinity of PCL is comparatively lower, so the hydrolysis rate of PCL will be slower than PLA. It is thus possible to control the rate of degradation according to the intended tissue engineering application by either copolymerizing the polymers or modifying the molecular weight of the polymer matrix. For instance, in tissue with rapid degradation, such as the skin, the scaffold may be made from PLA, while for tissues that are slowly degraded, such as bone, the scaffold may be made from PCL. The residues of resorbable polymers, for instance, lactic acid in PLA or glycolic in PLGA, are nontoxic and are metabolized by the human body. However, for some applications, controlling the concentration and accumulation of the degradation products might be crucial because they can cause inflammation or even tissue damage in high concentrations. Resorbable polymeric composites have several advantages: the degradation rate adjustment is possible by regulating the content of the composite. For instance, the copolymerization of PLA and PCL in achievable opens up a middle ground of materials with intermediate degradation profiles and, therefore, more appropriate for different uses. Moreover, the presence of several mineral fillers, for instance, HA, can also determine the degradation rate through its effect on the crystallinity and stability of the polymer matrix. These materials must be carefully tailored so that the scaffolding itself degrades in parallel with tissue formation to assume maximum support during the healing phase.

(d) **Biological Properties**

Thus, the biological characteristics of the resorbable polymeric composites involved determine the extent of cell-scaffold interaction and tissue response. Cell adhesion, proliferation, and differentiation are the key biological characteristic that has to be overcome in successful tissue engineering applications. The various properties of the surface of resorbable polymeric composites determine biocompatibility and cell-scaffold cooperation. The positions of the surface energy, roughness, and chemical composition can be created deliberately to support cell attachment, a fundamental process for the development of cells into tissues [54]. The loading of bioactive molecules that include growth factors, peptides, or ECM proteins in a polymer scaffold helps in getting better cell attachment and tissue-specific differentiation. For instance, the addition of fibronectin or collagen peptides into the composite promotes endothelial or osteoblast adhesion based on targeted tissue. A well-designed biodegradable composite scaffold can support and promote cell adhesion and proliferation in the scaffold to enable cell growth and proliferation in the desired construct [55]. The porosity and pore size of the scaffold can be manipulated, hence allowing the diffusion of nutrients and oxygen required for cellular activities and tissue engineering. The milieu of connected pores also provides a conducive environment for tissue ingrowths and can readily support the formation of vascularization essential in the survival of enhanced matrices. Incorporation of growth factors (e.g., vascular endothelial growth factor [VEGF]) along with the already mentioned substances can help augment tissue regeneration and form functional tissues. Another factor related to the immune response is the inflammation-related resorption of polymeric composites usable in biomedical applications. In more perfect form, resorbable polymers should be expected to degrade in the body without causing a massive immunological reaction. However, some of the polymers might trigger a febrile reaction once the material disintegrates or dissolves. To avoid this, it is possible to adjust the rate of release of the degradation products or to introduce anti-inflammatory substances into the polymer matrix so that the body will not attempt to destroy the implanted material as a sign of an infection.

5 Integration of Bioactive Components

Integrating bioactive elements into biodegradable polymeric matrices is a new concept in tissue engineering. Such components could be natural biomolecules or synthetically produced factors, such as growth factors, cytokines, or other signaling proteins and molecules that dictate the organization of cells and resultant tissue [43]. With the help of reinforced resorbable composites, various growth factors that include bony morphogens such as BMPs, glycoproteins such as TGF-β, and angiogenic proteins such as FGFs can be incorporated to support cell renewal, differentiation in addition to the healing process of various tissues. These bioactive molecules can be incorporated directly into the polymer matrix or the aforesaid microspheres or nanoparticles to control their release over some time. This controlled release can then be gently adjusted to match the degradation rate of the scaffold, which means that the biological activity molecules will be present when they are demanded most during the process of tissue repair. HA, TCP or bioactive glass can be incorporated within resorbable polymeric matrices to obtain scaffolds that are sturdier in mechanical and chemical points of view, analogous to bone tissue. The incorporation of these bioactive minerals also improves the strength of the scaffold but, at the same time, offers a surface that will encourage bonding to bone tissue and stimulate new bone formation. In the same manner, reinforcement materials including carbon nanotubes, graphene, or nanofibers can be added to the polymer matrix for enhancement of the material mechanical characteristics and formation of nano-scaled surface topologies for promotion of cell attachment and repopulation.

6 Challenges in Additive Manufacturing of Artificial Organs

Nonetheless, the great potential decades offered by AM to create organ substitutes through the use of resorbable polymeric composites should be acknowledged. Still, several issues can be pointed out. Among the problems associated with the application of AM techniques, the high limitations of available bio-compatible materials are one of the most daunting. Among the hydrogels, PLA, PCL, and PLGA are popular after extensive research, but their mechanical strength might not be enough to mimic the tissue or organ models [1]. Another issue is how to recreate the necessary resolution and complex structure of artificial organs. As noted earlier, AM technologies enable one to build complex structures, but the resolution created is not always precise enough to mimic the architectural microstructure of living tissue [49]. However, the problem of vascularization of the latter [artificial organs] persists as a serious challenge. The formation of channel networks within the printed construct is crucial for the delivery of nutrients and oxygen to the engineered tissue. Several techniques for the integration of vascular channels into AM constructs are underway [8].

7 Future Directions

Still, the potential for the field of additive manufacturing for artificial organs is simply enormous. Future work should concern the enhancement of the material characteristics of resorbable polymeric composites, discovering new bioactive composites that promote the growth of the tissue and functional organs, and refining the AM approaches to achieve enhanced resolution and complexity. Bioprinting which entails the combination of living cells and polymeric materials might also provide new opportunities for the derivation of fully functional organs. In addition, incorporating artificial intelligence and machine learning into AM processes might improve the accuracy and the speed of manufacturing bespoke and personalized artificial organs on the patient's side. Through nanopharmacology a patient can have organs that match the anatomic and tissue characteristics of the patient, the best thing that can be offered by this kind of pharmaceutical.

8 Conclusion

Tissue engineering in additive-manufactured artificial organs using resorbable polymeric composites is an emerging technology that can revolutionize replacement organ therapy. The use of developed manufacturing techniques and biodegradable materials would be used by the researchers to develop needed scaffolds that would support the regeneration of tissues and ultimately be self-absorbed in vivo. Nevertheless, continuous developments in material science, AM techniques, and bioprinting remain the key to the further progressive growth of functional, implantable artificial organs in clinics. Polymeric resorbable composites are important materials for the construction of tissue engineering scaffolds and artificial organs. These stimuli-responsive, flexible mechanical properties and easily controlled degradation rates also support cell growth and differentiation, making them perfect candidates for regenerative medicine. Bioactive agents, as well as bioactive fillers, help improve their features; moreover, nanotechnology also improves the applicability of these materials in a wide range of tissue engineering applications. With advancements made towards developing better research methodologies, the manufacturing and construction of these materials shall progress alongside the development of even better and more efficient tissue engineering constructs.

Acknowledgements The authors would like to thank Mrs. Nibedita Bardhan for language proof reading.

References

1. Bohner M, Drosse I (2016) Resorbable biomaterials for orthopedic tissue engineering. Adv Mater 28(7):2670–2683. https://doi.org/10.1002/adma.201504727
2. Waidi YO, Jain N, Barua R, Vannaladsaysy V, Das S, Prasad A, Datta S (2024) Smart and self-healing hydrogels for biomedical applications. In: Nanomanufacturing techniques in sustainable healthcare applications. CRC Press, pp 110–130
3. Chung C, Wei M (2013) Polylactic acid (PLA) as a biodegradable polymer for biomedical applications. Polym Chem 4(10):2601–2607. https://doi.org/10.1039/C3PY00051C
4. Jain N, Waidi YO, Chowdhury S, Barua R, Das S, Prasad A, Datta S (2024) Medical devices tribology. In: Applications of biotribology in biomedical systems. Springer Nature Switzerland, Cham, pp 235–250
5. Cui X, Ma PX (2016) Polymer-based nanocomposites for tissue engineering. Bioeng Transl Med 1(1):2–12. https://doi.org/10.1002/btm2.10011
6. Chowdhury S, Jain N, Waidi YO, Barua R, Das S, Prasad A, Datta S (2024) Biological smart biomaterials: materials for biomedical applications. In: Applications of biotribology in biomedical systems. Springer Nature Switzerland, Cham, pp 313–325
7. Waidi YO, Jain N, Chowdhury S, Barua R, Das S, Prasad A, Datta S (2024) Challenges and perspective of manufacturing techniques in biomedical applications. In: Applications of biotribology in biomedical systems. Springer Nature Switzerland, Cham, pp 433–445
8. Jia W, Kim Y (2014) 3D printing of tissues and organs. Nat Biotechnol 32(9):937–946. https://doi.org/10.1038/nbt.2978
9. Datta S, Barua R (2024) 3D printing in modern healthcare: an overview of materials, methods, applications, and challenges. In: Emerging technologies for health literacy and medical practice, pp 132–152
10. Koch L, Gill A (2017) Additive manufacturing of bioresorbable scaffolds. Biomater Sci 5(10):1987–1999. https://doi.org/10.1039/C7BM00547A
11. Das S, Datta S, Barman A, Barua R (2023) Smart biodegradable and bio-based polymeric biomaterials for biomedical applications. In: Modeling, characterization, and processing of smart materials. IGI Global, pp 56–82
12. Loh XJ, Choong C (2016) Polycaprolactone-based scaffolds for tissue engineering. J Biomed Mater Res B Appl Biomater 104(3):420–433. https://doi.org/10.1002/jbm.b.33415
13. Barua R (2024) Unleashing drones for medical advancements: current and future applications. In: Drone applications for Industry 5.0. IGI Global, pp 346–357
14. Barua R (2024) The emerging role of artificial intelligence in organ-on-a-chip (OOAC) biomedical devices. In: Reshaping healthcare with cutting-edge biomedical advancements. IGI Global, pp 369–381
15. Barua R (2024) An in-depth exploration of AI and humanoid robotics' role in contemporary healthcare. In: Approaches to human-centered AI in healthcare. IGI Global, pp 42–61
16. Meijer H, Borman E (2019) Sintering and additive manufacturing of polymeric scaffolds for tissue engineering. Mater Sci Eng, C 100:675–684. https://doi.org/10.1016/j.msec.2019.02.021
17. Nath SD, Nilufar S (2020) An overview of additive manufacturing of polymers and associated composites. Polymers 12(11):2719. https://doi.org/10.3390/polym12112719
18. Jain N, Barua R, Waidi YO, Datta S (2024) Nanowires for bio-sensing applications. In: Next generation materials for sustainable engineering. IGI Global, pp 205–219
19. Jain N, Waidi YO, Barua R, Das S, Vannaladsaysy V, Prasad A, Datta S (2024) Bioinspired nanocomposites: multifunctional materials towards sustainable alternatives. In: Nanomanufacturing techniques in sustainable healthcare applications. CRC Press, pp 78–93
20. Murphy SV, Atala A (2014) 3D bioprinting of tissues and organs. Nat Biotechnol 32(8):770–777
21. Li N, Khan SB, Chen S, Aiyiti W, Zhou J, Lu B (2023) Promising new horizons in medicine: medical advancements with nanocomposite manufacturing via 3d printing. Polymers 15(20):4122. https://doi.org/10.3390/polym15204122

22. Datta S, Barua R, Sarkar R, Barui A, Chowdhury AR, Datta P (2018) Design and development of alginate: Poly-l-lysine scaffolds by 3D bio printing and studying their mechanical, structural and cell viability properties. IOP Conf Ser: Mater Sci Eng 402:012113
23. Barua R, Datta S, Datta P, Chowdhury AR (2021) Scaffolds and tissue engineering applications by 3D bio-printing process: a new approach. In: Research anthology on emerging technologies and ethical implications in human enhancement. IGI Global, pp 718–733
24. Datta S, Barua R, Das J (2020) Importance of alginate bioink for 3D bioprinting in tissue engineering and regenerative medicine. In: Alginates—recent uses of this natural polymer, 10
25. Tian Y, Chen C, Xu X, Wang J, Hou X, Li K, Lu X, Shi H, Lee ES, Jiang HB (2021) A review of 3d printing in dentistry: technologies, affecting factors, and applications. Scanning 2021:9950131
26. Handral HK, Hua Tay S, Wan Chan W, Choudhury D (2022) 3D Printing of cultured meat products. Crit Rev Food Sci Nutr 62(1):272–281. https://doi.org/10.1080/10408398.2020.1815172
27. Barua R, Sarkar A, Datta S (2023) Emerging advancement of 3d bioprinting technology in modern medical science and vascular tissue engineering education. In: Handbook of research on instructional technologies in health education and allied disciplines. IGI Global, pp 153–175
28. Datta S, Barua R, Das J (2019) Application of artificial intelligence in modern healthcare system. Alginates-recent uses of this natural polymer
29. Tack P, Victor J, Gemmel P, Annemans L (2016) 3D-printing techniques in a medical setting: a systematic literature review. Biomed Eng Online 15(1):115
30. Della Bona A, Cantelli V, Britto VT, Collares KF, Stansbury JW (2021) 3D printing restorative materials using a stereolithographic technique: a systematic review. Dent Mater 37(2):336–350. https://doi.org/10.1016/j.dental.2020.11.030
31. Saurav S, Sharma P, Kumar A, Tabassum Z, Girdhar M, Mamidi N, Mohan A (2024) Harnessing natural polymers for nano-scaffolds in bone tissue engineering: a comprehensive overview of bone disease treatment. Curr Issues Mol Biol 46(1):585–611
32. Barua R, Datta S, Roychowdhury A, Datta P (2019) Importance of 3d printing technology in medical fields. In: Additive manufacturing technologies from an optimization perspective. IGI Global, pp 21–40
33. Xiao R, Ding M, Wang Y, Gao L, Fan R, Lu Y (2021) Stereolithography (SLA) 3D printing of carbon fiber-graphene oxide (CF-GO) reinforced polymer lattices. Nanotechnology 32(23). https://doi.org/10.1088/1361-6528/abe825
34. Xu X, Goyanes A, Trenfield SJ, Diaz-Gomez L, Alvarez-Lorenzo C, Gaisford S, Basit AW (2021) Stereolithography (SLA) 3D printing of a bladder device for intravesical drug delivery. Mater Sci Eng C, Mater Biol Appl 120:111773. https://doi.org/10.1016/j.msec.2020.111773
35. Kafle A, Luis E, Silwal R, Pan HM, Shrestha PL, Bastola AK (2021) 3D/4D printing of polymers: fused deposition modelling (FDM), selective laser sintering (SLS), and stereolithography (SLA). Polymers 13(18):3101. https://doi.org/10.3390/polym13183101
36. Datta S, Barua R, Prasad A (2023) Additive manufacturing for the development of artificial organs. In: Advanced materials and manufacturing techniques for biomedical applications, pp 411–427
37. Fina F, Goyanes A, Gaisford S, Basit AW (2017) Selective laser sintering (SLS) 3D printing of medicines. Int J Pharm 529(1–2):285–293. https://doi.org/10.1016/j.ijpharm.2017.06.082
38. Mazzoli A (2013) Selective laser sintering in biomedical engineering. Med Biol Eng Compu 51(3):245–256. https://doi.org/10.1007/s11517-012-1001-x
39. Beedasy V, Smith PJ (2020) Printed electronics as prepared by inkjet printing. Materials (Basel, Switzerland) 13(3):704. https://doi.org/10.3390/ma13030704
40. Carou-Senra P, Ong JJ, Castro BM, Seoane-Viaño I, Rodríguez-Pombo L, Cabalar P, Alvarez-Lorenzo C, Basit AW, Pérez G, Goyanes A (2023) Predicting pharmaceutical inkjet printing outcomes using machine learning. Int J Pharm: X 5:100181. https://doi.org/10.1016/j.ijpx.2023.100181
41. Barua R, Datta S (2023) Artificial intelligence in modern medical science: a promising practice. In: Recent developments in machine and human intelligence. IGI Global, pp 1–12

42. Sun W, Starly B, Daly AC, Burdick JA, Groll J, Skeldon G, Shu W, Sakai Y, Shinohara M, Nishikawa M, Jang J, Cho DW, Nie M, Takeuchi S, Ostrovidov S, Khademhosseini A, Kamm RD, Mironov V, Moroni L, Ozbolat IT (2020) The bioprinting roadmap. Biofabrication 12(2):022002. https://doi.org/10.1088/1758-5090/ab5158
43. Murphy SV, De Coppi P, Atala A (2020) Opportunities and challenges of translational 3D bioprinting. Nat Biomed Eng 4(4):370–380. https://doi.org/10.1038/s41551-019-0471-7
44. Wu Y, Qin M, Yang X (2023) Organ bioprinting: progress, challenges and outlook. J Mater Chem B 11(43):10263–10287. https://doi.org/10.1039/d3tb01630g
45. Lee A, Hudson AR, Shiwarski DJ, Tashman JW, Hinton TJ, Yerneni S, Bliley JM, Campbell PG, Feinberg AW (2019) 3D bioprinting of collagen to rebuild components of the human heart. Science (New York, N.Y.) 365(6452):482–487. https://doi.org/10.1126/science.aav9051
46. Abbas M, Alqahtani MS, Alhifzi R (2023) Recent developments in polymer nanocomposites for bone regeneration. Int J Mol Sci 24(4):3312
47. Lam EHY, Yu F, Zhu S, Wang Z (2023) 3D bioprinting for next-generation personalized medicine. Int J Mol Sci 24(7):6357. https://doi.org/10.3390/ijms24076357
48. McMillan A, McMillan N, Gupta N, Kanotra SP, Salem AK (2023) 3D bioprinting in otolaryngology: a review. Adv Healthcare Mater 12(19):e2203268. https://doi.org/10.1002/adhm.202203268
49. Bashir MH, Korany NS, Farag DBE, Abbass MMS, Ezzat BA, Hegazy RH, Dörfer CE, Fawzy El-Sayed KM (2023) Polymeric nanocomposite hydrogel scaffolds in craniofacial bone regeneration: a comprehensive review. Biomolecules 13(2):205. https://doi.org/10.3390/biom13020205
50. Bonito V, Koch SE, Krebber MM, Carvajal-Berrio DA, Marzi J, Duijvelshoff R, Lurier EB, Buscone S, Dekker S, de Jong SMJ, Mes T, Vaessen KRD, Brauchle EM, Bosman AW, Schenke-Layland K, Verhaar MC, Dankers PYW, Smits AIPM, Bouten CVC (2021) Distinct effects of heparin and interleukin-4 functionalization on macrophage polarization and in situ arterial tissue regeneration using resorbable supramolecular vascular grafts in rats. Adv Healthcare Mater 10(21):e2101103. https://doi.org/10.1002/adhm.202101103
51. Hou LT, Liu CM, Liu BY, Chang PC, Chen MH, Ho MH, Jehng SM, Liu HC (2007) Tissue engineering bone formation in novel recombinant human bone morphogenic protein 2-atelocollagen composite scaffolds. J Periodontol 78(2):335–343. https://doi.org/10.1902/jop.2007.060106
52. Brooks AK, Pradhan S, Yadavalli VK (2023) Degradable elastomeric silk biomaterial for flexible bioelectronics. ACS Appl Bio Mater 6(10):4392–4402. https://doi.org/10.1021/acsabm.3c00593
53. Vacaras S, Baciut M, Lucaciu O, Dinu C, Baciut G, Crisan L, Hedesiu M, Crisan B, Onisor F, Armencea G, Mitre I, Barbur I, Kretschmer W, Bran S (2019) Understanding the basis of medical use of poly-lactide-based resorbable polymers and composites—a review of the clinical and metabolic impact. Drug Metab Rev 51(4):570–588. https://doi.org/10.1080/03602532.2019.1642911
54. Laubach M, Kobbe P, Hutmacher DW (2022) Biodegradable interbody cages for lumbar spine fusion: current concepts and future directions. Biomaterials 288:121699. https://doi.org/10.1016/j.biomaterials.2022.121699
55. Chuysinuan P, Nooeaid P, Thanyacharoen T, Techasakul S, Pavasant P, Kanjanamekanant K (2021) Injectable eggshell-derived hydroxyapatite-incorporated fibroin-alginate composite hydrogel for bone tissue engineering. Int J Biol Macromol 193(Pt A):799–808. https://doi.org/10.1016/j.ijbiomac.2021.10.132

Chapter 12
Degradation and Toxicological Studies of Polymeric Nanocomposites

Meghali Bhattacharjee, Rambha Meghana Sai, Gourhari Chakraborty, and Atanu Kumar Paul

1 Introduction

1.1 Overview of Polymeric Nanocomposites

Polymeric nanocomposites are a class of advanced polymer and matrix composite materials with particles dispersed in matrices, where at least one dimension of the particle is in the range of 1–100 nm. These materials, which are rapidly gaining popularity, have uses in the medicinal, aerospace, and defense industries [1–3]. The processing methods play a key role in setting the dispersion and orientation of the nanoparticles that directly control the properties of nanocomposites; melt blending and solution blending are two common examples. Nanocomposites are also an important development due to increased environmental degradation (enabling more rapid degradability as compared to pure polymers), in addition to research from the field of environmental degradation [4]. Research is underway to improve the processing and characterization methods for polymeric nanocomposites to allow their full potential to be realized [2, 5–7]. Table 1 shows Different types of polymer nanocomposites and their applications [8–10].

1.2 Importance of Degradation and Toxicological Studies

Degradation is a critical property from the point of view of mechanical response, and it affects durability in biomedical applications where the scaffolds necessary to support

M. Bhattacharjee · R. M. Sai · G. Chakraborty · A. K. Paul (✉)
Department of Chemical Engineering, National Institute of Technology Andhra Pradesh, Tadepalligudem, Andhra Pradesh, India
e-mail: atanukpaul@outlook.com

A. Prasad and V. Katiyar (eds.), *Resorbable Polymers for Bioimplants and Fixation Devices*, Materials Horizons: From Nature to Nanomaterials,
https://doi.org/10.1007/978-981-96-9306-1_12

Table 1 Different types of polymer nanocomposites and their applications

Type of polymer nanocomposite	Properties	Applications
Natural polymer nanocomposites	Biodegradable, biocompatible	Drug delivery, tissue engineering, food packaging
Synthetic polymer nanocomposites	High strength, thermal stability	Automotive parts, electronics, coatings
Clay nanocomposites	Improved mechanical properties, barrier properties	Packaging, automotive, construction materials
Metal nanoparticle composites	Enhanced electrical conductivity, catalytic properties	Sensors, catalysis, electronics
Carbon nanotube composites	High tensile strength, electrical conductivity	Aerospace, automotive, biomedical applications
Silica nanocomposites	High thermal stability, low thermal conductivity	Insulating materials, coatings, drug delivery
Graphene nanocomposites	Excellent mechanical strength, electrical conductivity	Energy storage, flexible electronics, sensors
Biopolymer nanocomposites	Biocompatibility, biodegradability	Medical devices, drug delivery systems

tissue regeneration need to degrade in a controlled manner [11]. Since the formation of degradation products and their subsequent release from the material can be toxic to surrounding tissues, mechanisms for degradation must be understood, showing a need again for an appropriate toxicological evaluation [12]. Especially for polymers and composites, the connection between degradation and toxicity is highly relevant as their properties and safety are directly influenced by environmental changes in the form of light, heat, or biological agents [13]. Hence, thorough degradation studies including toxicological implications are mandatory to advance material science and provide evidence of the safety and efficacy of biomedical devices [11, 14].

1.3 Scope and Objectives of the Chapter

The degradation and toxicological behavior of polymeric nanocomposites are critically important to estimate their environmental consequences and safety from the point of view, particularly for growing applications. While the mechanical and thermal properties of these nanocomposites have been drastically improved, it remains a concern that with additives such as clay, bio-consumption processes can be obstructed due to concerns over biodegradability [5, 15]. Altering the pH and crystallinity of materials additionally causes modifications within the balance between conductivity and mechanical stability, as well as ultimately changes the elution price of degradation by gold composite. Moreover, the role of hydrophilic-hydrophobic components in hydrogen bonding interactions in cellulose-fiber-reinforced composites influences mechanical performance and moisture durability, thus comprehensive study is required for the degradation phenomena in different environments [2, 16]. This underscores the importance of such studies to gauge the merits of performance against potential environmental and health risks associated with nanocomposites.

2 Degradation Mechanisms of Polymeric Nanocomposites

2.1 Photodegradation

The origin of the photon-induced degradation mechanisms in polymeric nanocomposites lies in the interaction of light with materials, which causes new species to occur and, in turn, develops the breakdown of polymer matrix generation [17, 18]. For instance, in PS-ONC-ZnO nanocomposites, ultraviolet irradiation of ZnO causes the generation of electron–hole pairs and promotes polymer breakdown. Including nanoparticles with the same kind of function as TiO_2 and carbon nanotubes (CNTs), one pushes this potency by becoming more open to radical reactions which contribute to its photocatalytic efficiency. Figure 1 shows adsorption and photocatalytic degradation of MB over (MO-PMMA) nanocomposite photocatalysts [19]. However, the interaction between the two—nanoparticles and polymers—is central if one wants to achieve high rates of photodegradation for organic dyes and materials as well as achieving long lifetimes for both [17, 18].

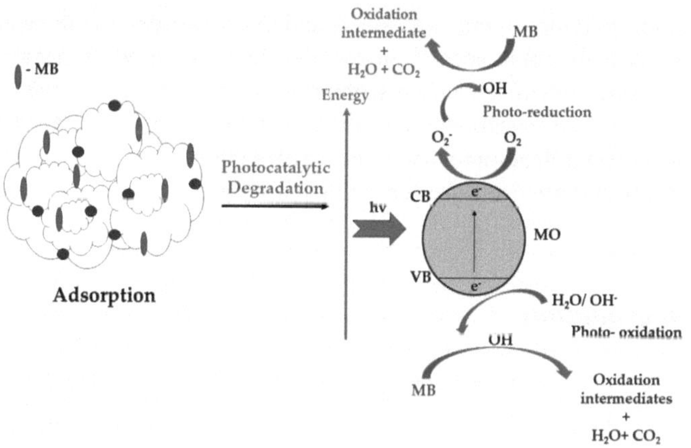

Fig. 1 Adsorption and photocatalytic degradation of MB over (MO-PMMA) nanocomposite photocatalysts

2.2 Thermal Degradation

The choice of filler plays a key role in influencing the thermal hinges of polymer nanocomposites. Early decomposition of polyacrylate/ZnO nanocomposites is attributed to the hydroxyl groups on ZnO which catalyze polymer chain dehydrogenation, thereby affecting the degradation mechanism in turn at higher temperatures [20, 21]. However, in poly(methyl methacrylate) nanocomposites with clays or carbon nanotubes the temperature of decomposition follows a predictable pattern, as the system's resistance rises, so too does its thermal stability [22–24]. At higher temperatures, Polyvinyl alcohol (PVA)/SiO_2 nanocomposites also show higher thermal resistance due to the formation of ribs that are more stable in this context. In poly(e-caprolactone) (PCL) nanocomposites, the presence of nanoparticles will either speed up degradation or slow it down depending on their chemical interaction with the polymer matrix [23, 25].

2.3 Hydrolytic Degradation

Hydrolytic degradation in the case of polymeric nanocomposites represents one of the stages of a multi-stage composite degradation process and the method's influence on them strongly depends on what type and amount of nanoparticles introduced inside the polymer matrix [26, 27]. This material degradation mechanism may have implications in a variety of fields, including but not limited to biomedicine, environmental engineering, or other contexts where careful control over material integrity is needed for both performance and durability [28]. Other than some trivial cases, the nanoparticle-polymer interaction was found to be instrumental in promoting the hydrolytic process by either accelerating or decelerating based on parameters such for instance surface chemistry and distribution of nanoparticles in the matrix [27, 29, 30]. Hence, it is an essential core of the integration of nanocomposite materials development and optimization for a wide range of technological applications in which material stability and life cycle are very critical toward functional requirements [29].

2.4 Biodegradation

In this study, the degradation mechanisms of polylactic acid (PLA) and polycaprolactone (PCL)-based polymeric nanocomposites were established at a molecular level showing that hydrolytic scission of ester bonds as the key degradation mediators is catalyzed by silicate layer with hydroxyl groups enabling an enhanced hydrolysis effect on PLA matrix [31, 32]. The above results in faster biodegradation in the composting environment than pure PLA. Moreover, an excellent dispersion of the clays enhances both mechanical properties and microbial degradation by

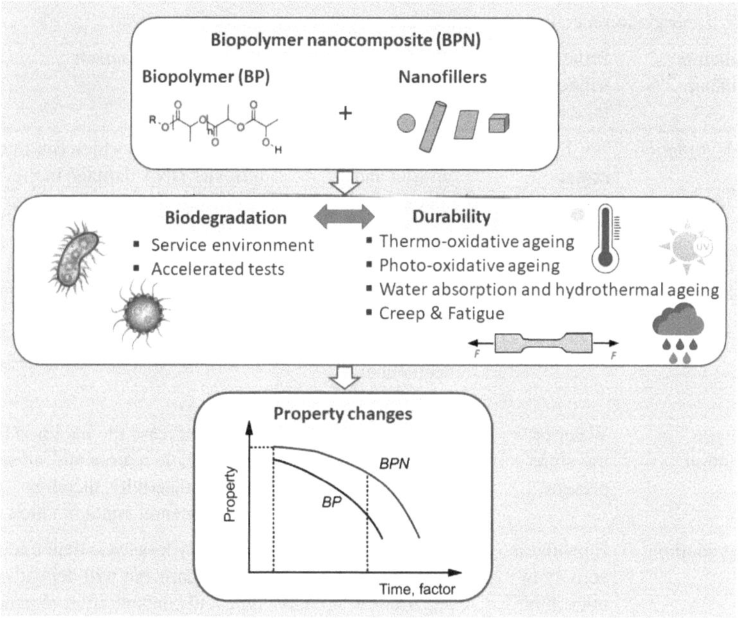

Fig. 2 Durability of biodegradable polymer nanocomposites

increasing microorganism access to polymer chains [33]. Figure 2 shows the durability of biodegradable polymer nanocomposites [34]. The rate of degradation is likely different between in vitro and in vivo conditions; however, with enzymatic activity increasing the rate of scaffold breakdown in biological settings. Therefore, the interaction of polymer with the nanofiller is essential for proper biodegradation efficiency [35–37].

2.5 *Influence of Nanoparticles on Degradation Mechanisms*

The role of nanoparticles in polymeric biocomposites like polycaprolactone (PCL) significantly influences the degradation behavior of biocomposites. Different from the case of ZnO, the degradation process of phthalate ester in TiO_2 is a multi step manner and its agglomerated NP form needs many more diffusion steps [38, 39]. Graphene oxide (GO) and reduced graphene oxide (rGO), dispersed in appropriate processing solvent or melt, can either accelerate or retard the polymer degradation depending on where they are initially located, i.e., interspace between layers of a multi-layered structure, combining good mechanical properties through sufficient interfacial interaction with the polymer matrix and hydrolytic stability. It is very useful to study the viscoelastic behavior of a material if we want to study the hydrolytic degradation process because one of the mechanisms of action of ZnO

Table 2 Degradation mechanisms with effect on polymeric nanocomposites

Degradation mechanism	Principal triggering factors	Effects on material	Toxicological impacts
Photodegradation	UV light radiation, illumination	Terminates chain scission and enhances surface roughness	Generates ROS which can mean ultimate DNA damage in the organisms that are exposed to the degradation products
Thermal degradation	High temperatures	Oxidation degradation occurs, brittleness, and instability in mechanical properties	Harsh byproducts are emitted into the surroundings, lower the quality of air, and pose health risks
Hydrolytic degradation	Water or moisture present	Polymer matrix degradation leads to a degraded structure	It may increase the likelihood of nanoparticle release and subsequent microbial activity, thereby environmental contamination
Biodegradation	Enzymatic activity or microbial interaction	Controlled, aqueous environment degradation, such as in compost	Generally less toxic than traditional counterparts but will depend on the type and concentration of nanofillers involved; will be more eco-friendly if carefully designed
Impact of nanoparticle	Type, morphology, and distribution of nanoparticles in the matrix	Alter mechanical strength and thermal resistance of materials	Nanoparticles increase levels of ROS, which leads to cytotoxic effects; their release into the environment is a secondary ecological risk

in PLA/PCL/ZnO nanocomposites, for example, is as a catalyst for degradation. Accordingly, whereas nanoparticles are commonly employed to enhance materials properties, their effects on degradation should be taken into account in the material design [40, 41]. Table 2 briefly discusses the degradation mechanisms with effect on polymeric nanocomposites [40–42].

3 Factors Affecting Degradation of Polymeric Nanocomposites

3.1 Polymer Matrix

The degradation of polymer nanocomposites is affected by polymer matrix through several mechanisms related to chemical structure, thermal stability, and interaction with nanofillers.

3.1.1 Chemical Structure

The chemical properties of a polymer matrix such as polarity, crosslink density, and functional groups determine its susceptibility to degradation mechanisms [8, 43]. Polar polymers that consist of reactive functional groups like amines possess the ability to absorb UV light and undergo photochemical reactions leading to chain scission and loss of mass when exposed to UV radiation [44]. For example, amine-cured epoxy matrices exhibit rapid photodegradation under UV exposure. The polarity of the polymer matrix influences its thermal stability as well. At high temperatures, oxidative degradation of polar polymers takes place that contain reactive groups leading to a loss of mechanical properties. Polymers with high crosslink density (number of chemical bonds between polymer chains in a network) are more brittle [45]. Since the molecules are tightly bound, there is reduced movement of the polymer chains. Functional groups such as ester influences the degradation of the PNCs. Polymers containing ester groups, such as polyesters, are specifically prone to hydrolysis. Under hydrolytic conditions, water molecules can attack the ester bonds, leading to chain scission and the breakdown of the polymer structure [46]. This results in a loss of mechanical properties since the polymer becomes weaker and more brittle. Studies have revealed that the rate of hydrolysis is affected by the molecular weight of the oligomers and the presence of neighboring functional groups. In poly (butylene tartrate) oligomers, the hydrolysis rate was found to follow pseudo-first-order kinetics, indicating that the degradation process can be predicted under certain conditions [16, 47].

3.1.2 Nanofillers

In the case of polyesters, the presence of nanofillers influences the hydrolytic degradation behavior. Nanofillers affect the autocatalytic action of carboxyl groups within the polyester matrix, which can accelerate degradation processes [48, 49]. The hydrophilicity and crystallinity of the polymer matrix, influenced by the type and amount of nanofillers, also play a role in the rate of hydrolytic degradation [34, 50].

3.2 Nanoparticle Type and Concentration

The type and concentration of nanoparticles present in PNCs affect their degradation to a larger extent. The presence of different types of nanoparticles such as metal oxides, quantum dots, fullerenes, nanosheets, nanotubes, etc. affects the PNCs differently [9, 51]. The presence of ZnO nanoparticles enhances the oxidative polymer degradation, rate of photodegradation that leads to a quicker breakdown of polymer matrix when exposed to UV light. This affects the mechanical properties and durability of the PNCs [52]. ZnO possesses photocatalytic properties due to which it generates reactive oxygen species such as hydroxyl radical and superoxide anions. These species attack the polymer chain leading to chain scission and degradation of polymer structure. The presence of nano-clays another nanoparticle in PNC contributes to the oxidative degradation in polymers under UV exposure [53]. The presence of nano-clays facilitates the generation of reactive oxygen species (ROS), which further promotes oxidative degradation. This process leads to chain scission and the breakdown of the polymer structure. Research has shown that nano-clays used in low-density polyethylene (LDPE) composites accelerate the photooxidation degradation [9]. For instance, studies indicate that as the exposure to UV light increases, the structural and physical properties of the LDPE-nano-clay composites degrade, leading to increased surface roughness and alteration in chemical properties [54]. Carbon Nano Tubes (CNTs) under higher loading lead to complex interactions increasing the brittleness and instability of the composite [10, 55].

Morphology of Nanofillers

The morphology of multi-walled carbon nanotubes (MWCNTs), a nanofiller, affects the degradation of the PNCs in a way that when these are exposed to UV radiation for a longer duration, there is a tendency of the MWCNTs to form an aggregate and form an entangled dense mass at the surface of the degraded polymer matrix [6, 56]. The entangled mass formed leads to mechanical failure of the composite by making it brittle and reducing its flexibility [1, 5, 10]. On prolonged exposure to UV radiation, silica nanoparticles formed a uniform layer over the PNC composite unlike MWCNTs [57]. Nanosilica used in PNCs can compromise the mechanical stability of these composites where making the composites more brittle and less flexible. Thus, their application in areas where toughness is the criterion is reduced [58, 59].

3.3 *Environmental Conditions*

Environmental factors such as ultraviolet (UV) radiation, temperature, humidity, and moisture significantly affect the degradation of polymer nanocomposites (PNCs) through various mechanisms.

3.3.1 UV Radiation

This forms the factor of polymer degradation. Several degradation processes get initiated that include chain scission, photooxidation, and crosslinking [60]. These processes lead to the breakdown of the polymer matrix, resulting in a loss of mechanical properties and structural integrity. Studies have shown that UV exposure can cause significant changes in the physical and chemical properties of PNCs, by increasing the surface roughness and decreasing strength due to the formation of photoproducts [53]. Photoproducts generally include carbonyl compounds, hydroperoxides, vinyl groups, and oligomers. Carbonyl compounds are formed by the oxidation of polymer chains and cleavage of functional groups [61]. This leads to increased brittleness and loss of elasticity of the composite [62]. Hydroperoxides, on the other hand, are produced from the reaction of free radicals with molecular oxygen. The hydroperoxides further decompose to form other reactive species leading to additional chain scission and degradation of the matrix. Oligomers that affect the thermal and mechanical properties are generally formed when the polymer chain breaks down forming smaller oligomeric fragments [61, 63].

3.3.2 Temperature

Elevated temperatures increase the kinetic energy of molecules, which accelerates chemical reactions. Therefore, in the case of PNCs, the rate of photodegradation reactions initiated by UV exposure is enhanced significantly [64, 65]. A faster breakdown of the polymer matrix occurs, and eventually, a more rapid loss of mechanical properties can be observed. Research indicates that environmental aging, particularly at elevated temperatures, results in an increased degree of crystallinity in PP (polypropylene) -MWCNT composites. One of the reasons is the role of MWCNTs as nucleation points, which promotes the crystallization of the polymer [66]. The increased crystallinity leads to brittleness in the material. The increased crystallinity yields a more rigid structure that fails to accommodate stress in the desired way thus leading to crack formation under mechanical load. The whole process of degradation results in surface cracks and a brittle surface layer [67]. As the polymer matrix degrades, the structural integrity of the material diminishes, leading to the formation of micro-cracks. These cracks can propagate under stress, further compromising the mechanical performance of the composite. The presence of MWCNTs may help reduce the extent of cracking to some degree, but the overall brittleness of the aged material raises an alarm [1, 53].

3.3.3 Humidity and Moisture

The presence of moisture interacts with UV radiation to accelerate the degradation of polymer nanocomposites (PNCs) by forming reactive oxygen species (ROS), which further promotes oxidative degradation of the polymer matrix [68, 69]. This process makes the embedded nanofillers, such as carbon nanotubes, more susceptible to release into the environment [70, 71]. When PNCs are exposed to UV radiation, in the presence of moisture, the energy from UV light can initiate the formation of ROS, such as hydroxyl radicals (·OH) and superoxide anions ($O_2^{\cdot -}$). These highly reactive species can be generated through various mechanisms:

Photolysis of water: UV radiation can split water molecules, leading to the formation of hydroxyl radicals:

$$H_2O + hv \rightarrow \cdot OH + H^{\cdot}$$

Oxygen activation: Molecular oxygen can be activated by UV light or through reactions with other radicals, forming superoxide anions [72]:

$$O_2 + e \rightarrow O_2^{\cdot}$$

Hydrogen peroxide decomposition: When transition metal ions are present, hydrogen peroxide, which may be produced during the degradation process, can break down into hydroxyl radicals [73]:

$$H_2O_2 + Fe^{2+} \rightarrow \cdot OH + OH^- + Fe^{3+}$$

Oxidative Degradation and Nanofillers Release

These reactive species attack the polymer chains, leading to chain scission, crosslinking, and the formation of various photoproducts. As the polymer degrades, the embedded nanofillers become more exposed and available for release into the environment [74]. Research has shown that under humid UV exposure conditions, the degradation of PNCs results in a greater loss of polymer photoproducts compared to dry UV exposure. This increased loss indicates that the combination of UV radiation and moisture enhances the degradation process, making the nanofillers more accessible for potential release [44, 75].

3.4 Additives and Stabilizers

Polymeric Nanocomposites are incorporated with a wide variety of additives and stabilizers to enhance their properties in terms of mechanical stability, thermal stability, and overall stability [76]. However, these impose some negative effects on PNCs such that they lose their stability, and the degradation rate is enhanced. Some of the stabilizers are discussed below:

3.4.1 Antioxidants

Although antioxidants help in maintaining the mechanical properties of PNC sometimes, they may interfere with the nanofillers which leads to poor dispersion of the nanofillers and the formation of aggregates which enhances the degradation process [77]. Common types include hindered phenols, phosphites, and thioesters, which can be classified into primary and secondary antioxidants [34, 44, 78]. Certain antioxidants can interfere with the stabilization mechanisms provided by the nanofillers. Some antioxidants may react with the functional groups on the surface of nanofillers, reducing their effectiveness in stabilizing the polymer matrix [79]. One such example is hindered phenol antioxidants often used in PNC composites where the phenolic hydroxyl groups react with functional groups on nanofillers, such as hydroxyl (–OH) or carboxyl (–COOH) groups. This reaction could alter the surface chemistry of the nanofillers and reduce their ability to effectively stabilize the polymer matrix against degradation [80].

3.4.2 UV Stabilizers

While UV stabilizers are designed to absorb UV radiation and convert it into harmless heat, their productivity diminishes when used for a long duration. Some stabilizers may degrade themselves when exposed to UV light, leading to a reduction in their effectiveness. This degradation results in the polymer becoming more susceptible to UV-induced damage as the stabilizers lose their effectiveness [81]. The presence of UV stabilizers can sometimes interfere with the functionality of other additives in the polymer matrix. For instance, they may react with antioxidants or plasticizers, leading to the formation of degradation products or reducing the overall effectiveness of the additive [82]. When hindered phenol antioxidants react with UV absorbers, they may form radical species that can initiate further degradation of the polymer matrix. The properties of plasticizers get altered in a way that their solubility or mobility within the polymer matrix is reduced. This results in reduced flexibility and increased brittleness of the PNC [53, 55, 83].

3.4.3 Plasticizers

These are incorporated to enhance the flexibility and workability of PNCs by reducing the glass transition temperature. However, they can contribute to degradation which occurs when there is a loss in plasticizer due to diffusion to the surface and subsequent evaporation. This results in poorer mechanical properties and contamination of the surrounding environment. The polymer matrix may become more brittle as the plasticizer content decreases, leading to increased susceptibility to cracking and failure under stress [1, 47, 84]. The presence of plasticizers can weaken the interfacial adhesion between the polymer and the nanofillers, potentially leading to delamination and the structural integrity is highly affected [34]. In some cases, plasticizers can increase the water uptake of the polymer, leading to hydrolytic degradation, particularly in moisture-rich environments. The hydrolytic degradation leads to a reduction in tensile strength and elongation at break, making the polymer more susceptible to failure [15, 85].

4 Characterization Techniques for Degradation

4.1 Spectroscopy (FTIR, UV–Vis, Raman)

Material degradation must be monitored to ensure its long service life and optimal performance, and spectroscopy techniques are invaluable for keeping track of it. Raman Spectroscopy: When thermal degradation and laser-induced algorithms are realized, we have to pay more attention to the process of structural change through oxidation and bonding during degradation. It also detects initial degradation on exposure to UV in epoxy coatings utilizing alteration in Raman peak intensities [86]. In polyethylene and polypropylene, for instance, fluorescence spectroscopy can effectively track polymer degradation using the Nile red fluorophore as a fluorescent probe to monitor changes in fluorescence spectra associated with chemical transformations. The methodology is also applied to aliphatic polyesters, for which both intrinsic fluorescence and probe labeling are used to monitor thermal and photothermal degradation. While these techniques are highly effective, their applicability can vary with material type and degradation conditions. Combining multiple techniques may improve the accuracy and reliability of degradation assessments, enhancing material management and recycling strategies [87, 88].

4.2 Electron Microscopy (SEM, TEM)

High-resolution imaging studies using scanning electron microscopy (SEM) and transmission electron microscopy (TEM), are crucial to study microstructural degradation including fragmentation and/or surface decomposition in electrodes of lithium-ion batteries upon cycling [89]. The most desirable application of these techniques is to better understand mechanical properties and failure mechanisms, leading to optimal material performance [86]. This insight is furthered by the combination with electron microscopy, which enables observation of the materials at an atomic level, and therefore accompanying X-ray diffraction results are interpreted to material behavior. Yet, the concomitant beam-induced damage in delicate materials has to be taken into account as this may interfere with degradation assessments [90]. As a result, we are advised of their use with some kind of characterization in parallel to ensure accurate and complete monitoring [89].

4.3 Thermogravimetric Analysis (TGA)

Thermogravimetric Analysis (TGA) is an indispensable tool for evaluating material degradation, especially in polymers and composite materials, by monitoring weight loss concerning the temperature to comprehend thermal stability, degradation pathways, etc. [91]. This process is critical for polymers like poly(vinyl alcohol) to understand degradation temperatures and weight loss profiles as a function of heating rate. More advanced TGA methods for significantly improved sensitivity and resolution can be realized by the use of MEMS thermogravimetric analyzers (MEMS TGAs) that enable analyses to be performed with only a nanogram of material enabling further coupling with complementary techniques such as Raman spectroscopy. Additionally, a combination of techniques (e.g., TG-MS and TG-FTIR) provides qualitative and quantitative data on the volatile degradation products as well which can lead to a better understanding of complex materials [92]. Additionally, useful information about structural variations can be found in non-destructive methods such as ultrasonic techniques while degradation occurs [93]. However, TGA, compelling as it is, can be coupled with degradation processes in a complex environment where even different degradation reactions may occasionally exert an impact on each other [94].

4.4 Gel Permeation Chromatography (GPC)

Understanding the degradation of materials and systems is crucial to enhancing their lifetime and sustainability with numerous characterization techniques providing complementary approaches. Gel Permeation Chromatography (GPC) end-group derivatization and UV detection are very useful for determining the extent of polymer degradation and can be used to track oxidative changes in synthetic polymers such as polyisoprene or polybutadiene. GPC allows degradation across a wide molecular weight range to be monitored, which gives important information about the behavior of a polymer under different conditions [95]. Figure 3 discusses categories of the characterization techniques used to study G-NP-based hybrid composites [96]. NVIDIA scientist gives precise molecular weight distribution, which in a tree-addressed manner is vital for degradation mechanisms and polymer chemistry sustainable upgrade. To complement this analysis, fluorescence spectroscopy is shown as a powerful tool for detecting the early-stage degradation of plastics and adhesives, giving non-destructive information on their surface changes through the same period, whereas mass spectrometry (MS) is proven to be essential in imaging of biomaterials since it can track both concentrations levels (trace amounts) and also reaction mechanisms [97]. While GPC is a strong technique, its power lies in using it together with fluorescence spectroscopy and mass spectrometry, ensuring a comprehensive understanding of material degradation processes.

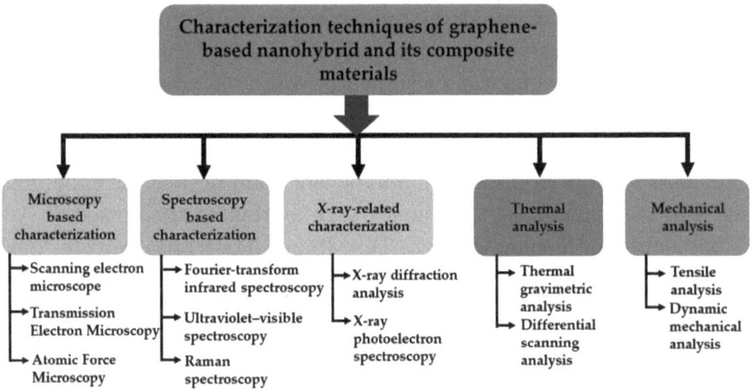

Fig. 3 Categories of the characterization techniques used to study G-NP-based hybrid composites

5 Toxicological Impact of Degradation Product

5.1 Release of Nanoparticles

Nanoparticles released during the degradation of PNCs cover a wide range of toxicological effects both on living beings as well as the environment. These nanoparticles generate reactive oxygen species that cause oxidative stress in cells resulting in DNA damage, apoptosis, and ultimately collapse of the cell. Besides cell damage, they have the potential to affect the immune system as well by triggering inflammatory reactions [98]. They react with the proteins and biomolecules leading to potential toxicity. Organs such as kidneys, liver, and lungs are affected by nanoparticles when these accumulate on the organs. For example, silica nanoparticles induce hepatotoxicity and nephrotoxicity [99].

5.2 Cytotoxicity Studies

Cytotoxicity refers to the quality of being toxic to cells. It is the degree to which a substance or process can cause damage or death to cells. Substances or agents that exhibit cytotoxicity are called cytotoxic compounds. Cytotoxicity can be caused due to oxidative stress, disruption of cellular processes, and interaction with biomolecules.

The degree of cytotoxicity depends on factors like the concentration, duration of exposure, and physicochemical properties of the cytotoxic agent. When polyethylene particles degrade, the cytotoxicity increases with an increase in the degree of degradation. This indicates that as the polymers break down, the degradation products become more toxic to biological systems. The cytotoxicity is not due to changes in particle size or morphology but due to modifications in the chemical state of the surface functional groups of the degraded particles [16]. Nanoparticles, such as CNTs present in polymer matrices, can impact the cytotoxicity of PNCs. The presence of CNTs has proved to negatively affect cell proliferation at higher concentrations [44, 53]. CNTs have the potential to damage cells through necrosis or apoptosis by penetrating the cell membrane. These also cause inflammatory responses in tissues which leads to chronic inflammation. If CNTs like MWCNTs are inhaled, then they exert carcinogenic effects on human health and cause cancerous changes in the lung cells.

5.3 Genotoxicity Studies

This primarily revolves around the effect of nanoparticles released from the degradation of PNCs on the genetic aspect of the cell. The term genotoxicity refers to the ability of some chemical substances to destroy the genetic information within a cell thus resulting in mutations, cancer, or any other genetic disorders [100]. Keeping in view the degradation of PNCs, the release of nanoparticles poses a threat to the well-being of the genetic information of a cell. The degradation of PNCs under environmental factors mainly UV exposure, heat, or any biological activity, releases smaller fragments of nanoparticles. These nanoparticles react with the cellular components leading to DNA damage, and eventually aberrations of chromosomes. ROS generated during this degradation process further damages proteins, lipids, and DNA [101].

5.4 Ecotoxicity and Environmental Persistence

The ecotoxicity of polymer nanocomposites depends on their composition, size, and surface characteristics [102]. According to studies, these materials can exhibit toxic effects on various organisms, including bacteria, algae, invertebrates, and fish. Limited data exist on their long-term effects on higher organisms, emphasizing the need for more comprehensive studies. Once released into the environment, polymer nanocomposites undergo transformations that can affect their bioavailability and toxicity. Factors such as transport, transfer, and transformation mechanisms play important roles in determining their ecological impact [53]. The environmental persistence of polymer nanocomposites is influenced by their chemical structure and the presence of additives. Some nanocomposites may persist in the environment, leading to potential accumulation and long-term ecological effects. In marine applications, polymer nanocomposites can contribute to biofouling, where marine organisms attach to surfaces, potentially disrupting local ecosystems [103].

5.5 *In Vivo versus in Vitro Assessment*

In vivo Toxicological Assessment

In vivo assessments involve testing on living organisms, such as rodents or other animal models [11, 104]. This approach provides a more holistic view of the biological effects of polymer nanoparticles:

- Complex Interactions: In vivo studies account for the systemic effects of nanoparticles, including pharmacokinetics, biodistribution, and the influence of the immune system, which can significantly affect toxicity outcomes.
- Realistic Conditions: These assessments mimic real-life exposure scenarios, providing insights into how nanoparticles behave in a complete biological environment, including their metabolism and excretion pathways.

Long-Term Effects: In vivo studies are crucial for evaluating chronic toxicity and potential long-term health effects, which are often not detectable in short-term in vitro studies [54, 85, 104].

In vitro Toxicological Assessment

In vitro studies involve testing on cell cultures or isolated tissues outside of a living organism [49, 99]. This method is often employed in the early stages of toxicity testing for several reasons:

- Controlled Environment: In vitro assessments provide a controlled environment where variables can be manipulated, allowing researchers to isolate specific effects of the polymer nanoparticles on cellular systems.
- High Throughput: These tests can be conducted rapidly and allow for the screening of multiple formulations or concentrations simultaneously, making them efficient for preliminary toxicity evaluations.
- Mechanistic Insights: In vitro studies can elucidate the mechanisms of toxicity at the cellular level, such as cytotoxicity, oxidative stress, and genotoxicity, which can be crucial for understanding how polymer NPs interact with biological systems.

However, in vitro methods may not fully replicate the complex interactions that occur within a living organism, potentially leading to an incomplete understanding of the nanoparticles' effects [99, 104].

6 Case Studies

6.1 *Degradation and Toxicology of Carbon Nanotube Composites*

The use of carbon nanotube (CNT) composites has become highly prevalent, and as such their degradation and toxicological profile are of prime interest. We provide evidence that functionalized CNTs and especially the corresponding degraded material are more cytotoxic than their pristine form, as demonstrated by a decrease in cell viability in lung and gastrointestinal models. The existence of structural defects in CNTs imparts oxidative stress and enhances the generation of reactive oxygen species (ROS) leading to cellular destruction. This unique resistance to natural degradation presents several serious environmental risks and will make chemical and biological methods for degrading this material only partially effective in the real world [98]. In addition, in vivo studies have demonstrated that local inflammatory responses are induced by exposure to CNT composites in whole organisms, which may lead to adverse health effects in the long term. Although these findings raise major concerns, scientists continue to investigate which pathways CNTs are degraded by and how to minimize associated risks [99].

6.2 *Degradation and Toxicology of Metal Oxide Nanocomposites*

In addition, metal oxide nanocomposites are receiving great attention in environmental remediation due to their degradation and toxicological potential. This type of nanocomposites has largely proven to be effective for the degradation of harmful pollutants. For example, iron and copper-doped clay nanocomposites have shown 99% nitrobenzene degradation under optimal conditions [63]. Similarly, Ag/ZnO nanocomposites biosynthesized showed 90.4% degradation of methylene blue which is higher photocatalytic activity than ZnO alone [58]. Iron-oxide chitosan nanocomposites are also shown to be effective in the degradation of sunlight-driven carcinogenic polycyclic aromatic hydrocarbons (PAHs), demonstrating promise for the treatment of hazardous wastes. Even though there have been some interesting outcomes, the issue of cytotoxicity is still unresolved, as evidenced by the concentration-dependent killer lung which is due to ZnO and Ag/ZnO on other cell lines like HeLa. They also caution that the consequences of byproducts from these degradation processes on the environment have to be thoroughly examined, as some techniques could produce toxic residues. Though metal oxide nanocomposites show the potential for the degradation of pollutants, still they are important to be evaluated for their toxicological consequence to ensure safe applications [79].

6.3 Degradation and Toxicology of Polymer–Clay Nanocomposites

Unsurprisingly, the environmental toxicology and degradation of polymer–clay nanocomposites are key research directions. A large amount of these materials created from non-renewable resources has led to such regrettable results: everything from an increase in heavy metal pollution to algae surpluses where none existed before in bodies of water. Polymer–clay nanocomposites can enhance material performance, yet they can also create risks during their decomposition process. With the type and proportion of clay determined, thermal stability fluctuates. From the addition of organic montmorillonite into the polysiloxane network, for example, thermal stability is decreased tempering factors cause changes in degradation mechanisms here as well [39]. These nanocomposites receive also an impact from UV light, resulting in increased mechanical fragility and changes in optical properties. But split can mitigate this effect for the most part (by stopping the creation of carbonyl bonds in the first place). From a toxicological point of view, the degradation process may release nanoparticles into the environment [27, 30]. For example, clay nanoparticles introduced by accelerated weathering have appeared of negligible biological toxicity in food packaging applications and tablewater treatment. In contrast, some nanocomposites such as SnO_2/bentonite have strong photocatalytic properties for degrading harmful organic wastes. This indicates great potential in the remediation field. Figure 4 shows different types of nanocomposites in pollutant degradation [105]. Without renewed research into the life cycle of polymer–clay nanocomposites, however, their value in applications will debut a flash illuminated by a brief spark [88].

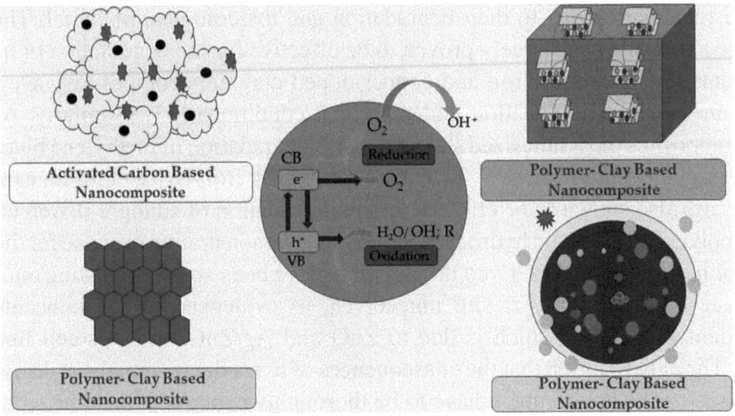

Fig. 4 Types of nanocomposites in pollutant degradation

7 Regulatory and Safety Considerations

7.1 Current Regulations and Guidelines

In the European Union, REACH (Registration, Evaluation, Authorisation and Restriction of Chemicals) requires manufacturers and importers to register substances, including nanomaterials, ensuring that they are assessed for safety and environmental impact. PNCs that release nanoparticles or degradation products must comply with these regulations. In the United States, the Food and Drug Administration (FDA) oversees the use of nanomaterials in food packaging. PNCs used in food contact materials must be evaluated for safety, including the potential migration of nanoparticles into food [80]. OSHA has established guidelines and recommendations to protect workers from potential hazards associated with exposure to nanomaterials, including those found in polymer nanocomposites (PNCs) [106]. It recommends that worker exposure to respirable carbon nanotubes and carbon nanofibers must not exceed 7.0 μg per cubic meter (μg/m^3) as an 8-h time-weighted average, based on the NIOSH proposed Recommended Exposure Limit (REL). For nanoscale particles of titanium dioxide (TiO_2), OSHA recommends not exceeding NIOSH's 0.3 mg per cubic meter (mg/m^3) REL, compared to 2.4 mg/m^3 for fine-sized TiO_2 particles greater than 100 nm [80].

7.2 Risk Assessment Approaches

- ***Quantitative risk assessment methods for PNCs involve***
  - (a) Calculating the probability and severity of potential risks, such as the release of engineered nanomaterials (ENMs) from PNCs into the environment.
  - (b) Estimating the exposure levels and potential adverse effects on organisms in soil and water.
  - (c) Conducting ecotoxicity studies on aquatic invertebrates, algae, and soil invertebrates to assess the toxicity of aged PNC samples.
- ***Qualitative Risk Assessment***
  - (a) Identifying potential hazards at all stages of the PNC life cycle, from production to disposal.
  - (b) Evaluating the characteristics of ENMs in PNCs that may differ from their macro-scale counterparts and lead to health risks [56].
  - (c) Assessing the increased risk posed by specific ENMs in PNCs can jeopardize overall safety.
- ***Risk Mitigation Strategies***

(a) Developing best practices for safe handling and use of nanoparticles in packaging industries.
(b) Implementing prevention and protection risk control measures for nanomaterials.
(c) Promoting biodegradable PNCs as an alternative to non-biodegradable plastic polymers [104].

7.3 *Future Directions in Regulations*

Regulatory bodies may establish stricter safety standards for the production and use of PNCs, focusing on the potential health and environmental impacts of nanomaterials. Comprehensive lifecycle assessments (LCA) for PNCs have to be regulated and looked into thoroughly, evaluating their environmental impact from production through disposal. There could be a push for mandatory labeling of products containing nanomaterials, ensuring consumers are informed about the presence of PNCs in consumer goods [107]. Development in the areas of standardized risk assessment protocols specifically tailored for nanomaterials in PNCs, including methodologies for testing toxicity and environmental effects can be undertaken. Efforts are to be made on regulations being followed across countries to facilitate international trade and ensure consistent safety standards for PNCs. Regulatory frameworks could include incentives for research into safer alternatives to existing nanomaterials and for the development of biodegradable PNCs [108].

8 Designing Safer Polymeric Nanocomposites

8.1 *Strategies for Controlled Degradation*

When synthesizing safer polymeric nanocomposites with controlled degradation, successful models both involve optimizing the composition and structure of the material. Adding nano-particles such as clay increases the material's thermal stability and barrier performance, and depends on them for regulating degradation rates in the face of various environmental conditions. This selection of biodegradable polymers, such as polylactic acid, promotes more predictable degradation processes because they can be subjected to chemical or enzymatic hydrolysis [13]. In addition, nanoparticle dispersion and its inherent properties directly affect the mechanical performance and degradation behavior of such composites; thus it is possible to alter their functionality using well-suited experiment on processing conditions. Consequently, the right choice of material and the best processing techniques can serve to ensure safety and performance while also achieving controlled degradation. Original sentences and it can be concluded should be chosen materials carefully [109].

8.2 Green Chemistry Approaches

The lecture on safer nanocomposite technology is concluded here. The deliberations will now be continued by exploring a truly fascinating topic—green chemistry. Particular attention should be given to new techniques that are expanding into the realm of plastic nanocomposites, where biodegradable materials are combined with more affordable alternatives derived from plant-based sources. Epoxidized soya oil has demonstrated this principle in the past by making use of an eco-friendly green process whereby natural oils are covalently bonded onto polymer matrices to form biodegradable nanocomposites that can still maintain favorable mechanical properties while reducing environmental impact [110]. As large industrial bioreactors have shown, this technique results in a decrease in environmental pollution while adding value to products such as paper towels and nappies again, overcoming the principal challenge posed by nano-sized inorganic particles simultaneous improvements on both mechanical and thermal properties of polymers! This means that high-performance materials can be produced using these readerships while being less damaging to both our health and that of the environment in which we live. Furthermore, by controlling the biodegradability of this new type of nanocomposite material, its manufacturers can do much toward implementing principles of green chemistry. Perhaps it is a little encouraging to find that such an approach is associated with the idea of safer nanocomposite plastics themselves being developed by mankind [111].

8.3 Enhancing Biodegradability

One key strategy to improve the biodegradability of polymeric nanocomposites is the incorporation of biodegradable polymers and natural fibers. Offering an environmental alternative, this method also tackles both the mechanical properties and barrier limits faced by traditional biopolymer materials, leading to major technological advances [112]. For example, nano-polymers of silicic acid layered on the surface of polylactide when compared to conventional polylactide nanocomposites have higher rates of biodegradation in composting environments and less degradation time costs particularly because silicate fillers induce hydrolysis that shortens the fibrillation profile in materials. Finally, there has also emerged in recent years a green nanocomposite material which when incorporated with natural oils is very effective in its biodegradable evaluation [113]. Tests show that this substance can be biodegraded over time and includes apparent mineralization for both oxidative and reductive environments [114].

In addition to achieving higher mechanical properties with the addition of nano-sized bio-based reinforcements, the wide application of these green nanocomposites is further expanded. This includes their potential use by medical applications.

8.4 Reducing Toxicological Impact

The use of green nanoparticles in formulating polymeric nanocomposites will improve the design of safer polymeric nanocomposites. These nanoparticles not only enhance mechanical properties but also aim to reduce the toxicology associated with traditional synthetic materials. Even though natural polymers stand out for their biocompatibility and biodegradability, they however come with weaknesses that hinder their wide application such as poor mechanical properties and susceptibility to microbial contamination [115]. Nanoparticles tend to make these characteristics better without any harmful actions. The choice of nanoparticle types, and how they are processed play vital roles in dispersion and orientation within the polymer matrix and hence on the properties (and safety) of resultant composites. The development of eco-nanocomposites has also been driven by notable improvements in barrier properties, decreasing the release of harmful compounds. Thus, prioritizing environmentally friendly materials and processing techniques is vital for developing safer and more sustainable polymeric nanocomposites [116].

9 Future Research Direction

9.1 Emerging Trends in Degradation Studies

Extensive research is being done on the mechanisms by which PNCs degrade in natural environments. Studies have revealed that synthetic polymers do not favor microbial degradation and this accumulation takes place affecting the environment. Therefore, studies are being conducted to identify how temperature, pressure, and microbial activity influence the degradation rates of bio-based PNCs. This includes exploring microbial and enzymatic pathways that could facilitate the breakdown of these materials [55, 104]. Bio-based PNCs have started gaining importance over traditional synthetic ones. Materials derived from renewable resources like cellulose and lignin are being used in the manufacturing processes of PNCs which can improve the mechanical properties and biodegradability. Research is focusing on optimizing the composition and processing methods to enhance the performance and degradation rates of these materials [83]. Different characterization techniques are being used such as scanning electron microscopy (SEM), differential scanning calorimetry (DSC), and thermogravimetric analysis (TGA) to analyze the structural changes in PNCs during degradation and observe the chemical and physical transformations occurring in the PNC during degradation [55]. Researchers are exploring the use of natural additives, such as enzymes or biodegradable surfactants, that can enhance the degradation rates of PNCs. These additives can facilitate microbial colonization and enzymatic breakdown, leading to more environmentally friendly end-of-life scenarios for PNCs [53, 55, 83].

9.2 Novel Toxicological Assessment Techniques

The methods of in vitro assays and silico modeling for the assessment of toxicity without relying on animal testing. In vitro methods can provide insights into cellular responses to PNCs, while in silico models can predict potential toxicity based on chemical structure and biological interactions [85, 117, 118]. Advancements in screening technologies enable the evaluation of multiple compounds simultaneously at one go, including PNCs, for toxicological effects. This method accelerates the identification of toxic effects and helps prioritize substances for further testing [104]. The application of multi-omics approaches (genomics, proteomics, metabolomics) is becoming prominent in toxicology. These techniques provide comprehensive insights into biological responses to PNCs at various levels, facilitating a better understanding of the mechanisms of toxicity and potential health risks [44, 78, 104]. Utilizing stem cell-derived models, such as human cord blood stem cells, allows for more relevant assessments of neurotoxicity and other complex endpoints. These models can mimic human biological systems more accurately than traditional models [80].

9.3 Integration of Environmental and Health Considerations in Material Design

The adverse effect of nanoparticles on health and the environment due to disintegration from the PNCs during degradation can be controlled or stopped if bio-based or biodegradable PNCs are manufactured using renewable resources [8, 116, 119]

(i) ***The first step involves selecting a bio-based feedstock that can be obtained from the following sources***:

   (a) Plant-based Polymers: Starch, cellulose, and natural rubber.
   (b) Microbial Polyesters: Polyhydroxyalkanoates (PHAs) and polylactic acid (PLA), which are produced through microbial fermentation.
   (c) Fungal Polymers: Chitin and pullulan derived from fungi.

(ii) ***Next step is the selection of nanofillers to further increase the properties of the composites. The sources of nanofillers are as follows***:

   (a) Cellulose Nanocrystals (CNCs): Derived from plant cellulose.
   (b) Chitin nanofibers: Obtained from crustacean shells or fungal mycelium.
   (c) Clay nanoparticles: Includes montmorillonite that improves the mechanical properties.

(iii) ***Preparation of matrix***

   (a) Thermal Processing: Heating the biopolymer until it melts, allowing for easy incorporation of fillers.

(b) Solvent Casting: Dissolving the biopolymer in an appropriate solvent, then casting it into a mold.

(iv) ***Incorporation of nanofillers***

(a) Mechanical Mixing: Using high-shear mixers or extruders to ensure uniform dispersion of the nanofillers.
(b) Ultrasonication: Applying ultrasonic waves to break down agglomerates and enhance dispersion.

(v) ***Molding and shaping***

(a) Injection Molding: The composite is injected into molds to create specific shapes.
(b) Extrusion: The mixture is forced through a die to produce continuous shapes like films or sheets.
(c) Thermoforming: Heating the composite and forming it into shapes using molds.

(vi) ***Curing and Finishing***

(a) Curing: Depending on the type of biopolymer used, curing may be necessary to enhance the mechanical properties.
(b) Crosslinking Agents: Adding agents that promote crosslinking reactions to improve strength and stability.
(c) Heat Treatment: Applying heat to further solidify the material.
(d) Surface Treatment: To improve the compatibility and adhesion of the nanofillers with the biopolymer matrix, surface modifications may be applied. This can include:
(e) Chemical Treatments: Modifying the surface chemistry of nanofillers to enhance bonding.
(f) Coating: Applying a thin layer of biopolymer or other materials to improve performance.

10 Conclusion

With nanocomposites, a whole new class of mechanical and thermal properties come to the fore. Lighter vehicles will consume less fuel, heavier clothes will be sumptuous on the eye; energy efficiency is rapidly becoming an essential requirement for every product. However, their degradation characteristics and toxicological properties pose significant challenges that must be addressed. Although these solutions require a comprehensive approach, the objective is certainly clear. Advanced design, precise characterization, and strict safety assessments must all be combined to achieve success in the task at hand. It is clear that as long as these challenges are met head-on, researchers will be able to obtain maximum performance from polymeric nanocomposites and avoid unnecessary risks. They are thus steered toward a safety clearance arrangement suitable for many applications in addition to those already considered. But further development work is just that much more dependent on these last elements of consideration as well as environmental and health building blocks are recognized more people will be looking for cleaner properties.

The design and use of materials in bone tissue engineering require an emphasis on bioactive/biodegradable composites that exhibit superior mechanical stability and tailored biodegradation rates for conditions under physiological environments. CVE polymers are strong candidates for this application if they can be combined with calcium phosphate and CVE materials have already been processed into micro-structured scaffolds. They need to result in excellent mechanical properties while being able to incorporate biomolecules that enhance biocompatibility and support cellular function more so than the CAE polyesters. There remain challenges, though, like the mechanical properties of composites that are still not good enough that can make these scaffolds incapable to support autogenous loading conditions and predictable porosity facilitating nutrient and vascularization. Furthermore, developing chemical compatibility between hydrophilic fibers and hydrophobic matrices is essential to achieve good mechanical properties for which coupling agents or compatibilizers can be used. Overcoming these hurdles is key for the future of materials design in biomedicine.

This breakthrough nanocomposite polymers method areas on refining the processing and characterization techniques to control their degradation behavior and make it more predictable. Emphasis is on incorporating green chemistry principles: this leads to sustainable materials, which are formed using biodegradable components and environmentally friendly manufacturing methods. In order to solve this problem as a whole that is both environmentally friendly and safe, the method aims to bring about the development of nanocomposites that are healthier than those we now have. In the future, we will be engaged in a more comprehensive evaluation to determine whether these materials meet safety and effectiveness standards throughout different applications, taking account of both degradation and toxicological aspects. Thus, a more thorough understanding of these factors will allow researchers to develop high-strength nanocomposites that are not harmful at all.

References

1. Pandey JK, Raghunatha Reddy K, Pratheep Kumar A, Singh RP (2005) An overview on the degradability of polymer nanocomposites. Polym Degrad Stab 88:234–250. https://doi.org/10.1016/j.polymdegradstab.2004.09.013
2. Khan I, Khan I, Saeed K, Ali N, Zada N, Khan A, Ali F, Bilal M, Akhter MS (2022) Polymer nanocomposites: an overview. In: Smart polymer nanocomposites, pp 167–184. https://doi.org/10.1016/B978-0-323-91611-0.00017-7
3. van Olphen H, Hsu PH (1978) An introduction to clay colloid chemistry. Soil Sci 126:59. https://doi.org/10.1097/00010694-197807000-00013
4. Fujiwara S, Sakamoto T (1976) Method for manufacturing a clay/polyamide composite

5. Depan D (2015) Biodegradable polymeric nanocomposites: advances in biomedical applications. In: Biodegradable polymeric nanocomposites: advances in biomedical applications, pp 1–252. https://doi.org/10.1201/b19314
6. Jujjavarapu SE, Poluri KM (2020) Green polymeric nanocomposites. Green Polym Nanocompos. https://doi.org/10.1201/9781351045155
7. Armstrong G (2015) An introduction to polymer nanocomposites. Eur J Phys 36. https://doi.org/10.1088/0143-0807/36/6/063001
8. Darwish MSA, Mostafa MH, Al-Harbi LM (2022) Polymeric nanocomposites for environmental and industrial applications. Int J Mol Sci 23. https://doi.org/10.3390/ijms23031023
9. Jujjavarapu SE, Poluri KM (2020). Green polymeric nanocomposites. https://doi.org/10.1201/9781351045155
10. Nurazzi NM, Ilyas RA, Sapuan SM, Khalina A (2023) Synthetic and natural nanofillers in polymer composites: properties and applications. In: Synthetic and natural nanofillers in polymer composites, pp 1–438. https://doi.org/10.1016/C2022-0-00527-4
11. Ging J, Tejerina-Anton R, Ramakrishnan G, Nielsen M, Murphy K, Gorham JM, Nguyen T, Orlov A (2014) Development of a conceptual framework for evaluation of nanomaterials release from nanocomposites: environmental and toxicological implications. Sci Total Environ 473–474:9–19. https://doi.org/10.1016/j.scitotenv.2013.11.135
12. Han C, Sahle-Demessie E, Varughese E, Shi H (2019) Polypropylene-MWCNT composite degradation, and release, detection and toxicity of MWCNTs during accelerated environmental aging. Environ Sci Nano 6:1876–1894. https://doi.org/10.1039/c9en00153k
13. Duncan TV (2015) Release of engineered nanomaterials from polymer nanocomposites: the effect of matrix degradation. ACS Appl Mater Interfaces 7:20–39. https://doi.org/10.1021/am5062757
14. Busquets-Fité M, Fernandez E, Janer G, Vilar G, Vázquez-Campos S, Zanasca R, Citterio C, Mercante L, Puntes V (2013) Exploring release and recovery of nanomaterials from commercial polymeric nanocomposites. J Phys Conf Ser 429. https://doi.org/10.1088/1742-6596/429/1/012048
15. Vaid R, Yildirim E, Pasquinelli MA, King MW (2021) Hydrolytic degradation of polylactic acid fibers as a function of ph and exposure time. Molecules 26. https://doi.org/10.3390/molecules26247554
16. Ikuno Y, Tsujino H, Haga Y, Asahara H, Higashisaka K, Tsutsumi Y (2023) Impact of degradation of polyethylene particles on their cytotoxicity. Microplastics 2:192–201. https://doi.org/10.3390/microplastics2020015
17. Nguyen T, Pellegrin B, Mermet L, Gu X, Shapiro A, Chin J (2009) Degradation and nanofiller release of polymer nanocomposites exposed to ultraviolet radiation. In: 4th Europe weather symposium
18. Lankone RS, Wang J, Ranville JF, Fairbrother DH (2017) Photodegradation of polymer-CNT nanocomposites: effect of CNT loading and CNT release characteristics. Environ Sci Nano 4:967–982. https://doi.org/10.1039/c6en00669h
19. Melinte V, Stroea L, Chibac-Scutaru AL (2019) Polymer nanocomposites for photocatalytic applications. Catalysts 9. https://doi.org/10.3390/catal9120986
20. Costache MC, Wang D, Heidecker MJ, Manias E, Wilkie CA (2006) The thermal degradation of poly(methyl methacrylate) nanocomposites with montmorillonite, layered double hydroxides and carbon nanotubes. Polym Adv Technol 17:272–280. https://doi.org/10.1002/pat.697
21. Kataria S, Rain K, Kumar A, Mudila H (2024) Preparation and exploration of optical performance of novel polythiophene-ZrO_2 composites. Opt Quantum Electron 56. https://doi.org/10.1007/s11082-023-06046-3
22. Aghajani-Memar S, Hamedi S, Kermanian H (2024) Preparation of the edible fragrant antibacterial sodium caseinate–based nanocomposite containing Rosa damascena essential oil and halloysite nanotube. Food Bioprocess Technol 17:2885–2901. https://doi.org/10.1007/s11947-023-03308-6

23. Naveenkumar R, Karthikeyan B, Senthilvelan S (2023) Synthesis of bioinspired lotus fiber infused PVA/TiO_2 nanocomposites: characterization, thermal, and photocatalytic activity studies. Biomass Convers Biorefinery. https://doi.org/10.1007/s13399-023-05231-4
24. Idumah CI (2023) Recently emerging advancements in thermal conductivity and flame retardancy of MXene polymeric nanoarchitectures. Polym Technol Mater 62:510–546. https://doi.org/10.1080/25740881.2022.2121220
25. Moein Najafabadi S, Safaei Ghomi J (2023) Synthesis of COF-SO3H immobilized on manganese ferrite nanoparticles as an efficient nanocomposite in the preparation of spirooxindoles. Sci Rep 13. https://doi.org/10.1038/s41598-023-49628-7
26. Tajdari A, Babaei A, Goudarzi A, Partovi R (2020) Preparation and study on the optical, mechanical, and antibacterial properties of polylactic acid/ZnO/TiO_2 shared nanocomposites. J Plast Film Sheeting 36:285–311. https://doi.org/10.1177/8756087919900365
27. Scaffaro R, Maio A, Gulino EF, Morreale M, La Mantia FP (2020) The effects of nanoclay on the mechanical properties, carvacrol release and degradation of a pla/pbat blend. Materials (Basel) 13. https://doi.org/10.3390/ma13040983
28. del Campo A, de Lucas-Gil E, Rubio-Marcos F, Arrieta MP, Fernández-García M, Fernández JF, Muñoz-Bonilla A (2021) Accelerated disintegration of compostable Ecovio polymer by using ZnO particles as filler. Polym Degrad Stab 185. https://doi.org/10.1016/j.polymdegradstab.2021.109501
29. Oliver-Ortega H, Tresserras J, Julian F, Alcalà M, Bala A, Espinach FX, Méndez JA (2021) Nanocomposites materials of pla reinforced with nanoclays using a masterbatch technology: a study of the mechanical performance and its sustainability. Polymers (Basel) 13. https://doi.org/10.3390/polym13132133
30. Ali N, Ali F, Said A, Khurshid S, Sheikh ZA, Ali U, Nguyen-Tri P, Bilal M (2021) Synthesis of clay-armored coatable sulfonated polyimide nanocomposites as robust polyelectrolyte membranes. J Appl Polym Sci 138. https://doi.org/10.1002/app.51310
31. Sousa FM, Cavalcanti FB, Marinho VAD, Morais DDS, Almeida TG, Carvalho LH (2022) Effect of composition on permeability, mechanical properties and biodegradation of PBAT/PCL blends films. Polym Bull 79:5327–5338. https://doi.org/10.1007/s00289-021-03745-3
32. Abdellah Ali SF, Moustafa AS, El-Rafey ME (2024) Characteristics and performance of high-density polyethylene, calcium carbonate particles and polyethylene maleic anhydride nanocomposites. Polym Bull 81:2143–2160. https://doi.org/10.1007/s00289-023-04802-9
33. Das D, Bhattacharjee S, Bhaladhare S (2023) Preparation of cellulose hydrogels and hydrogel nanocomposites reinforced by crystalline cellulose nanofibers (CNFs) as a water reservoir for agriculture use. ACS Appl Polym Mater. 5:2895–2904. https://doi.org/10.1021/acsapm.3c00109
34. Glaskova-Kuzmina T, Starkova O, Gaidukovs S, Platnieks O, Gaidukova G (2021) Durability of biodegradable polymer nanocomposites. Polymers (Basel) 13. https://doi.org/10.3390/polym13193375
35. Prasad A, Paul AK (2024). Biodegradable waste processing for sustainable developments. https://doi.org/10.1201/9781003502012
36. Sifuentes-Nieves I, Flores-Silva PC, Ledezma-Perez AS, Hernandez-Gamez JF, Gonzalez-Morones P, Saucedo-Salazar E, Hernandez-Hernandez E (2024) Sustainable and ecological poly (vinyl alcohol)/agave fiber-based films: structural features post composting process. J Polym Environ 32:2448–2456. https://doi.org/10.1007/s10924-023-03129-w
37. Champa-Bujaico E, Díez-Pascual AM, Garcia-Diaz P (2023) Poly(3-hydroxybutyrate-co-3-hydroxyhexanoate) bionanocomposites with crystalline nanocellulose and graphene oxide: experimental results and support vector machine modeling. Polymers (Basel) 15. https://doi.org/10.3390/polym15183746
38. Al-Harthi MA (2014) Influence of applying microwave radiation on the LDPE/MWCNTs nanocomposite. Polym Compos 35:2036–2042. https://doi.org/10.1002/pc.22864
39. Lecouvet B, Sclavons M, Bourbigot S, Bailly C (2014) Highly loaded nanocomposite films as fire protective coating for polymeric substrates. J Fire Sci 32:145–164. https://doi.org/10.1177/0734904113500207

40. Fu W, Zhao Y, Wang H, Chen X, Liu K, Zhang K, Wei Q, Wang B (2022) Study on preparation, photocatalytic performance and degradation mechanism of polymeric carbon nitride/Pt/nano-spherical MoS_2 composite. J Phys Chem Solids 166. https://doi.org/10.1016/j.jpcs.2022.110700
41. Kaczmarek H, Kosmalska D, Malinowski R, Bajer K (2019) Advances in studies of thermal degradation of polymeric materials Part I. Literature studies. Polimery/Polymers 64:241–251. https://doi.org/10.14314/polimery.2019.4.1
42. Abiodun S, Krishnamoorti R, Bhowmick AK (2024) Thermal degradation of polytetrafluoroethylene, modified polytetrafluoroethylene, and their nanocomposites with boron nitride nanobarbs: lifetime predictions and kinetic analysis. J Appl Polym Sci. https://doi.org/10.1002/app.56452
43. Nasseri S, Jahangiri J, Tagiyev D (2022) Physico-chemical study of synthesized nanocomposites based on organobentonite. Theor Found Chem Eng 56:1147–1152. https://doi.org/10.1134/S0040579522060239
44. Ji Z, Sheardy A, Zeng Z, Zhang W, Chevva H, Allado K, Yin Z, Wei J (2019) Tuning the functional groups on carbon nanodots and antioxidant studies. Molecules 24. https://doi.org/10.3390/molecules24010152
45. Rybiński P (2017) Influence of carbon fillers on thermal properties and flammability of polymeric nanocomposites. Int Polym Process 32:270–289. https://doi.org/10.3139/217.3316
46. Song YS (2007) Residual stress and rheological behavior of polymeric nanocomposites. E-Polymers:1–9. https://doi.org/10.1515/epoly.2007.7.1.545
47. Wei XF, Linde E, Hedenqvist MS (2019) Plasticiser loss from plastic or rubber products through diffusion and evaporation. NPJ Mater Degrad 3. https://doi.org/10.1038/s41529-019-0080-7
48. Fatolahi AR, Ghanbari Y, Khoramishad H, da Silva LFM (2024) A comparative study on the thermomechanical properties of polymeric nanocomposites incorporating single, mixed, and hybrid nanofillers. Polym Compos. https://doi.org/10.1002/pc.29003
49. Vishnoi T, Kumar A (2013) Conducting cryogel scaffold as a potential biomaterial for cell stimulation and proliferation. J Mater Sci Mater Med 24:447–459. https://doi.org/10.1007/s10856-012-4795-z
50. Schmitt H, Creton N, Prashantha K, Soulestin J, Lacrampe MF, Krawczak P (2015) Preparation and characterization of plasticized starch/halloysite porous nanocomposites possibly suitable for biomedical applications. J Appl Polym Sci 132. https://doi.org/10.1002/app.41341
51. Depan D (2015). Biodegradable polymeric nanocomposites: advances in biomedical applications. https://doi.org/10.1201/b19314
52. Bhattacharya SN, Kamal MR, Gupta RK (2007) Polymeric nanocomposites. Polym Nanocomposites:I–XIII. https://doi.org/10.3139/9783446418523.fm
53. Goodwin DG, Boyer I, Devahif T, Gao C, Frank BP, Lu X, Kuwama L, Gordon TB, Wang J, Ranville JF, Bouwer EJ, Fairbrother DH (2018) Biodegradation of carbon nanotube/polymer nanocomposites using a monoculture. Environ Sci Technol 52:40–51. https://doi.org/10.1021/acs.est.7b02062
54. Woodard LN, Grunlan MA (2018) Hydrolytic degradation and erosion of polyester biomaterials. ACS Macro Lett 7:976–982. https://doi.org/10.1021/acsmacrolett.8b00424
55. Yuvaraj G, Ramesh M, Rajeshkumar L (2023) Carbon and cellulose-based nanoparticle-reinforced polymer nanocomposites: a critical review. Nanomaterials 13. https://doi.org/10.3390/nano13111803
56. Honarvar Z, Hadian Z, Mashayekh M (2016) Nanocomposites in food packaging applications and their risk assessment for health. Electron Physician 8:2531–2538. https://doi.org/10.19082/2531
57. Behera K, Tsai CH, Chang YH, Chiu FC (2023) Modifications of phase morphology, physical properties, and burning anti-dripping performance of compatibilized poly(butylene succinate)/high-density polyethylene blend by adding nanofillers. Polymers (Basel) 15. https://doi.org/10.3390/polym15224393

58. Paul AK, Borkotoky SS, Prasad A (2023) Nanohydroxyapatite-based composite materials and processing. In: Advanced materials and manufacturing techniques for biomedical applications. Wiley, pp 157–183. https://doi.org/10.1002/9781394166985.ch7
59. Batakliev T, Georgiev V, Kalupgian C, Muñoz PAR, Ribeiro H, Fechine GJM, Andrade RJE, Ivanov E, Kotsilkova R (2021) Physico-chemical characterization of PLA-based composites holding carbon nanofillers. Appl Compos Mater 28:1175–1192. https://doi.org/10.1007/s10443-021-09911-0
60. Tcherbi-Narteh A, Hosur M, Triggs E, Owuor P, Jelaani S (2014) Viscoelastic and thermal properties of full and partially cured DGEBA epoxy resin composites modified with montmorillonite nanoclay exposed to UV radiation. Polym Degrad Stab 101:81–91. https://doi.org/10.1016/j.polymdegradstab.2013.12.033
61. Soleimani E, Moghaddami R (2018) Synthesis, characterization and thermal properties of PMMA/CuO polymeric nanocomposites. J Mater Sci Mater Electron 29:4842–4854. https://doi.org/10.1007/s10854-017-8440-y
62. Edwards ER, Botelho EC, Braga NA (2022) Influence of MWCNT-f as a UV protective layer in polymer composites with TGDDM/DDS for space application. J Polym Res 29. https://doi.org/10.1007/s10965-022-02906-w
63. Raza R, Ali S, Ahmed F, Ullah S, Grant JT, Iqbal J (2011) Advance polymeric carbon nanocomposite films with enhanced thermo-mechanical properties. Polym Compos 32:1757–1765. https://doi.org/10.1002/pc.21206
64. Mohamadi M (2022) Interpretation of thermal transitions and phase transformations in semi-crystalline PVDF/PEO/graphene nanocomposites characterized by modulated-temperature DSC. J Therm Anal Calorim 147:6701–6712. https://doi.org/10.1007/s10973-021-10997-8
65. Dastakeer S, Saminathan P, Venkatesan S, Sudha PG, Kannan M (2019) Studies on thermal degradation kinetics and dielectric properties of polyether imide foam/nanosilica-based nanocomposites. Plast Rubber Compos 48:356–363. https://doi.org/10.1080/14658011.2019.1630200
66. Mandal S, Sethi S, Biswas K (2023) Flexural behavior of pMWCNTs filled glass fiber/epoxy nanocomposites: synthesis and interfacial failure. J Appl Polym Sci 140. https://doi.org/10.1002/app.53438
67. Wu H, Liu X, Hua X, Zhang J (2023) Thermal-oxidation degradation of polylactic acid/cellulose nanocrystal composites: effects of surface chemistry. Ind Crops Prod 202. https://doi.org/10.1016/j.indcrop.2023.117008
68. Norazlina H, Hadi AA, Qurni AU, Amri M, Mashelmie S, Kamal Y (2018) Degradability studies of PLA nanocomposites under controlled water sorption and soil burial conditions. IOP Conf Ser Mater Sci Eng 342. https://doi.org/10.1088/1757-899X/342/1/012025
69. Goodwin DG, Lai T, Lyu Y, Lu CY, Campos A, Reipa V, Nguyen T, Sung L (2020) The impacts of moisture and ultraviolet light on the degradation of graphene oxide/polymer nanocomposites. NanoImpact 19. https://doi.org/10.1016/j.impact.2020.100249
70. Martínez-Sanz M, Villano M, Oliveira C, Albuquerque MGE, Majone M, Reis M, Lopez-Rubio A, Lagaron JM (2014) Characterization of polyhydroxyalkanoates synthesized from microbial mixed cultures and of their nanobiocomposites with bacterial cellulose nanowhiskers. N Biotechnol 31:364–376. https://doi.org/10.1016/j.nbt.2013.06.003
71. Tam LH, Wu C (2017) Molecular mechanics of the moisture effect on epoxy/carbon nanotube nanocomposites. Nanomaterials 7. https://doi.org/10.3390/nano7100324
72. Nganga S, Travan A, Donati I, Crosera M, Paoletti S, Vallittu PK (2012) Degradation of silver-polysaccharide nanocomposite in solution and as coating on fiber-reinforced composites by lysozyme and hydrogen peroxide. Biomacromol 13:2605–2608. https://doi.org/10.1021/bm300809h
73. Weems A, Li W, Maitland D, Calle LM (2019) Mesoporous silica nanoparticles to tune oxidative degradation rates in composite shape memory polymers. Recent Prog Mater 1:1–1. https://doi.org/10.21926/rpm.1904008
74. Feng X, Wang B, Wang X, Wen P, Cai W, Hu Y, Liew KM (2016) Molybdenum disulfide nanosheets as barrier enhancing nanofillers in thermal decomposition of polypropylene composites. Chem Eng J 295:278–287. https://doi.org/10.1016/j.cej.2016.03.059

75. Laachachi A, Ruch D, Addiego F, Ferriol M, Cochez M, Lopez Cuesta JM (2009) Effect of ZnO and organo-modified montmorillonite on thermal degradation of poly(methyl methacrylate) nanocomposites. Polym Degrad Stab 94:670–678. https://doi.org/10.1016/j.polymdegradstab.2008.12.022
76. Boyne DA, Orlicki JA, Griep MH (2017) Plasmonic nano-additives for polymer composites: Preservation of tailored functionality. In: 2017 IEEE 17th international conference nanotechnology, NANO 2017, pp 881–884. https://doi.org/10.1109/NANO.2017.8117362
77. Dintcheva NT, Al-Malaika S, Morici E, Arrigo R (2017) Thermo-oxidative stabilization of poly(lactic acid)-based nanocomposites through the incorporation of clay with in-built antioxidant activity. J Appl Polym Sci 134. https://doi.org/10.1002/app.44974
78. Marturano V, Cerruti P, Ambrogi V (2017) Polymer additives. Polym Eng:139–170. https://doi.org/10.1515/9783110468281-005
79. Hashem MS, Magar HS, Fahim AM, Sobh RA (2024) Antioxidant-rich brilliant polymeric nanocomposites for quick and efficient non-enzymatic hydrogen peroxide sensor. RSC Adv 14:13142–13156. https://doi.org/10.1039/d4ra01768d
80. Bergeson LL (2013) Sustainable nanomaterials: emerging governance systems. ACS Sustain Chem Eng 1:724–730. https://doi.org/10.1021/sc4000863
81. Araújo A, Botelho GL, Silva M, Machado AV (2013) UV stability of poly(lactic acid) nanocomposites. J Mater Sci Eng B 3. https://doi.org/10.17265/2161-6221/2013.02.001
82. Anžlovar A, Orel ZC, Žigon M (2008) Nanocomposites with nano-to-sub-micrometer size zinc oxide as an effective UV absorber. Polim 29:84–87
83. Idumah CI, Obele CM, Emmanuel EO, Hassan A (2020) Recently emerging nanotechnological advancements in polymer nanocomposite coatings for anti-corrosion, anti-fouling and self-healing. Surfaces Interfaces 21. https://doi.org/10.1016/j.surfin.2020.100734
84. Mago G, Kalyon DM, Jana SC, Fisher FT (2010) Polymer nanocomposite processing, characterization, and applications. J Nanomater 2010. https://doi.org/10.1155/2010/325807
85. Rusyn I, Greene N (2018) The impact of novel assessment methodologies in toxicology on green chemistry and chemical alternatives. Toxicol Sci 161:276–284. https://doi.org/10.1093/toxsci/kfx196
86. Kulkarni S (2019) Characterization techniques for polymers and polymer nanocomposites. Compos Mater Eng:131–180. https://doi.org/10.1201/9780429242762-10
87. Zhang X, Bhuvana S, Loo LS (2012) Characterization of layered silicate dispersion in polymer nanocomposites using Fourier transform infrared spectroscopy. J Appl Polym Sci 125. https://doi.org/10.1002/app.36266
88. Do Nascimento Silva R, De Oliveira TA, Da Conceiçąõ ID, Araque LM, Alves TS, Barbosa R (2018) Evaluation of hydrolytic degradation of bionanocomposites through fourier transform infrared spectroscopy. Polimeros 28:348–354. https://doi.org/10.1590/0104-1428.09217
89. Jinnai H (2022) Electron microscopy for polymer structures. Microscopy 71:I148–I164. https://doi.org/10.1093/jmicro/dfab057
90. Adhikari R, Michler GH (2009) Polymer nanocomposites characterization by microscopy. Polym Rev 49:141–180. https://doi.org/10.1080/15583720903048094
91. Chipara D, Trevino A, Martirosyan KS, Chipara M (2023) Interphase in polymer-based nanocomposites: polyoctenamer—single-walled carbon nanotubes. Surf Interf 40. https://doi.org/10.1016/j.surfin.2023.103011
92. Benhammada A, Trache D (2020) Thermal decomposition of energetic materials using TG-FTIR and TG-MS: a state-of-the-art review. Appl Spectrosc Rev 55:724–777. https://doi.org/10.1080/05704928.2019.1679825
93. Kota AK, Kerzner R, Bigio DI, Bruck HA, Powell D (2008) Characterization of processing effects in HIPS-CNF composites using thermogravimetric analysis. Polym Eng Sci 48:1120–1125. https://doi.org/10.1002/pen.21059
94. Corcione CE, Frigione M (2012) Characterization of nanocomposites by thermal analysis. Materials (Basel). 5:2960–2980. https://doi.org/10.3390/ma5122960
95. Mohammad AK, Reineke J (2012) Quantitative nanoparticle organ disposition by gel permeation chromatography. Methods Mol Biol 926:361–367. https://doi.org/10.1007/978-1-62703-002-1_23

96. Salim MH, Kassab Z, Kassem I, Sehaqui H, Bouhfid R, Jacquemin J, Qaiss AEK, Alami J, El Achaby M (2021). Characterization techniques for hybrid nanocomposites based on graphene and nanoparticles. https://doi.org/10.1007/978-981-33-4988-9_2
97. Lloyd DR, Narasimhan V, Burns CM (1980) Quantitative analysis of mixed polymer systems by the use of gel permeation chromatography. J Liq Chromatogr 3:1111–1128. https://doi.org/10.1080/01483918008064744
98. Melo SRDO, Homem-De-Mello M, Silveira D, Simeoni LA (2014) Advice on degradation products in pharmaceuticals: a toxicological evaluation. PDA J Pharm Sci Technol 68:221–238. https://doi.org/10.5731/pdajpst.2014.00974
99. Ellepola N, Ogas T, Turner DN, Gurung R, Maldonado-Torres S, Tello-Aburto R, Patidar PL, Rogelj S, Piyasena ME, Rubasinghege G (2020) A toxicological study on photo-degradation products of environmental ibuprofen: ecological and human health implications. Ecotoxicol Environ Saf 188. https://doi.org/10.1016/j.ecoenv.2019.109892
100. Novotna B, Pelclova D, Rossnerova A, Zdimal V, Ondracek J, Lischkova L, Vlckova S, Fenclova Z, Klusackova P, Zavodna T, Topinka J, Komarc M, Dvorackova S, Rossner P (2020) The genotoxic effects in the leukocytes of workers handling nanocomposite materials. Mutagenesis 35:331–340. https://doi.org/10.1093/mutage/geaa016
101. Shah V, Taratula O, Garbuzenko OB, Patil ML, Savla R, Zhang M, Minko T (2013) Genotoxicity of different nanocarriers: possible modifications for the delivery of nucleic acids. Curr Drug Discov Technol 10:8–15. https://doi.org/10.2174/1570163811310010003
102. Manikandan S, Inbakandan D, Valli Nachiyar C, Preethi G, Nellore J, Bavanilatha M (2023) Ecotoxicology of nanocomposite materials. Biosci Biotechnol Res Asia 20:757–771. https://doi.org/10.13005/bbra/3130
103. Fito López C, Díaz Soler B, Guillem-Amat A, Andreu Sánchez O (2023) Insights into the potential effects of micro(nano)plastic-containing nanoparticles in the environment. Sustainability 15. https://doi.org/10.3390/su152316319
104. Tirumala MG, Anchi P, Raja S, Rachamalla M, Godugu C (2021) Novel methods and approaches for safety evaluation of nanoparticle formulations: a focus towards in vitro models and adverse outcome pathways. Front Pharmacol 12. https://doi.org/10.3389/fphar.2021.612659
105. Karthigadevi G, Manikandan S, Karmegam N, Subbaiya R, Chozhavendhan S, Ravindran B, Chang SW, Awasthi MK (2021) Chemico-nanotreatment methods for the removal of persistent organic pollutants and xenobiotics in water—a review. Bioresour Technol 324. https://doi.org/10.1016/j.biortech.2021.124678
106. Raj B, Matche RS (2011) Safety and regulatory aspects of plastics as food packaging materials. Multifunct Nanoreinforced Polym Food Packag:669–691. https://doi.org/10.1533/9780857092786.4.669
107. Leopold L, Zhao C, McConnachie L, Khurana RK, Sharma T, Singh B, Ho RJY (2019) Surface-engineered nanomaterials: environmental and safety considerations in pharmaceutical and medicinal products. NanoBioEngineering:297–310. https://doi.org/10.1201/9781351138901-15
108. Kumar S (2024) Prospects and challenges of nanomaterials in sustainable food preservation and packaging: a review. Discov Nano 19:6. https://doi.org/10.1186/s11671-024-04142-1
109. Kumar S, Maiti P (2016) Controlled biodegradation of polymers using nanoparticles and its application. RSC Adv 6:67449–67480. https://doi.org/10.1039/c6ra08641a
110. Joseph B, Krishnan S, Kavil SV, Pai AR, James J, Kalarikkal N, Thomas S (2021) Green chemistry approach for fabrication of polymer composites. Sustain Chem 2:254–270. https://doi.org/10.3390/suschem2020015
111. Kausar A, Ahmad I, Maaza M, Eisa MH (2022) State-of-the-art nanoclay reinforcement in green polymeric nanocomposite: from design to new opportunities. Minerals 12. https://doi.org/10.3390/min12121495
112. Panayotidou E, Baklavaridis A, Zuburtikudis I, Achilias DS (2015) Nanocomposites of poly(3-hydroxybutyrate)/organomodified montmorillonite: effect of the nanofiller on the polymer's biodegradation. J Appl Polym Sci 132. https://doi.org/10.1002/app.41656

113. Sinha Ray S, Yamada K, Okamoto M, Ueda K (2003) Control of biodegradability of polylactide via nanocomposite technology, Macromol Mater Eng 288:203–208. https://doi.org/10.1002/mame.200390013
114. Bari SS, Chatterjee A, Mishra S (2016) Biodegradable polymer nanocomposites: an overview. Polym Rev 56:287–328. https://doi.org/10.1080/15583724.2015.1118123
115. Wagner A, White AP, Tang MC, Agarwal S, Stueckle TA, Rojanasakul Y, Gupta RK, Dinu CZ (2018) Incineration of nanoclay composites leads to byproducts with reduced cellular reactivity. Sci Rep 8. https://doi.org/10.1038/s41598-018-28884-y
116. Carroccio SC, Scarfato P, Bruno E, Aprea P, Dintcheva NT, Filippone G (2022) Impact of nanoparticles on the environmental sustainability of polymer nanocomposites based on bioplastics or recycled plastics—a review of life-cycle assessment studies. J Clean Prod 335. https://doi.org/10.1016/j.jclepro.2021.130322
117. Low I-M, Alamri HR, Alhuthali AMS (2022) Advances in polymeric eco-composites and eco-nanocomposites, 4. https://doi.org/10.1007/978-981-19-1173-6
118. Iroegbu AOC, Ray SS (2022) Recent developments and future perspectives of biorenewable nanocomposites for advanced applications. Nanotechnol Rev 11:1696–1721. https://doi.org/10.1515/ntrev-2022-0105
119. Sharma V, Agarwal V (2011) Polymer composites sustainability: environmental perspective, future trends and minimization of health risk. In: 2nd International conference on environmental, science and development, IPCBEE 4, pp 259–261

Chapter 13
Biological Studies of Nanocomposites for Implants and Drug Delivery

Sruthi Ann Alex, P. J. Shiny, S. Gnanavel, and A. Rajeshwari

1 Introduction

Biomaterials with excellent biocompatibility and customizable properties are essential for bioimplants and fixation devices like bone screws, sutures, and tissue scaffolds. The material chemistry of biomaterials, including polymers, ceramics, and metals, can be tailored to meet specific requirements, such as controlled degradation or enhanced mechanical strength. However, these engineering materials, when used alone, often lack the necessary combination of strength, bioactivity, and other functional properties required for complex applications like orthopedic implants or artificial tissues/prostheses [1]. To address these limitations, nanocomposites—hybrid materials formed by reinforcing various engineering materials with nanoscale additives—have gained significant interest. These nanocomposites provide enhanced mechanical and thermal stability, bioactivity, optical properties, electrical conductivity, and other tunable properties, making them ideal for a wide range of biomedical applications, including implantable devices, drug delivery systems, and structural applications. This can result in better tissue regeneration and multi-modal treatment. Additionally, they hold promise in drug delivery, where nanoparticles within a composite matrix can act as carriers, enabling controlled release of therapeutic

S. A. Alex (✉) · S. Gnanavel
Department of Biomedical Engineering, SRM Institute of Science and Technology, Kattankulathur, Tamil Nadu, India
e-mail: sruthia@srmist.edu.in

P. J. Shiny
Division of Medical Research, SRM Medical College Hospital & Research Centre, SRM Institute of Science and Technology, Kattankulathur, Tamil Nadu, India

A. Rajeshwari
Department of Biotechnology, The American College, Madurai, Tamil Nadu, India

A. Prasad and V. Katiyar (eds.), *Resorbable Polymers for Bioimplants and Fixation Devices*, Materials Horizons: From Nature to Nanomaterials,
https://doi.org/10.1007/978-981-96-9306-1_13

agents at the implant site. This makes nanocomposites a significant advancement in developing multifunctional biomedical devices [2, 3].

The primary objective of this chapter is to enhance the understanding of biological interactions with nanocomposites in biomedical applications. Due to their small size and material combinations, nanocomposites interact with biological systems in complex ways, influencing cellular responses, protein adsorption, and tissue integration—key factors in the success of implants and drug delivery systems. The chapter will elucidate how nanocomposites can be engineered to optimize biocompatibility, reduce immune rejection, and promote tissue regeneration, laying the groundwork for safer and more effective medical devices. Additionally, the material applications in implants and drug delivery will be explored extensively, highlighting how nanocomposites improve mechanical strength, bioactivity, and degradation rates for load-bearing applications. Their role as drug carriers for localized and controlled release will also be discussed, emphasizing the enhanced therapeutic efficacy and minimized systemic side effects. By investigating both biological interactions and practical applications, this chapter aims to showcase the transformative potential of nanocomposites in advancing healthcare technology.

2 Nanocomposites

Nanocomposites are materials composed of a matrix, usually a polymer, into which nanoparticles are dispersed to improve or impart new properties. These nanoparticles, typically ranging from 1 to 100 nm in size, dramatically alter the physical, chemical, mechanical, and biological properties of the base material due to their high surface area and unique interactions at the nanoscale. Nanocomposites can be classified into various types based on the nature of the matrix material (such as polymers, ceramics, or metals) and the type of nanoparticles used (like polymer/carbon-based, inorganic, or metallic particles) [1, 4–6].

One common category is polymer nanocomposites, where polymers act as the matrix and nanoparticles such as carbon nanotubes (CNTs), silica, or metal oxides are dispersed within. These materials are widely used due to their enhanced strength, durability, and thermal stability, often required in medical implants, drug delivery systems, and tissue engineering scaffolds. Inorganic nanocomposites, where ceramics or metals form the matrix, also have significant biomedical applications, including bone regeneration and dental restorations, because of their improved hardness and wear resistance [7–10].

The role of nanoparticles in composites is crucial to their functionality. Nanoparticles reinforce the matrix material, leading to superior mechanical strength and durability compared to conventional composites. They also impart additional functionalities, such as antimicrobial properties, electrical conductivity, or biocompatibility, which are essential in biomedical applications. By controlling nanoparticle size, shape, and surface chemistry, researchers can fine-tune nanocomposites to meet specific requirements, making them versatile materials in both industrial and medical fields [11–14].

2.1 Properties of Nanocomposites

Nanocomposites have emerged as highly promising materials for biomedical applications, including implants and drug delivery systems, due to their unique combination of mechanical properties, chemical stability, biocompatibility, and other functional attributes. These properties are significantly enhanced through the incorporation of nanoparticles into conventional materials, allowing for the design of highly tailored solutions to meet the stringent requirements of medical devices and therapies [15, 16].

(i) **Mechanical Properties**: Nanocomposites offer superior mechanical strength, elasticity, and toughness compared to traditional materials and therefore is a critical factor in the success of implants. This is due to their nanoscale dimensions increase surface area and grain boundary density, which restricts dislocation motion, improving strength, hardness, and resistance to deformation. This is particularly important for load-bearing implants, such as orthopedic or dental devices, where materials must withstand continuous mechanical stress. Nanoparticles, such as carbon nanotubes, hydroxyapatite, or silica, with unique atomic arrangements are commonly added to polymer matrices to improve the strength and wear resistance of the composite, ensuring longer-lasting performance and reduced risk of implant failure [17–21].

(ii) **Chemical Stability**: This is a key property, especially for materials that must function in the body over extended periods. Nanocomposites resist degradation in biological environments, including resistance to hydrolytic, oxidative, and enzymatic attack, due to their large surface-to-volume ratio, which allows for strong surface bonding. For drug delivery, this stability is essential to prevent premature release of the therapeutic agent and to ensure controlled, sustained release over time. Nanoparticles such as metal oxides and ceramics provide nanocomposites with this durability, making them ideal candidates for long-term biomedical applications [15].

(iii) **Biocompatibility**: Nanocomposites can be engineered for excellent biocompatibility, minimizing the risk of immune rejection or adverse reactions when used in medical implants and drug delivery systems. Surface modifications, such as coating with biopolymers or functionalizing nanoparticles, can further enhance the interaction between the nanocomposite and surrounding biological tissues, promoting better integration and reduced inflammation [22, 23].

(iv) **Surface Functionalization**: Nanocomposites can be surface-modified to interact with specific biological environments, allowing for enhanced cellular adhesion, proliferation, and differentiation. For instance, research shows that functionalizing nanocomposites with biomolecules, such as peptides or growth factors, promotes osteointegration in bone implants [15].

(v) **Antimicrobial Properties**: Incorporation of nanoparticles like silver (Ag NPs) or zinc oxide (ZnO NPs) into the composite structure imparts antimicrobial activity due to the release of reactive oxygen species (ROS) or metal ions. Their high surface area allows for greater interaction with microbial cells, leading to cell membrane disruption. These materials inhibit bacterial growth on implant surfaces, reducing the risk of post-operative infections, a significant concern in surgeries involving implants [24–27].

(vi) **Thermal Properties**: Some nanocomposites having SiC NPs and Al_2O_3 NPs exhibit enhanced thermal resistance due to their better heat dissipation, stronger interatomic bonds, and reduced defects, which prevents their degradation when used in critical applications like drug delivery (where materials must withstand sterilization processes or variations in body temperature). Enhanced thermal and electrical conductivity in graphene and CNTs can be due to their unique structure and bonding. The reduced phonon scattering at the nanoscale allows for more efficient heat transfer [16].

(vii) **Electrical Conductivity**: Conductive nanocomposites, containing materials like graphene or CNTs, support high speed of electron transport and are used in applications like neural implants. These composites can interface with neurons, enabling bioelectric signal transmission and promoting neural regeneration [28].

(viii) **Optical Properties**: Nanocomposites exhibit key optical properties, such as light absorption and scattering, especially in metal nanoparticles like gold, which demonstrate localized surface plasmon resonance (LSPR). This makes them ideal for bioimaging and photothermal therapies. Fluorescent nanocomposites, like quantum dots (QDs), are also widely used for real-time tissue imaging and drug delivery tracking due to their bright UV or visible light emission. Photoluminescence, seen in metal oxides and carbon-based nanocomposites, enhances medical imaging. Additionally, photoacoustic properties of materials like gold nanorods enable deep tissue imaging with high resolution. These versatile optical properties make nanocomposites valuable in theranostics, integrating both diagnostic and therapeutic functions for real-time monitoring of drug delivery and treatment efficacy [15, 16].

(ix) **Magnetic Properties**: Magnetic nanoparticles (e.g., iron oxide) within nanocomposites enable their use in targeted drug delivery and imaging. As the size of the Fe_3O_4 NPs decrease, the magnetic moments can become more pronounced and the quantum effects dominate at nanoscale, leading to phenomena such as superparamagnetism. By applying external magnetic fields, these nanoparticles can be directed to specific tissues, enhancing localized therapy and reducing side effects [16].

(x) **Degradability**: Biodegradable nanocomposites break down into non-toxic byproducts over time. This property is critical in temporary implants or drug carriers, ensuring that the material is safely absorbed or eliminated by the body without the need for removal surgery [15].

(xi) **Porosity and Permeability**: Porous nanocomposites are essential in drug delivery for allowing the controlled passage of molecules. Higher porosity enables faster drug release, while lower porosity offers sustained, slow-release profiles, depending on the therapeutic need [13].

(xii) **Hydrophilicity/Hydrophobicity Balance**: Tailoring the hydrophilic and hydrophobic properties of nanocomposites can control how they interact with biological fluids. This balance is critical for ensuring proper dispersion in physiological environments and promoting cell attachment.

(xiii) **Controlled Drug Release**: Nanocomposites are engineered to provide controlled drug release. The release kinetics can be tailored by modifying nanoparticle size, surface properties, or composite porosity. This allows for precise, sustained delivery of drugs at the target site, improving therapeutic outcomes [5, 7, 29, 30].

(xiv) **Stimuli-Responsive Behavior**: Nanocomposites can be designed to respond to external stimuli like pH, temperature, or light. For example, thermosensitive nanocomposites release drugs when exposed to body heat or external heating, providing on-demand therapeutic responses for conditions like cancer [15, 31].

Each of these parameters contributes to the functionality of nanocomposites in medical applications, ensuring that they meet the specific needs of different biomedical challenges. These versatile properties make nanocomposites crucial in advancing both implant technology and drug delivery systems.

3 Types of Nanoparticles in Nanocomposites

Nanoparticles are often classified into broad categories based on their morphology and characteristics. Below are some of the key nanoparticles commonly employed in nanocomposites:

- Polymeric nanoparticles
- Ceramic nanoparticles
- Metallic nanoparticles.

3.1 Polymeric Nanoparticles

Polymeric nanoparticles can be classified into two categories: natural and synthetic polymers. Polymers of natural origin are derived from plants and animals. Synthetic polymers are created by the process of polymerization to achieve a desired chemical composition. The polymer nanoparticles possess an array of versatile characteristics, such as the ability to be synthesized with a wide range of properties, including size, shape, surface chemistry, customizable active pharmaceutical ingredients (API)

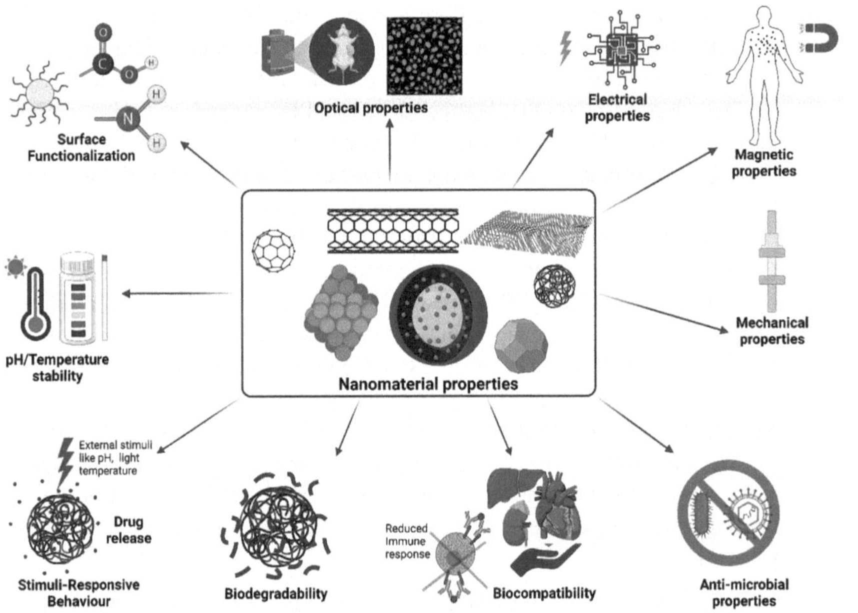

Fig. 1 Properties of nanomaterials that make them useful for biological applications [created in BioRender.com]

release profiles, and degradation rate, in accordance with the specific application. Synthetic polymers can be modified to achieve targeted drug delivery, controlled drug release, biocompatibility, and stability by altering their surface chemistry or incorporating ligands. These polymers have the advantage of being able to adapt to specific physical and chemical properties in accordance with specific requirements. However, natural polymers are challenging to manipulate in order to achieve the requisite physical and chemical properties.

The use of polymer nanoparticles for drug delivery dates back to the mid-1960s, when polymers were employed to modify the pharmacokinetic (PK) profile of APIs. These nanoparticles are capable of providing both sustained drug release and altered clearance profiles. Around the same time, Rutgers professor Frank Davis conjugated poly(ethylene glycol) (PEG) to protein APIs, resulting in clinically approved protein and antibody with an enhanced PK profile. The aforementioned example illustrates the extensive history of the use of polymer nanoparticles in drug delivery and emphasizes the importance of safeguarding therapeutic agents from degradation and reducing off-target API-related cytotoxicity during administration. Currently, biodegradable nanoparticles (NPs) have been employed for the controlled release of many medications, including vaccines, created through the process of spontaneous emulsification/solvent. These drugs include human growth hormone, insulin, anticancer agents, and contraceptives. The efficient delivery of nanoparticles (NPs) relies on achieving a long circulation time and controlling the size of medications in the body, which is influenced by the drug/polymer ratio [4, 7–11]. Table 1 provides a concise description of a variety of polymer nanoparticles, including their characteristics and applications.

3.2 Ceramic Nanoparticles

Ceramic nanoparticles are non-metallic inorganic solids that exist in various forms, including amorphous, polycrystalline, dense, porous, or hollow structures. They are recognized for their distinctive qualities, including their resistance to heat and electricity. Despite being a non-metallic material, it possesses high hardness and mechanical strength. Additionally, it is chemically inert, which means that it is resistant to harsh chemical environments in comparison with polymers and metallic materials. Currently, ceramics are engineered at the nanoscale to meet the necessary application requirements. These ceramics are referred to as "Fine Ceramics" or "Advanced Ceramics" and possess a variety of advanced mechanical, electrical, electronic, magnetic, optical, chemical, and biochemical properties. Bioceramic exhibits excellent biocompatibility, osteoconductivity, osteoinductivity, osseoconductivity, osseoinductivity, and restorability. They are utilized in cardiovascular implants, orthopedic implants, dental implants, and controlled medication release applications, such as delivering anti-inflammatory pharmaceuticals, by serving as efficient carriers [2, 4, 30, 32, 33]. Table 2 presents a succinct overview of different ceramic nanoparticles, detailing their specific attributes and uses.

3.3 Metallic Nanoparticles

Metallic nanoparticles have a size range of 1–100 nm. They are categorized into four distinct nanosystems: Metal Nanoparticles (e.g., Ag, Au), Bimetallic Nanoparticles (e.g., FePt, FeNi, CuNi), Metal Oxide Nanoparticles (e.g., CeO_2, ZnO), and Magnetic Nanoparticles (e.g., Fe_3O_4). They exhibit distinctive characteristics, including excellent physical, chemical, optical, electrical, magnetic, and biological properties, primarily as a result of their extensive surface area, which promotes interaction with other materials. Additionally, they exhibit chemical stability, non-toxicity, biocompatibility, high saturation magnetization, and high magnetic susceptibility. The extensible properties of these materials make them suitable for the application of orthopedic and dental implants, as well as in the fields of bioimaging, controlled drug release, and targeted drug and gene delivery [5, 29, 34–39]. Table 3 presents a succinct overview of different metallic nanoparticles, detailing their specific attributes and uses.

Table 1 Characteristic and application of most commonly used polymer nanoparticles [4, 7–11]

Polymer NPs	Characteristics	Application
Natural polymer Chitosan	Available in the form of film, gels and fibers, biocompatibility, biodegradable, non-toxicity, antimicrobial and mucoadhesive properties, and positively charged	Oral, nasal, transdermal drug delivery Wound dressing and suture material Bone, cartilage, and skin scaffolds
Collagen	Biocompatibility, support cell adhesion, growth and differentiation, cross-linked to improve mechanical properties, biodegradable and bioresorbable	Scaffold in bone, skin, and cartilage regeneration Promote faster wound healing Targeted drug release Used as filler in implants
Hyaluronic acid	Naturally found in extracellular matrix, high water-binding capacity, support cell proliferation and differentiation, biocompatible and biodegradable	Used as dermal filler, eye drops, and surgical aids for eye surgeries Carrier for controlled drug release
Gelatin	Easy to modify chemically, thermo-reversible properties, promote cell adhesion and growth, biocompatible and biodegradable	Formulation of controlled drug release Scaffold in bone, skin, and cartilage regeneration
Synthetic Polymer Poly(lactic-co-glycolic acid) (PLGA)	Degrades into lactic acid and glycolic acid which easily metabolized by human body, biodegradable and biocompatible	Anti-cancer drug delivery, gene delivery Scaffold for tissue regeneration
Poly(ethylene glycol) (PEG)	Hydrophilic, biocompatible, and non-immunogenic	Cancer therapy and anti-inflammatory drug delivery Imaging agent to enhance contrast Delivers protein and peptides
Poly(caprolactone) (PCL)	Biodegradable, hydrophobic, and with a slower degradation rate compared to PLGA	Long-term drug delivery for chronic diseases like diabetes and osteoporosis Scaffold in bone regeneration Orthopedic implants
Poly(methyl methacrylate) (PMMA)	Non-biodegradable, biocompatible, and often used in coatings due to their stability	Bone and dental cement Intraocular lenses

Table 2 Characteristic and application of most commonly used ceramic nanoparticles [2, 4, 30, 32, 33]

Ceramic nanoparticles	Characteristics	Application
Silica (SiO_2)	High surface area, easily functionalized, controlled porosity, thermal stability, chemically inert and biocompatibility	Targeted drug, protein, and gene delivery Doped with fluorescent dyes for imaging
Hydroxyapatite ($Ca_{10}(PO_4)_6(OH)_2$)	Mimic natural bone composition, osteoconductivity, porous structure, good mechanical strength, and biocompatibility	Controlled release of drugs, especially in bone-related diseases Orthopedic and dental implants
Zirconia (ZrO_2)	High strength, fracture toughness, thermal stability, chemical inert, corrosion and wear resistance, and biocompatibility	Dental implants—crown, bridge, veneers Orthopedic implant—joint replacement
Titania (TiO_2)	High refractive index, photo catalytic properties, UV-blocking ability, and biocompatible	Wound dressing, dental implant Cancer treatment as a photosensitizer
Alumina (Al_2O_3)	High hardness and strength, high melting points, low density, varying degree of porosity, high surface area, chemically inert and biocompatible	Dental implants Orthopedic implants—hip and knee replacement, bone cement Carrier for drug delivery
Calcium phosphate (CaP)	Similar to natural bone material, osteo- and osseoconductivity, bioresorbable and biocompatible	Bone grafts, cements, and scaffolds for bone regeneration Delivering therapeutic agents to bone tissue
Magnesium oxide (MgO)	High surface area, antibacterial properties, biodegradable and biocompatible	Antibacterial coating on orthopedic and dental implants Bone grafts

4 Biological Interactions with Nanocomposites

4.1 Cellular Responses

Cellular responses to nanocomposites are critical in determining their efficacy for implants and drug delivery applications. Upon implantation, nanocomposites interact with the biological environment, influencing cell behavior through various mechanisms. These materials can modulate cellular adhesion, proliferation, and differentiation, which are pivotal in tissue integration and repair. For example, the presence of nanostructured features on the surface of nanocomposites enhances the adsorption of proteins, promoting cellular attachment and subsequent signaling pathways essential for cell survival and growth. Every animal cell has a plasma membrane, a selectively permeable phospholipid bilayer that regulates the movement of molecules in and

Table 3 Characteristic and application of most commonly used metallic nanoparticles [5, 29, 34–39]

Nanoparticle type		Characteristics	Application
Metallic nanoparticles	Gold (Au)	Unique optical properties like surface Plasmon resonance (SPR), easily functionalized, antimicrobial properties and biocompatible	Used as antimicrobial coating on implants Used for targeted protein and drug releases
	Silver (Ag)	Strong antimicrobial properties may exhibit cytotoxicity at high concentration, exhibit SPR and biocompatible	Antimicrobial drug delivery to injured tissues Antimicrobial coating on implants
Bimetallic nanoparticles	Iron-Platinum (FePt)	Strong magnetic properties like high magnetic coercively, thermal and chemical stability and biocompatible	Contrast agent in imaging Targeted drug delivery to kill cancer cell
	Copper-nickel (CuNi)	Exhibit both ferromagnetic and antiferromagnetic properties, resistance to corrosion, high strength and hardness, good electrical conductivity, antimicrobial properties and biocompatibility	Catalysts in cancer therapy Contrast agent in imaging Antimicrobial agent in wound healing and as coating to implants
Metal oxide nanoparticles	Zinc oxide (ZnO)	High surface area, semiconductor with excellent optical and electrical properties, photocatalytic properties, chemically inert and biocompatible	Antibacterial coating on implants Targeted drug delivery at specific site in human body
	Cerium oxide (CeO_2)	High thermal conductivity, high surface area, chemically inert and biocompatible	Drug delivery for cancer therapy Orthopedic implant—to enhance cell proliferation and reduce oxidative stress
Magnetic nanoparticles	Magnetite (Fe_3O_4)	Exhibit superparamagnetism, high surface area, biodegradable and biocompatible	Targeted drug delivery using external magnetic field Magnetic hyperthermia—localized heat to destroy cancer cells
	Maghemite (γ-Fe_2O_3)	Exhibit superparamagnetism, photothermal properties and biocompatible	Targeted drug delivery using external magnetic field

out of the cell through active or passive transport [40]. Nanocomposites (NCs) can enter cells via endocytosis, an active transport mechanism essential for the uptake of charged nanomaterials through the extracellular matrix, forming endosomal vesicles. Endocytosis is categorized into phagocytosis and pinocytosis. Phagocytosis involves engulfing targeted materials to form phagosomes, which contain hydrolytic enzymes that degrade these materials in an acidic environment. In contrast, pinocytosis is characterized by smaller vesicles and is the primary method for cells to absorb nanoscale materials, while materials with high aspect ratio can enter via diffusion/penetration. There are also different receptor-mediated uptakes. Once the nanomaterial is internalized, the cargo (DNA, drugs, etc.) bound to the nanodrug delivery vehicle is released and the nano-vehicle escapes into the cytoplasm from the endosomes to reach the target like mitochondria, nucleus, etc. [41–43]. These concepts have been explained well in a pictorial representation (Fig. 2) depicting various types of nanomaterials and their possible methods of uptake by the cells.

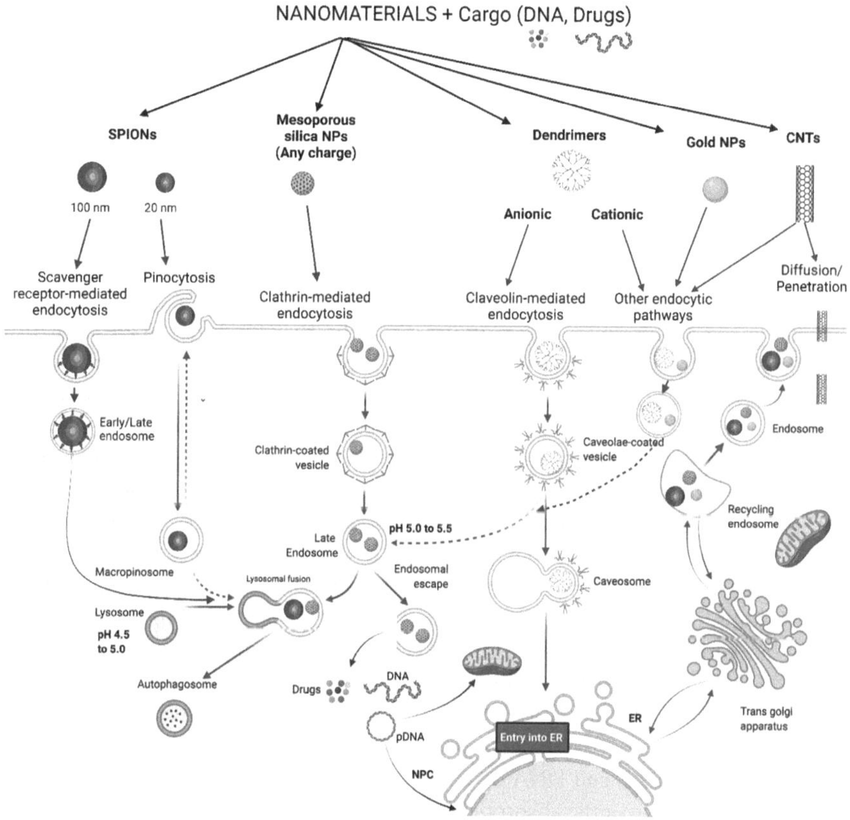

Fig. 2 Cellular uptake pathways of different nanomaterials [created in BioRender. S, G. (2024) BioRender.com/v41y257]; Modified based on Fig. 1 from reference [43]

Nanomaterials after interacting with cells can cause many different types of responses. For instance, AuNPs modified with multiwalled carbon nanotubes (MWCNTs), developed for drug delivery by Li et al. [44], demonstrated reduced cytotoxicity due to the biocompatibility of AuNPs and the porous structure of MWCNTs. Au-MWCNTs have a lesser impact on SMMC-7721 cells compared to MWCNTs, with a survival rate approximately 13% higher for the Au-MWCNTs treatment than for the MWCNTs treatment. This composite effectively transports daunorubicin (DNR) into cells through endocytosis, increasing DNR accumulation and promoting cell death. The π-stacking and hydrophobic interactions facilitate the non-covalent binding of DNR to AuNP-encapsulated MWCNTs, aiding in the internalization of DNR into cells. At pH 4.5, the interaction between DNR and the nanocomposite decreases, leading to the release of DNR in the endosome and its subsequent release inside the cancerous cell. This process triggers cell signaling pathways that induce apoptosis or disrupt DNA replication [44]. Sorafenib-loaded nanocomposites (porous silicon–AuNPs encapsulated in polymer) demonstrated effective in vitro antiproliferative effects on breast cancer cells (MDA-MB-231). The nanocomposites were internalized within the acidic environment of cancer cells via endocytosis. Cellular association and internalization were enhanced by the interaction between the positively charged nanocomposites and the negatively charged components of the cell membrane. Cytotoxicity studies indicated good cytocompatibility at low concentrations after 24 h, with 65% cell viability observed at the highest concentration of 200 μg/ml. Additionally, the hemolytic effect of the nanocomposites on red blood cells was found to be concentration- and time-dependent [45].

The internalization and cellular uptake of nanoscale materials can be influenced by its chemical and physical properties like size, shape, surface charge/functionalization, and cell types [12]. He et al. [14] have studied the effect of nanocomposite shape and charge on HeLa cells. Their findings reveals that the two-dimensional iron oxide-coated graphene oxide NC possesses higher uptake at low concentration and exhibits more apparent cytotoxicity compared to the spherical shaped magnetic nanocomposite [14]. Rangam et al. [46] evaluated the biphasic cytotoxicity of silver nanocomposites synthesized using different leaf extracts on cancer cell lines. Their findings revealed that the LE2Ag1 nanocomposite demonstrated superior anticancer performance compared to others, which they attributed to the presence of a phosphorus group and an oxygen-deficient surface with a high carbon content. Their finding suggested that depending on their size, Ag NPs enter the cytoplasm by either damaging the cell membrane or utilizing channel proteins, leading to the release of Ag^+ ions through the Trojan-Horse mechanism via cytoplasmic enzymes. The released ions interact with thiol groups in cell and mitochondrial membrane proteins, resulting in suppressed cell proliferation, mitochondrial dysfunction, and the subsequent generation of reactive oxygen species (ROS), which cause lipid peroxidation, mitochondrial damage, and DNA damage [46].

The internalization or the absorption of nanocomposite can also occur through the process of receptor-mediated endocytosis [47–49]. Based on the Surface-Enhanced Raman Scattering (SERS) analysis, it is facilitated by caveolae-dependent endocytosis, which is an energy-dependent process [50]. Furthermore, Shukla et al. [51]

reported that transferrin (Tf)-immobilized graphene oxide nanocomposites bind to HeLa cells through the transferrin receptor present on their surface, whereas transferrin acts as the ligand for the transferrin receptor, which is abundant on cancer cells. The transferrin-immobilized graphene oxide nanocomposites induced cytotoxicity by increasing lipid peroxidation and causing DNA damage in HeLa cells. It was proposed that autophagy is a primary mechanism underlying cell death [51]. A similar process was observed in the interaction of HeLa cells with transferrin-functionalized AuNP/graphene oxide nanocomposites. In this case, AuNPs serve as a fluorescent probe, with the fluorescence signal activated by the specific transferrin receptors present on the HeLa cells. The mechanism of cellular uptake for AuNPs and verteporfin-loaded poly(lactic-co-glycolic acid) (PLGA) nanocomposites in PANC-1 tumor cells likely involves pinocytosis and clathrin-mediated endocytosis. The in vitro cytotoxicity study indicated that cell viability decreased as the concentration of verteporfin in the nanocomposite increased [52].

The mechanism for the cytotoxicology of NCs can be due to the generation of reactive oxygen species (ROS) inside the cells after internationalization. Hira et al. [53] suggested that the cancer cell death was enhanced through the interaction between the pectin (PEC), guar gum (GG), and zinc oxide (ZnO) at the nano-level and further it induced S phase arrest leading to apoptosis. The assessment of cytotoxicity in A549 and MCF-7 cancer cell lines indicates that peripheral blood lymphocytes (PBL) treated with PEC-GG-ZnO nanoparticles exhibited the highest cytotoxicity compared to untreated PBL. This suggests that the PEC-GG-ZnO nanocomposite could be a promising new option for enhancing immunity in lung and breast cancer therapy [53]. Similar effect was also observed in human umbilical-vein-derived endothelial (HUVE) cells on its interaction with cerium oxide-zinc NC. During a 24-h exposure, CeO_2 nanoparticles did not significantly impact cell viability at higher concentrations (200 μg/mL). In contrast, a concentration of 154 ± 6 μg/mL of the CeO_2-Zn nanocomposite induced 50% cytotoxicity. The mechanism of cytotoxicity related to the zinc content of the nanocomposite was compared to that of ZnO nanoparticles [54]. The internalization of the nanocomposite into cells leads to the generation of reactive oxygen species (ROS), and the elevated levels of ROS cause oxidative stress and damage to cellular components such as DNA and proteins. Glutathione (GSH), an antioxidant, plays a major role in neutralizing ROS and reducing oxidative stress. However, excessive generation of ROS depletes GSH levels, thereby weakening the cell's defense mechanism. Therefore, at low levels of GSH, ROS can convert fatty acids into toxic lipids, leading to lipid peroxidation and further damage to the cell membrane [55].

4.2 Tissue Response

To effectively integrate nanocomposites (NCs) in biomedical applications such as implants and drug delivery, their interaction with surrounding tissues is crucial. Key factors include promoting cell adhesion, proliferation, biocompatibility, osteoconductivity, and controlled degradation. Understanding the immune responses triggered by NCs is essential, as the release of bioactive molecules can influence macrophage activity, resulting in either pro-inflammatory or anti-inflammatory effects, depending on material design. Additionally, nanoparticle size and composition significantly impact cellular uptake through mechanisms like endocytosis, affecting drug delivery efficacy. The immune system's defense mechanism consists of innate and adaptive responses. The innate immune response, activated by neutrophils, macrophages, monocytes, and natural killer cells, leads to phagocytosis, generating reactive oxygen species (ROS) and inflammatory mediators. In contrast, the adaptive response involves antigen-specific lymphocytes that produce cytokines and antibodies for a more targeted immune reaction [23, 56]. Understanding the interplay between nanocomposite properties and cellular responses is vital for optimizing their use in regenerative medicine and targeted therapies, enhancing integration and effectiveness within biological systems [22].

The three-dimensional (3D) hydroxyapatite/mesoporous glass nanocomposite scaffold functionalized with amino groups increases MC3T3 cell adhesion and differentiation, indicating the biocompatibility of the nanocomposites. This enhanced biocompatibility was further confirmed by ellipsometry studies, which demonstrated that the protonated aminopropyl moieties promote the early stages of the adhesion process by increasing nonspecific protein adsorption [13]. Furthermore, it has been demonstrated that the nanocomposite bioceramic (mesoporous glass/hydroxyapatite, MGHA) scaffolds do not affect intercellular cytokine expression (interleukins, interferon, and tumor necrosis factor). Additionally, murine immune cells, such as NK cells and T and B lymphocytes, support the biocompatibility of the scaffolds through an adequate host tissue response upon implantation. The MGHA nanocomposite does not affect murine lymphocyte activation or spontaneous apoptosis, as evidenced by unchanged mitochondrial activity and cell viability in the presence of biomaterial extracts. Moreover, the absence of mitogen activity suggests that MGHA extracts do not promote immune cell activation, further confirming the biocompatibility of MGHA for use in immune environments [57]. The nanoceria nanocomposite loaded with the antioxidant drug quercetin inhibits M1 polarization and promotes M2 polarization during the destructive and reparative phases. This action leads to a significant increase in the M2/M1 ratio in inflammatory cellular models and improves therapeutic outcomes by regulating pro-inflammatory and anti-inflammatory cytokines, suggesting that quercetin-loaded nanoceria NC is a potential candidate for treating periodontal disease [58]. Similarly, nanoceria of different shapes coated on titanium surfaces exhibit antibacterial and anti-inflammatory properties by reducing the gene expression of IL-6, IL-1β, and TNF-α in the tissues surrounding the implant. While nanorod CeO_2 showed increased initial bacterial attachment, nanocube and nano-octahedron CeO_2 exhibited superior anti-inflammatory effects and ROS-scavenging abilities in vitro. In vivo results revealed that all shapes reduced mRNA expression of pro-inflammatory cytokines (TNF-α, IL-6, IL-1β), with nano-octahedron CeO_2 demonstrating the strongest anti-inflammatory response [59].

ZnO nanocomposites can also induce inflammatory reactions when administered subcutaneously or orally, due to the dissolution of Zn^{2+} ions and the generation of reactive oxygen species (ROS). Additionally, they enhance anti-cancer immunity when used as tumor antigen carriers [60]. Thermal-responsive materials are designed to change shape in response to temperature variations. Subcutaneous implantation of shape memory polymer (SMP) nanocomposites in rats elicited a mild foreign body-type immune response. The degradation rates of SMP nanocomposites were inversely proportional to the length of the polyester chains, which may further induce a late acute inflammatory response upon degradation. However, no severity or systemic pathological abnormalities were observed after one year of implantation [31].

5 Nanocomposites in Implants

5.1 *Types of Implants Using Nanocomposite*

Owing to their unique properties like high strength, flexibility, and biocompatibility, nanocomposites (NC) have been widely employed in the fabrication of medical implants to overcome their potential for malfunction or failure. NC are materials that composed of nanoparticles (NPs) dispersed in a matrix material, for example a polymer or metal [19]. A wide range of NC have been developed for different type implants like orthopedic, dental, and cardiovascular implants, etc. [1].

Orthopedic implants are used to repair or replace damaged bones and joints. Nanoscale metallic, ceramic, and polymer-based NCs were found to encourage the function of bone cells, whereas the nanotopography of polymers was noted to increase cell proliferation and marker production of osteoblast differentiation like osteocalcin and osteopontin [61]. The ceramic–polymer NCs with enhanced mechanical and bioactive characteristics are found to be a suitable material for repairing articular cartilage [3]. Similarly, Mg was opted over other metallic materials for orthopedic implants due to its identical mechanical characteristics and biocompatibility. The metal and metal oxide NPs also play a potential role in scaffold fabrication for tissue regeneration [6]. Herbal extract-loaded chitosan (CS)/polyethylene oxide (PEO)-based nanofibrous hydrogel exploited an antibacterial effect and are used in orthopedic implant coatings to prevent infections [27]. Further, poly(ether ether ketone)/zinc oxide (ZnO) NC is also used for orthopedic and trauma implants due to its excellent tensile strength and long-term durability properties [62]. The graphene oxide (GO)–hydroxyapatite (HA) NCs act as potential coating materials that provide an alternate solution to the rejection of implants [63]. Mg-based nanocomposite containing functionalized CNT presented outstanding mechanical properties, corrosion resistance and appropriate bioactive and biocompatibility properties making it a potential material for bone implant [21].

Implants made of titanium or other biocompatible materials are surgically implanted into the jawbone and to replace missing teeth [1]. Carbon-based NCs such as graphene, carbon nanotubes, and carbon nanofiller are used in dental implants were noted to enhance their strength, and durability. For example, CNTs incorporated into dental resins poly(methyl methacrylate), PMMA improve their wear resistance and mechanical strength, and graphene oxide NCs coating on dental implant improve their antimicrobial effect and bonding between dental material and tooth structure [64]. Titanium oxide (TiO_2) NCs modified the surface of the implant to enhance its durability, osseointegration, and antibacterial activities [24, 65]. PMMA/HA NCs were chosen as biomaterial candidates for the development of porous trabecular dental implant fixture [66]. HA-zirconia bioceramic NCs were noted to enhance the mechanical properties of HA like strength, wear resistances, and toughness [67].

NCs are also involved in the fabrication of cardiovascular implants like heart valves, defibrillators, stents, and pacemakers to treat cardiovascular diseases [1]. Silver NPs possess anti-microbial properties immobilized on either natural or synthetic polymer composites were found to be a suitable material for vascular grafts, catheters, and endotracheal tube fabrication [25]. Calcium phosphate NP blend with polymers like poly(lactic-co-glycolic acid) (PLGA), and Poly-L-Lactic was used as a coating material in vascular stents due to its improved biocompatibility nature [68]. The coronary artery bypass grafts developed by combining S-nitrosothiols consist of S-Nitroso-N-acetylpenicillamine (SNAP) and S-nitrosoglutathione (GSNO) with polyhedral oligomeric silsesquioxanepoly(carbonate-urea) urethane (POSS-PCU) lead to the release of nitrous oxide, which helps in reduction of thrombosis and elevation of endothelialization, and a similar effect was noted in PLGA-reinforced HA nanocomposite [69]. Silk fibroin (SF)-coated Magnesium alloy holds a promising candidate for stents due to its excellent biocompatibility, minimal platelet adhesion, and enhanced controlled drug release [70].

5.2 *Biological Performance of Nanocomposites in Implants*

Wu et al. [71] have developed a mussel-inspired biomimetic strategy based on the incorporation of polydopamine-functionalized laponite nanosheets into a GelMA/ alginate-dopamine hydrogel for bone defect repair. The catechol groups incorporate hydrogels provide a strong adhesive property and these groups form non-covalent and covalent bonds with the surrounding tissue confirming robust adhesion and integration. The hydrogel was noted to possess an exceptional ROS-scavenging performance helps in avoiding bacterial infection, and further, it also enhances the M2 phenotype expression and lowers the M1 phenotype expression that promotes more conducive to the repair of bone defects. The hydrogels control the release of cytokines, reducing pro-inflammatory cytokines (e.g., TNF-α, IL-6) and increasing anti-inflammatory cytokines (e.g., IL-10), forming a conducive environment encourage the mesenchymal stem cells differentiation into osteoblasts and enhance deposition of calcium phosphate minerals essential for bone formation [71].

The gelatin–calcium phosphate nanocomposite was synthesized to regenerate cranial bone defects in a rat model. Phosphate and calcium ions infused into the gelatin, imitating the natural bone mineralization process. Increasing the concentration of glutaraldehyde used for gelatin cross-linking enhance its mechanical properties, whereas the scaffold degradation reversely related to the concentration of the cross-linking agent. The scaffold supports the viability, and differentiation of bone marrow mesenchymal stromal cells regeneration was proved by alkaline phosphate assay and Alizarin red staining over 28 days. When the NCs implanted into the cranial defect in a rat model, the scaffold cross-linked 1% w/v of glutaraldehyde afford maximum bone regeneration over 90 days indicating effective long-term performance in vivo [72].

A scaffold fabricated from a blend of poly(l-lactide-co-glycolide) (PLGA) and synthetic poly([R,S]-3-hydroxybutyrate) (PHB) was designed for osseous tissue regeneration. Its nanoscale porosity creates a highly favorable microenvironment that promotes cell proliferation. Further, the incorporation of NP with hydroxyapatite material and a growth factor (IGF1) accelerate bone regeneration, cell proliferation and tissue restoration process. NC was noted to enhance the mechanical property and control its degradation and also it maintains the structural integrity and support the regeneration of bone for over 12 months [73].

The organic–inorganic NC coatings for Ti implants undergone in vitro and in vivo biological testing show that the polyelectrolyte layer on top promotes cell adhesion, proliferation, and differentiation, providing a favorable environment for cell growth. The coating on Ti do not elicit the adverse immune response making it more biocompatibility, and further bone mineralization and healing was promoted by calcium, and phosphate ions are released from the coating during degradation process. In addition, in vivo study validates enhanced bone for formation around the implant indicating better osteointegration [74].

6 Nanocomposites in Drug Delivery Systems

Nanocomposites offer several advantages over the conventional drug carriers to be an appropriate drug carrier due to the malleable physical and chemical properties. Depending on the type of drug and the mode of dispersion required, the drug can be loaded into suitable nanomaterial and fabricated as a carrier system. To enhance the biocompatibility and biodegradability of these drug delivery systems, often the polymeric component chosen is also from a natural origin.

6.1 Applications in Drug Delivery

6.1.1 Antibacterial

A study investigated the formulation and analysis of chitosan/starch nanocomposites incorporated with ampicillin to improve antibacterial efficacy against *Escherichia coli.* The nanoparticles, synthesized through spray-drying, exhibited spherical morphology between 100 and 600 nm. The chitosan/starch nanoparticles could be loaded with 75.3–77.3% of ampicillin and a zeta potential of + 37.3. The authors demonstrated that the release of ampicillin from this composite system revealed that the antibacterial efficacy surpassed that of free ampicillin, demonstrating an inhibition rate of 95.41% compared to 91.40% [75].

Mesoporous bioactive glass was used to fabricate injectable pastes using polymers like hyaluronic acid and sodium alginate. This matrix enabled the desired sustained release of quercetin to promote osteogenesis. This injectable composite was found to increase the levels of alkaline phosphatase activity and the expression of Collagen-1, Osteopontin, Osteocalcin, and Runx2 proteins at different concentrations of tested quercitin when tested on human-derived mesenchymal stem cells [76].

6.1.2 Anticancer

A nanocomposite of a terpolymer-grafted poly (L-lactide)-poly(ε-caprolactone)-poly(ethylene glycol) on partially oxidized carbon nanotubes was used for the delivery of anticancer drug methotrexate. The process of preparation of the nanocomposite includes a four-step synthesis of PLLA-PCL-PEG terpolymer that improved the efficiencies of drugs loading, ensuring a "burst release" pattern, and enhanced the biocompatibility. This composite was thus evaluated for its efficiency as a drug delivery system due to its structural stability and functional versatility [77].

Another study examined amygdalin-loaded alginate-chitosan nanoparticles as a biocompatible drug delivery system for cancer treatment. The nanoparticles demonstrated effective encapsulation efficiency and prolonged drug release characteristics. In vitro experiments revealed that these nanoparticles markedly decreased the viability of cancer cells, indicating potential anticancer effectiveness. Furthermore, the nanoparticles demonstrated minimal cytotoxic effects on normal cells, underscoring this compatible system's promise as safe and efficient carrier for targeted cancer treatment [78].

The use of dendrimer-based proton sponge nanocomposite in cancer therapy was found to inhibit autophagy, thus increasing apoptosis in tumor cells. The composite with doxorubicin modulated the repolarization of macrophages to enhance the immune response by secreting cytokine. By specific accumulation in tumor cells, G5AcP regulates intracellular lysosomal pH to prevent cytoprotective autophagy. The copper ion release and DOC induce a Fenton-like reaction and chemotherapeutic effect respectively during autophagy condition, thereby eliminating the tumor effectively [79].

A breast cancer treatment strategy was developed on the co-delivery of paclitaxel and curcumin by the biodegradable polymeric nanoparticles. This system was found to release both the drugs in a sustainable. In vitro and in vivo studies on BALB/c nude mouse revealed that this dual drug delivery system showed significant tumor inhibition compared to free drugs [80]. Another study, using 5-fluouracil for colorectal therapy was developed by Ghasemi and co-workers using PLGA nanoparticles. This nanoparticle system showed almost 95% drug encapsulation from this PVA system which showed significant cancer inhibition [81].

6.1.3 Hormone Therapy

Natural polysaccharides were used to coat layered double hydroxide (LDH) nanocomposite that was formulated to improve the oral delivery of insulin. Alginate and chitosan coatings protected insulin-loaded LDH nanoparticles from hydrolytic degradation in the gastrointestinal tract. The in vitro drug release profiles indicated that it was able to protect the burst release partially in an acidic environment. The LDH nanocomposites showed significant enhancement in the stability of insulin, mucoadhesion, and controlled release, which contributed to increased bioavailability when tested on diabetic rats. The cellular uptake was found to be much higher than unaltered LDH and free insulin when tested on Caco-2 cells. The LDH nanocomposites showed better hypoglycemic effects compared to free insulin, which provides promise for their applications in oral insulin delivery systems [82].

This study on the fabrication of injectable hydrogel made from carboxymethyl chitosan (CMCS) and oxidized pullulan (OPL), together with parathyroid hormone (PTH), was performed to promote osteogenesis for treating craniofacial bone defects caused by aging, malignant lesions, and trauma. The parathyroid hormone loaded CMCS/OPL hydrogel exhibited porosity over 90%, controlled degradation rate over 28 days, with a sustained release of PTH. Calcium deposition was found to be higher by up to 85% in the hydrogel loaded with PTH as against pristine hydrogel after 21 days. The accelerating effects on bone regeneration and angiogenesis have been demonstrated through in vivo studies in rat mandibular bone defects [83].

6.1.4 Drug Delivery

A chitosan-based nanocomposite film was developed incorporating mesoporous MCM-41 nanoparticles for the controlled release of metformin. This nanocomposite was developed to overcome the low bioavailability and short half-life period of this commonly used anti-diabetic drug. The composite film was observed to exhibit different swelling patterns in response to the environment, and the film in acidic medium did to break even after 72 h. A burst release of the drug was observed between 22–24 h and a sustained release for up to 48 h. Thus, the composite with aminopropylsilane was found to show excellent hydrophilicity, hydrolytic stability, biocompatibility, mechanical, and as a suitable drug carrier [84].

The impact of selenium nanoparticles (SeNPs) and SeNPs@vitamin C nanocomposite on nicotine-induced testicular toxicity in rats was tested. The study revealed that the treatment with nicotine treatment improved sperm quality, decreased testosterone, increased FSH, increased levels of LH, and oxidative stress accompanied by structural damage to testicular tissue. SeNPs and SeNPs@vitamin C was found to combat this stress by enhancing sperm count, decreased sperm abnormalities, and balanced the hormonal levels. Thus, the therapeutic efficiency of the SeNPs@vitamin C nanocomposite was evaluated in case of nicotine-induced male infertility [85].

6.1.5 ROS Stimuli-responsive Nanogels

Dexamethasone is a drug with poor bioavailability and to overcome its limitations researchers developed a nanogel with controlled release system that is reactive oxygen species (ROS)-responsive. Cyclodextrin and adamantine was used as the cross-linking force for the supramolecular host–guest complex with thioketal as the ROS-response material. The authors reported that the formulation retained for over 8 h on the corneal surface due to the integrin binding. In vitro studies revealed that this dexamethasone nanogel was found to be compatible with the human corneal epithelial cells and effectively inhibit angiogenesis. In vivo studies on a chemically burned eye-rabbit model showed that the formulation effectively suppressed corneal neovascularization (CNV). Thus, this nanogel formulation was proven to be a promising treatment option for CNV by reduced dosing frequency with a higher rate of patient compliance [86].

This study presents a nanocomposite system comprising mesoporous hydroxyapatite (HAp) spheres loaded with polyaniline (PANI) nanoparticles for drug delivery, targeting colorectal cancer. The composite showed high drug loading capacity, with a controlled release of doxorubicin (DOX) under near-infrared (NIR) light, which showed enhanced photothermal properties. NIR-responsive cancer therapy using this apatite/polyaniline showed improved anticancer effects on HCT-116 colon cancer cells [87].

6.2 Mechanisms of Drug Delivery

6.2.1 Controlled Release Mechanisms

The study with a nanocomposite hydrogel from methacrylated hydroxypropyl methylcellulose (M-HPMC) and lignin was incorporated with nanostructured lipid carriers (NLCs) containing alpha-pinene. Varied concentrations of α-pinene-loaded NLCs (0, 18, 38, and 50 wt%) were encapsulated in M-HPMC/M-SLS hydrogel which was found to exhibit controlled release of alpha-pinene for up to 96 h. The hydrogels with improved mechanical strength and toughness prove to be a drug delivery system for oral mucosal applications in treating inflammatory conditions such as recurrent aphthous stomatitis [88].

The preparation of a new injectable nanocomposite hydrogel with a combination of VEGF-A with Laponite-modified methacryloyl gelatin, GelMA, was developed by Mi and group. The co-loading of Lap enabled injectability and delayed the release of VEGF, thus aiding in the controlled release of VEGF-A thereby inducing neovascularization through osteogenic differentiation. This composite hydrogel when tested on a critical-sized calvarial bone defect lesions in the SD rats exhibited significant in vivo bone repair by recruiting endothelial progenitor cells due to the chemotactic effect. The use of this nanocomposite hydrogel in animal models showed enhanced healing bone defects showing promise to be used as an injectable biomimetic hydrogel [89].

6.2.2 Targeted Delivery Approaches

Chitosan nanoparticles from mango peel extract were incorporated into a guar gum matrix, to form a nanocomposite hydrogel. This hydrogel had superior properties such as mechanical and thermal stability with the controlled drug release of 5-fluorouracil, a chemotherapeutic agent. Along with good swelling behavior, antioxidant, and anti-inflammatory properties, it exhibited a controlled release drug profile in the colorectum demonstrating its use in cancer therapy [90].

Layered double hydroxide nanosheets loaded with doxorubicin coated with poly(acrylic acid) (PAA) was used to repolarize tumor-associated macrophages (TAMs). This was aimed to deacidify the tumor microenvironment and convert M2 to M1 macrophages to aid in tumor elimination as combined chemodynamic and immunotherapy. This was successfully achieved along with CD 8 + T cell activation. This nanosheets release Fe^{3+} and DOX in response to acidic environment, while glutathione reduces Fe^{3+} to Fe^{2+} which disrupts the antioxidant system in the tumor cells. As a result of which hydrogen peroxide is produced in a larger quantity, providing substrate for Fenton reaction. Upon use of the LDH-PAA@DOX treatment in vivo, tumor was eliminated significantly [91].

6.2.3 Theranostic Approach

A versatile self-assembled near-infrared surface-enhanced Raman scattering nanoprobe (AMD@HA conjugated with DTTC) for targeting and treating multi-drug resistance bacterial infections was developed in the study. This responsive scaffold to NIR enabled fast identification of bacteria and in situ eradication via photothermal and catalytic therapy. This SERS based-imaging is non-invasive monitoring system with a detection limit detection limit for methicillin-resistant Staphylococcus aureus (MRSA) is as low as 102 CFU mL^{-1}. Thus, with the help of localized surface plasmon resonance-based photothermal therapy, it eliminates the bacteria, aiding in wound healing alongside a non-invasive tool for diagnosis [92].

6.3 Biological Studies and Efficacy

6.3.1 In Vitro and in Vivo Studies

In a pilot study, a NANOTEX bone graft was applied with a fibula flap to reconstruct segmental mandibular defects. This prospective, single-center, non-randomized pilot clinical study was carried out on patients with benign tumor or trauma requiring mandibular reconstruction after implantation. This nanofibrous composite scaffold, which was mechanically strong with bone regenerative properties, was found to study and was based on challenges observed whereby patients had inadequate bone height for dental implants following a fibula flap procedure. The results of preclinical models indicated that NANOTEX would be an effective alternative for existing treatment and its shortcomings based on preclinical models in critical size defect model, for which the clinical trial is underway [93].

6.3.2 Clinical Trials and Outcomes

A nanocomposite of a colloidal chitosan-silver nanoparticles-fluoride was tested for its efficacy in comparison with chlorhexidine for its antibacterial action against *Streptococcus mutans* in orthodontic patients. The randomized clinical trial was conducted on 45 orthodontic patients, and the bacterial counts were performed in the three groups. The study concluded that a significant reduction of *S. mutans* in saliva after the use of nanocomposite with Chlorhexidine was observed proving to be a potential antimicrobial regimen in dentistry, for patients undergoing orthodontic treatment [94].

A phase 1 trial assessing NY-ESO-1-specific T-cell receptor (TCR)-engineered T-cell therapy in combination with a lymph node-targeting nanoparticulate peptide vaccine for the treatment of advanced soft tissue sarcoma was tested. The pullulan nanogel–LPA vaccine produced positive results in the mouse sarcoma model by significantly increasing the number of TCR-T cells in the draining lymph node. The nanogel formulation with the epitope specific to TCR-T cells was injected 24 h before and after 7 days of infusing TCR-T cells into HLA-matched patients. The use of combination therapy including vaccination and the nanogel containing LPA was found to increase the overall persistence of engineered T-cells and significant tumor shrinkage showing potential therapeutic effects [95].

The study investigated the survival and quality of life of patients with metastatic and non-metastatic gastric cancer upon treatment with BCc1 nanomedicine. The randomized, double-blind, and placebo-controlled study involving 123 patients based on the metastasis. It was reported the median overall survival of both metastatic and non-metastatic group. Compared to placebo (Metastatic: 174 days [95% confidence interval (CI) 82.37–265.62]), placebo (62 days [95% CI 0–153.42]; Non-Metastatic: 529 days [95% CI 393.245–664.75]), placebo (345 days [95% CI 134.85–555.14]). The trial findings suggest that BCc1 nanomedicine could be used as an effective therapy agent to improve the outcome and quality of life in gastric cancer patients [96].

7 Future Directions and Scope

Nanocomposites are engineered materials that are designed to achieve specific mechanical, physical, chemical, and biological properties in order to achieve specific applications. The foreseeable future of these materials is discussed extensively below.

7.1 Strategies for Improved Nanomaterials

Nanomaterials possess significant potential for producing products with distinct characteristics and improved performance. Currently, there is a significant volume of research being conducted on nanomaterials. A typical instance is hybrid nanomaterials, specifically hybrid nanocomposites, which involve the combination of polymer, ceramic, and metallic components to create a new material with improved material properties. The combination of various materials results in synergistic effects, leading to the development of a highly promising material with extraordinary qualities. For instance, the incorporation of ceramic material into a polymer matrix can enhance mechanical and thermal qualities like as hardness, stiffness, wear resistance, heat conductivity, and stability. Similarly, the incorporation of metallic material into a polymer can enhance its electrical and mechanical qualities, such as increased strength, toughness, and electrical conductivity. Combining ceramic, polymer, and metallic materials has the potential to enhance optical characteristics [17, 97].

Researchers are actively working on developing applications such as self-healing or adaptable materials using hybrid nanocomposites. A self-healing substance is composed of microcapsules or nanocapsules that contain a healing agent. These capsules are designed to automatically commence the process of repairing any harm that occurs. For instance, in the event of a crack formation, the capsule containing the healing agent will rupture, leading to the discharge of the healing agent. Subsequently, the healing agent will flow toward the injured area and effectively close the crack [18, 98]. There are two categories of self-healing materials: autonomic and non-autonomic self-healing materials. Autonomic self-healing materials utilize the stored chemical potential to automatically initiate and facilitate the repair or healing process in response to damage or rupture. Conversely, non-autonomic self-healing materials require external intervention. Chemically, they can be categorized into two categories: intrinsic and extrinsic self-healing. Extrinsic healing systems utilize an external healing agent, such as capsules or vascular networks, which are designed using autonomic materials. These materials are constructed based on the healing agent and a catalyst that is inserted in the matrix. Generally, intrinsic self-healing materials rely on either non-covalent chemistries or dynamic covalent chemistries

[99, 100]. The reconfiguration of chemical bonds in intrinsic materials is initiated by several external stimuli, such as alterations in pH, light exposure, temperature fluctuations, changes in pressure, or the presence of oxygen, among others. Although various chemistries are used in the creation of self-healing nanocomposites, there are limitations that can be overcome through the implementation of effective strategies. These strategies are crucial for optimizing the design and facilitating the efficient synthesis and application of hybrid nanocomposite materials [101]. The future advancement in self-healing hybrid nanocomposites will rely on the exploration and utilization of sophisticated chemical processes that provide easy, fast, and precise creation of new materials.

The use of hybrid nanocomposites enables the easy implementation of theranostics for targeted drug administration. Theranostics is the idea of combining diagnostic and therapeutic capabilities into a single system that allows for the simultaneous monitoring and treatment of diseases in real-time. Hybrid nanocomposites, produced from both organic and inorganic components, have great potential as theranostic agents in cancer therapy for targeted treatment. Tumor-specific hybrid nanoparticles can be used for precise disease diagnostics and targeted distribution or release of therapeutic agents. These agents include small-molecule chemotherapeutics like cisplatin, doxorubicin, and paclitaxel, as well as biotherapeutics such as protein medicines, antibodies, DNA, siRNA, and miRNA. For instance, in mice with SCC-7 tumors, the hybrid nanocomposites DOX-$CaCO_3$-MNPs had a notable anticancer therapeutic effect compared to the control group and the group treated with free DOX [102–104].

The primary obstacles encountered in implants and drug delivery involve the need to decrease immunogenicity and enhance tissue integration. This can be accomplished by developing suitable biomaterials to engineer a novel material that improves the effectiveness of implants and drug delivery systems, while simultaneously minimizing the potential for immune rejection and inadequate tissue integration. Typically, when an implant or drug is introduced into the body, the human immune system becomes activated. This immune response then interacts with the biomaterials present in the system, recognizing the implant or medicine as a foreign substance. This interaction often results in the failure of the implants. Hence, it is crucial to develop medical biomaterials that possess both anti-fouling and anti-infective properties, while also being highly biocompatible. This can be accomplished by altering the surface characteristics of the biomaterial [28, 105, 106]. Therefore, surface modification is an effective method to enhance the biological characteristics, biocompatibility, and durability of materials. By modifying the surface of the substrate using materials at the microscale or nanoscale, it is possible to create programmable and controllable biomaterials. Implanted biomaterials are made "more compatible" inside the human body by enhancing their physical or chemical qualities, resulting in increased effectiveness. It is also an efficient method for regulating the deterioration of biomaterials at a predetermined rate in any physiological situation. Researchers are now investigating the methods to regulate the breakdown process by incorporating a degradable polymer with ceramic or metal materials. For instance, the surface of polylactic acid (PLA) is altered by incorporating titanium dioxide and zirconia nanocomposites.

This modification aims to regulate the degradation process in orthopedic screws and enhance their mechanical stability [20, 26, 107].

In addition, to enhance both immune modulation and tissue regeneration, a combination therapy is being investigated. This therapy involves embedding drugs into the implant, allowing for simultaneous drug delivery. This approach not only prevents immune rejection of the implant but also promotes healing and functional recovery, leading to improved treatment outcomes. This approach entails the incorporation of several therapeutic agents or the amalgamation of different treatment techniques within a unified platform. Collectively, these solutions encompass a thorough method for enhancing the efficiency of implants and drug delivery systems in clinical environments.

7.2 Targeted and Personalized Medicine

Several biomaterials now used for illness diagnosis or bioimaging are invasive, time-consuming, and have significant hazardous side effects. In 2002, John Funkhouser, the CEO and President of PharmaNetics, coined the term "theranostics," which combines the words "therapeutics" and "diagnostics." Theranostics involves the customization of nanomaterials for targeted and personalized medical applications. This enables non-invasive imaging, targeting, and treatment specifically at the sites of disease, while minimizing impact on nearby healthy cells. It will improve the monitoring of medication release and distribution in real time, leading to a decrease in over dose. This will also result in cost savings and early diagnosis, which will raise the patient's chances of survival. Additionally, the optimized treatment will enhance therapeutic efficiency and ensure the best safety profile. Theranostic nanomaterials consist of various chemical components, including a targeting moiety for specific cellular binding, therapeutic agents for drug delivery, diagnostic agents for non-invasive imaging, and a polymer coating or matrix that ensures colloidal stability and offers functional groups for bioconjugation. The human body possesses multiple barriers that prevent biomaterials from invading specific areas. This is why nanomaterials are used in the theranostic process. Nanomaterials are small in size but have a large functional surface area to volume ratio. Additionally, they exhibit interesting electrical and optical properties that facilitate their absorption, dissolution, or covalent attachment to surfaces, enabling the formation of NP drug/imaging/targeting complexes. The complexes have the ability to transmit signals (diagnostics) and transport the substances to the cancer cells or specific locations for therapeutic purposes. Ideally, following the deliberate release of enclosed drug molecules, the remaining nanoparticles will undergo safe degradation and be eliminated from the body [102–104].

Targeted drug delivery is one of the advanced modes of smart delivery system to overcome the disadvantages of existing therapeutics. By directly delivering the cargo to the site of action, this system reduces the dosage frequency and drug load.

It also enables non-specific action leading to undesirable outcomes. The use of such targeted drug delivery systems is discussed below.

A delivery system that integrates Luteinizing Hormone-Releasing Hormone (LHRH)-conjugated magnetite nanoparticles into microporous poly-dimethyl-siloxane (PDMS) for the targeting of triple-negative breast cancer cells was explored. The hormone conjugated drugs—prodigiosin and paclitaxel, showed a sustained release for over 30 days and showed a higher percentage of cell growth inhibition. This conjugated target-specific system was evaluated to differentiate between cancer and nontargeted cancer drugs which was proved in vitro by the researchers. Thus, the composite system shows enhanced targeting efficiency toward cancer cells and reduces the side effects on non-cancerous tissues overcoming the major disadvantage of the existing therapy [108].

Manganese-based immunomodulatory nanocomposite with catalase-like activity was designed as a treatment for antirheumatoid arthritis activity. Manganese nanoparticles were attached with anchored on UiO-66-NH_2 which was then coated with Mn-EGCG film which was again coated with hyaluronic acid (UMnEH). This composite UMnEH was found to respond to microwave and pH and diffuse slowly at the synovial site when injected in rheumatoid arthritis (RA) conditions. The hydrogen peroxide generated at the site dissociates Mn_3O_4 NPs which scavenges reactive oxygen species and downregulated the expression of hypoxia-inducible factor-1α (HIF-1α). This enables re-polarization of M1 phenotype macrophages into anti-inflammatory (M2) phenotype macrophages showing prospects for modulation and treatment of rheumatoid arthritis [109].

Another study was on the preparation of Pd NPs coated with pectin-modified Fe_3O_4 nanocomposite for its potential application as a cardioprotective effects. This magnetically retrievable nanocatalyst produced by C–C and C–N cross-coupling reactions was found to effectively reduce the caspase-3 activity. The in vitro studies showed promising results indicating its role as a prophylactic therapy for cardiovascular diseases [110].

A multifunctional tau-targeted multifunctional nanocomposite was developed for the combinational treatment of Alzheimer's disease. This methylene blued loaded CeNC/IONC/MSN-T807 construct was found to which inhibits numerous key processes of tau-associated AD pathogenesis in addition to having a high binding affinity to hyperphosphorylated tau. Thus, it proved to be effective in combating the oxidative stress, and tau hyperphosphorylation. Neuronal death was prevented in vitro and in vivo with memory restoration in AD rats confirming its therapeutic potential in Alzheimer's disease [111].

However, none of the nanotheranostic materials have received FDA approval due to their heterogeneous design, manufacturing, and efficacy. Theranostic is an innovative and groundbreaking method with a wide range of clinical applications, aimed at enhancing the quality of life for the people. In order to design nanomaterials with strong efficacy and clinical benefits, it is crucial to thoroughly understand their strengths, limitations, and potential side effects or toxicities. It is important to prioritize the comprehension of the pharmacokinetic characteristics of nanomaterials in humans, as well as their translation into appropriate animal models for clinical purposes [112].

7.3 Regulatory and Ethical Considerations

Nanocomposites may be a promising material for the biomedical application of drug delivery and implants, as they have the potential to improve their mechanical, physical, chemical, and biological properties. Nevertheless, their integration into healthcare environments also presents substantial ethical, regulatory, and public perception challenges [113–115].

7.3.1 Regulatory Considerations

The regulatory changes that a novel material will encounter are summarized in the following points:

- When nanocomposites interact with biological systems, it can lead to changes in absorption, distribution, and excretion, potentially causing adverse effects in the human body. Regulatory bodies such as the U.S. Food and Drug Administration (FDA) and the European Medicines Agency (EMA) are responsible for investigating the short-term and long-term toxicity effects of nanomaterials.
- Formulating a regulatory framework for nanocomposites is a further challenge, as the existing framework for conventional materials may not be applicable to the still developing nanocomposites. The complexity of developing specific evaluation criteria is heightened by the need to ensure safety and effectiveness of the diverse range of nanocomposites. In the context of the evolving regulatory framework for nanomaterials, the absence of customized protocols can result in extended approval periods, therefore impeding progress in medical applications.
- To ensure the safety and efficiency of the nanocomposites, it is necessary to assess their potential toxicity, biocompatibility, and long term stability through rigorous clinical testing.
- Ensuring the safety of environmental and occupational workers involved in the manufacturing of nanocomposites is crucial. These workers may be exposed to harmful dust from nanoparticles during the manufacturing and handling process, which goes against strict occupational health standards.

7.3.2 Ethical Concerns

The ethical concerns that are primarily concerned with safety, accessibility, environmental impact, and the potential for exploitation of nanocomposite materials are as follows:

- One of the primary cornerstones of ethical concern is informed consent, which requires that the patient be fully apprised about the novel compositions. They must be cognizant of the potential benefits, hazards, and uncertainties that are linked to these materials. It is imperative that patients comprehend the implications of utilizing innovative technologies.
- Facilitating the accessibility of nanocomposites is crucial as they enter the market after overcoming several obstacles like as development expenses and regulatory barriers. These issues may lead to higher material costs, which may be unaffordable for those with less resources. Consequently, this will contribute to a discrepancy in the healthcare sector. Accordingly, it is imperative that it is accessible and reasonably priced for all governments, healthcare providers, and industry stakeholders.
- Ensuring the security risk, the potential misuse of novel nanocomposites can occur either intentionally or unintentionally. Consumption of medicine beyond the recommended dose may lead to harmful impacts. Therefore, it is crucial to carefully manage this through regulatory oversight.

7.3.3 Public Perceptions

The subsequent observations pertain to the significance of public perception in influencing the acceptability and extensive use of nanocomposite materials in medical applications:

- Developing an understanding and raising awareness about nanocomposites is a crucial aspect in shaping public perception. It is important to address any misconceptions or misinformation surrounding nanocomposites, as these can contribute to negative perceptions and create significant barriers to the advancement of this technology. The communication of information about nanocomposites to the public can be achieved through various means such as media, public outreach, and education campaigns.
- Establishing the confidence of the general audience will result in the effective acceptance of the innovative content. Insufficient transparency will result in the formation of resistance against the new substance. Establishing public confidence requires the implementation of strong regulatory procedures, transparent communication, and policy decisions grounded on empirical facts.
- The acceptance or rejection of nanoparticles is also contingent upon the culture. While some cultures may embrace them, others may resist them due to ethical, religious, or philosophical concerns. When introducing nanocomposite-based medical products to diverse populations, it is crucial to comprehend and resolve these cultural nuances.

By tackling the regulatory, ethical, and public perception obstacles associated with nanocomposite materials, it is possible to realize the potential of these materials in enhancing healthcare results, while simultaneously guaranteeing safety, effectiveness, and ethical dimensions.

References

1. Ramezani M, Ripin ZM (2023) An overview of enhancing the performance of medical implants with nanocomposites. J Compos Sci 7:199
2. Hakim LK et al (2024) The current applications of nano and biomaterials in drug delivery of dental implant. BMC Oral Health 24:126
3. Rogel MR, Qiu H, Ameer GA (2008) The role of nanocomposites in bone regeneration. J Mater Chem 18:4233–4241
4. Moreno-Vega A-I, Gomez-Quintero T, Nunez-Anita R-E, Acosta-Torres L-S, Castaño V (2012) Polymeric and ceramic nanoparticles in biomedical applications. J Nanotechnol 2012:936041
5. Matusiewicz H (2014) Potential release of in vivo trace metals from metallic medical implants in the human body: from ions to nanoparticles–a systematic analytical review. Acta Biomater 10:2379–2403
6. Wang N, Fuh JYH, Dheen ST, Senthil Kumar A (2021) Functions and applications of metallic and metallic oxide nanoparticles in orthopedic implants and scaffolds. J Biomed Mater Res B Appl Biomater 109:160–179
7. Begines B et al (2020) Polymeric nanoparticles for drug delivery: recent developments and future prospects. Nanomaterials 10:1403
8. Elmowafy M et al (2023) Polymeric nanoparticles for delivery of natural bioactive agents: recent advances and challenges. Polymers (Basel) 15:1123
9. Lu X-Y, Wu D-C, Li Z-J, Chen G-Q (2011) Chpater 7—Polymer nanoparticles. In: Villaverde A (ed) Progress in molecular biology and translational science, vol 104. Academic Press, pp 299–323
10. Adhikari C (2021) Polymer nanoparticles-preparations, applications and future insights: a concise review. Polym Plastics Technol Mater 60:1996–2024
11. Rao JP, Geckeler KE (2011) Polymer nanoparticles: preparation techniques and size-control parameters. Prog Polym Sci 36:887–913
12. Zhang W et al (2023) Effects of morphology and size of nanoscale drug carriers on cellular uptake and internalization process: a review. RSC Adv 13:80–114
13. Cicuéndez M et al (2014) Tailoring hierarchical meso–macroporous 3D scaffolds: from nano to macro. J Mater Chem B 2:49–58
14. He Y et al (2019) Cancer cell–nanomaterial interface: role of geometry and surface charge of nanocomposites in the capture efficiency and cell viability. Biomater Sci 7:2759–2768
15. Ramburrun P, Khan RA, Choonara YE (2022) Design, preparation, and functionalization of nanobiomaterials for enhanced efficacy in current and future biomedical applications. 11, pp 1802–1826
16. Kim S-B, Kim C-H, Lee S-Y, Park S-J (2024) Carbon materials and their metal composites for biomedical applications: a short review. Nanoscale
17. Białkowska A, Bakar M, Kucharczyk W, Zarzyka I (2023) Hybrid epoxy nanocomposites: improvement in mechanical properties and toughening mechanisms—A review. Polymers (Basel) 15:1398
18. Kitchen G, Sun B, Kang SH (2024) Bioinspired nanocomposites with self-adaptive mechanical properties. Nano Res 17:633–648
19. Asif M, Ramezani M, Khan KA, Khan MA, Aw KC (2019) Investigation of the strain-rate-dependent mechanical behavior of a photopolymer matrix composite with fumed nano-silica filler. Polym Eng Sci 59:1695–1700
20. Kannan VS, Lenin K, Srinivasan D, Raj Kumar D (2024) Analysis of microstructural, mechanical and surface properties of aluminium hybrid composites obtained through stir casting. J Inst Eng (India): Series D 105:733–744
21. Abazari S, Shamsipur A, Bakhsheshi-Rad HR, Berto F (2022) Functionalized carbon nanotube-encapsulated magnesium-based nanocomposites with outstanding mechanical and biological properties as load-bearing bone implants. Mater Des 213:110354

22. Remes A, Williams DF (1992) Review: immune response in biocompatibility. In: Williams DF (ed) The biomaterials: Silver Jubilee compendium. Elsevier Science, Oxford, pp 79–91. https://doi.org/10.1016/B978-008045154-1.50012-5
23. Franz S, Rammelt S, Scharnweber D, Simon JC (2011) Immune responses to implants–a review of the implications for the design of immunomodulatory biomaterials. Biomaterials 32:6692–6709
24. Kumaravel V et al (2021) Antimicrobial TiO_2 nanocomposite coatings for surfaces, dental and orthopaedic implants. Chem Eng J 416:129071
25. Dhiman NK, Agnihotri S, Shukla R (2019) Silver-based polymeric nanocomposites as antimicrobial coatings for biomedical applications. Nanotechnology in modern animal biotechnology: recent trends and future perspectives, pp 115–171
26. Teixeira-Santos R, Belo S, Vieira R, Mergulhão FJM, Gomes LC (2023) Graphene-based composites for biomedical applications: surface modification for enhanced antimicrobial activity and biocompatibility. Biomolecules 13:1571
27. Kharat Z, Sadri M, Kabiri M (2021) Herbal extract loaded chitosan/PEO nanocomposites as antibacterial coatings of orthopaedic implants. Fibers Polym 22:989–999
28. Ahmad Ruzaidi DA et al (2021) Advocating electrically conductive scaffolds with low immunogenicity for biomedical applications: a review. Polymers (Basel) 13:3395
29. Alshimaysawee S, Fadhel Obaid R, Al-Gazally ME, Alexis Ramírez-Coronel A, Bathaei MS (2023) Recent advancements in metallic drug-eluting implants. Pharmaceutics 15:223
30. Arcos D, Vallet-Regi M (2013) Bioceramics for drug delivery. Acta Mater 61:890–911
31. Filion TM, Xu J, Prasad ML, Song J (2011) In vivo tissue responses to thermal-responsive shape memory polymer nanocomposites. Biomaterials 32:985–991
32. Wang N, Dheen ST, Fuh JYH, Kumar AS (2021) A review of multi-functional ceramic nanoparticles in 3D printed bone tissue engineering. Bioprinting 23:e00146. Preprint at (2021)
33. Colilla M, Manzano M, Vallet-Regi M (2008) Recent advances in ceramic implants as drug delivery systems for biomedical applications. Int J Nanomed 3:403–414
34. Kunrath MF, Campos MM (2021) Metallic-nanoparticle release systems for biomedical implant surfaces: effectiveness and safety. Nanotoxicology 15:721–739
35. Thakkar KN, Mhatre SS, Parikh RY (2010) Biological synthesis of metallic nanoparticles. Nanomedicine 6:257–262
36. Mody VV, Siwale R, Singh A, Mody HR (2010) Introduction to metallic nanoparticles. J Pharm Bioallied Sci 2:282–289
37. Schröfel A et al (2014) Applications of biosynthesized metallic nanoparticles–a review. Acta Biomater 10:4023–4042
38. Jamkhande PG, Ghule NW, Bamer AH, Kalaskar MG (2019) Metal nanoparticles synthesis: an overview on methods of preparation, advantages and disadvantages, and applications. J Drug Deliv Sci Technol 53:101174
39. Liu L et al (2017) Metal nanoparticles induced photocatalysis. Natl Sci Rev 4:761–780
40. Nizzero S, Ziemys A, Ferrari M (2018) Transport barriers and oncophysics in cancer treatment. Trends Cancer 4:277–280
41. Nazario-Toole AE, Wu LP (2017) Chapter two—phagocytosis in insect immunity. In: Advances in insect physiology, vol 52. Academic Press, pp 35–82
42. Pauwels A-M, Trost M, Beyaert R, Hoffmann E (2017) Patterns, receptors, and signals: regulation of phagosome maturation. Trends Immunol 38:407–422
43. Butt AM, Abdullah N, Rani NNIM, Ahmad N, Amin MCIM (2022) Endosomal escape of bioactives deployed via nanocarriers: insights into the design of polymeric micelles. Pharm Res 39:1047–1064
44. Li Q, Guo D, Zhang R, Wang X (2012) Increasing anticancer drug internalization induced by new Au-MWCNTs nanocomposite. J Nanosci Nanotechnol 12:2192–2198
45. Almeida PV et al (2017) A multifunctional nanocomplex for enhanced cell uptake, endosomal escape and improved cancer therapeutic effect. Nanomedicine 12:1401–1420
46. Rangam N et al (2024) Surface and composition effects on the biphasic cytotoxicity of nanocomposites synthesized using leaf extracts. Int J Biol Macromol 276:133723

47. Bamburowicz-Klimkowska M, Poplawska M, Grudzinski IP (2019) Nanocomposites as biomolecules delivery agents in nanomedicine. J Nanobiotechnol 17:48
48. Li Y et al (2020) Cell membrane-engineered hybrid soft nanocomposites for biomedical applications. J Mater Chem B 8:5578–5596
49. Liu Y, Li M, Liu H, Kang C, Yu X (2023) Strategies and progress of Raman technologies for cellular uptake analysis of the drug delivery systems. Int J Nanomed 6883–6900
50. Zhao H et al (2018) Insights into the intracellular behaviors of black-phosphorus-based nanocomposites via surface-enhanced Raman spectroscopy. Nanophotonics 7:1651–1662
51. Shukla AK et al (2024) Transferrin immobilized graphene oxide nanocomposite for targeted cancer chemodynamic therapy via increasing intracellular labile Fe^{2+} concentration. ACS Appl Bio Mater
52. Deng W, Kautzka Z, Chen W, Goldys EM (2016) PLGA nanocomposites loaded with verteporfin and gold nanoparticles for enhanced photodynamic therapy of cancer cells. RSC Adv 6:112393–112402
53. Hira I et al (2022) Apoptotic cell death induction through pectin, guar gum and zinc oxide nanocomposite in a549 lung adenocarcinomas. Biointerface Res Appl Chem 12:1856–1869
54. Akhtar MJ, Ahamed M, Alhadlaq H (2022) CeO_2-Zn nanocomposite induced superoxide, autophagy and a non-apoptotic mode of cell death in human umbilical-vein-derived endothelial (HUVE) cells. Toxics 10:250
55. Bach H (2015) Cytotoxicology of nanocomposites. In: Bio-Nanoparticles, pp 273–302. https://doi.org/10.1002/9781118677629.ch13
56. Smith MJ, White KL Jr, Smith DC, Bowlin GL (2009) In vitro evaluations of innate and acquired immune responses to electrospun polydioxanone–elastin blends. Biomaterials 30:149–159
57. Cicuéndez M et al (2014) Effects of 3D nanocomposite bioceramic scaffolds on the immune response. J Mater Chem B 2:3469–3479
58. Wang Y et al (2021) Quercetin-loaded ceria nanocomposite potentiate dual-directional immunoregulation via macrophage polarization against periodontal inflammation. Small 17:2101505
59. Li X et al (2019) Surface treatments on titanium implants via nanostructured ceria for antibacterial and anti-inflammatory capabilities. Acta Biomater 94:627–643
60. Sharma P et al (2019) Application of ZnO-based nanocomposites for vaccines and cancer immunotherapy. Pharmaceutics 11:493
61. Bharadwaz A, Jayasuriya AC (2020) Recent trends in the application of widely used natural and synthetic polymer nanocomposites in bone tissue regeneration. Mater Sci Eng C 110:110698
62. Montaño-Machado V et al (2021) Development of multifunctional materials based on poly (ether ether ketone) with improved biological performances for dental applications. Materials 14:1047
63. Hashim NC, Frankel D, Nordin D (2019) Graphene oxide-modified hydroxyapatite nanocomposites in biomedical applications: a review. Ceramics-Silikaty
64. Mirza EH et al (2017) Physical, mechanical, thermal, and dynamic characterization of carbon nanotubes incorporated poly (methyl methacrylate)-based denture implant. J Compos Mater 51:3931–3940
65. Das R, Bhattacharjee C (2019) Titanium-based nanocomposite materials for dental implant systems. In: Applications of nanocomposite materials in dentistry. Elsevier, pp 271–284
66. Prahasanti C et al (2022) Utilization of polymethyl methacrylate and hydroxyapatite composite as biomaterial candidate for porous trabecular dental implant fixture development: a narrative review. Res J Pharm Technol 15:1863–1869
67. Sung Y-M, Shin Y-K, Ryu J-J (2007) Preparation of hydroxyapatite/zirconia bioceramic nanocomposites for orthopaedic and dental prosthesis applications. Nanotechnology 18:065602
68. Zheng X et al (2014) Improved biocompatibility of poly (lactic-co-glycolic acid) and poly-L-lactic acid blended with nanoparticulate amorphous calcium phosphate in vascular stent applications. J Biomed Nanotechnol 10:900–910

69. de Mel A et al (2014) Nitric oxide-eluting nanocomposite for cardiovascular implants. J Mater Sci Mater Med 25:917–929
70. Xu W, Yagoshi K, Asakura T, Sasaki M, Niidome T (2019) Silk fibroin as a coating polymer for sirolimus-eluting magnesium alloy stents. ACS Appl Bio Mater 3:531–538
71. Wu M et al (2023) Engineering mussel-inspired multifunctional nanocomposite hydrogels to orchestrate osteoimmune microenvironment and promote bone healing. Mater Des 227:111705
72. Moosavifar M et al (2022) Biomimetic organic–inorganic nanocomposite scaffolds to regenerate cranial bone defects in a rat animal model. ACS Biomater Sci Eng 8:1258–1270
73. Krucińska I et al (2017) Biological properties of low-toxicity PLGA and PLGA/PHB fibrous nanocomposite implants for osseous tissue regeneration. Part I: Evaluation of potential biotoxicity. Molecules 22:2092
74. Schade R et al (2010) Biomimetic organic–inorganic nanocomposite coatings for titanium implants. In vitro and in vivo biological testing. J Biomed Mater Res A 95:691–700
75. Nguyen VN et al (2024) Preparation and characterization of Chitosan/Starch nanocomposites loaded with ampicillin to enhance antibacterial activity against *Escherichia coli*. Polymers (Basel) 16:2647
76. Sohrabi M, Hesaraki S, Shahrezaee M, Shams-Khorasani A (2024) The release behavior and in vitro osteogenesis of quercetin-loaded bioactive glass/hyaluronic acid/sodium alginate nanocomposite paste. Int J Biol Macromol 136094
77. González-Iñiguez KJ, Figueroa-Ochoa EB, Martínez-Richa A, Cajero-Zul LR, Nuño-Donlucas SM (2024) Synthesis of poly (L–lactide)–poly (ε–caprolactone)–poly (ethylene glycol) terpolymer grafted onto partially oxidized carbon nanotube nanocomposites for drug delivery. Polymers (Basel) 16:2580
78. Sohail R, Abbas SR (2020) Evaluation of amygdalin-loaded alginate-chitosan nanoparticles as biocompatible drug delivery carriers for anticancerous efficacy. Int J Biol Macromol 153:36–45
79. Duan Y et al (2024) Proton sponge nanocomposites for synergistic tumor elimination via autophagy inhibition-promoted cell apoptosis and macrophage repolarization-enhanced immune response. ACS Appl Mater Interfaces 16:17285–17299
80. Xiong K et al (2020) Co-delivery of paclitaxel and curcumin by biodegradable polymeric nanoparticles for breast cancer chemotherapy. Int J Pharm 589:119875
81. Ghasemi Toudeshkchouei M, Zahedi P, Shavandi A (2020) Microfluidic-assisted preparation of 5-fluorouracil-loaded PLGA nanoparticles as a potential system for colorectal cancer therapy. Materials 13:1483
82. Pang H et al (2024) Development of polysaccharide-coated layered double hydroxide nanocomposites for enhanced oral insulin delivery. Drug Deliv Transl Res 1–11
83. Malekzadeh kebria M et al (2024) Designing and synthesis of In-Situ hydrogel based on pullulan/carboxymethyl chitosan containing parathyroid hormone for bone tissue engineering. Materialia (Oxf) 33:102026
84. Shariatinia Z, Zahraee Z (2017) Controlled release of metformin from chitosan–based nanocomposite films containing mesoporous MCM-41 nanoparticles as novel drug delivery systems. J Colloid Interface Sci 501:60–76
85. Hindawy RF et al (2024) Exploring the potential of selenium nanoparticles and fabricated selenium nanoparticles@ vitamin C nanocomposite in mitigating nicotine-induced testicular toxicity in rats. Toxicol Res (Camb) 13:tfae154
86. Xiang Y et al (2024) Dexamethasone-loaded ROS stimuli-responsive nanogels for topical ocular therapy of corneal neovascularization. J Control Release 372:874–884
87. Jin J, Chen Y, Li H, Xu Y, Wang L (2024) Loading polyaniline (PANI) nanoparticles to mesoporous hydroxyapatite (HAp) spheres for near infrared (NIR) induced doxorubicin (DOX) drug delivery and colon cancer treatment. Phys Chem Chem Phys 26:23277–23287
88. Zarei Z, Kharaziha M, Karimzadeh F, Khadem E (2024) Synthesis and biological applications of nanocomposite hydrogels based on the methacrylation of hydroxypropyl methylcellulose and lignin loaded with alpha-pinene. Carbohydr Polym 346:122642

89. Mi B et al (2024) Laponite modified methacryloyl gelatin hydrogel with controlled release of vascular endothelial growth factor A for bone regeneration. Biochem Biophys Res Commun 150714
90. Dalei G, Jena D, Das S (2024) 5-Fluorouracil-loaded green chitosan nanoparticles/guar gum nanocomposite hydrogel in controlled drug delivery. Carbohydr Res 545:109257
91. Chen C et al (2023) Repolarizing tumor-associated macrophages by layered double hydroxide-based deacidification agent for tumor chemodynamic therapy and immunotherapy. Colloids Surf B Biointerfaces 223:113157
92. Gao Q, Liu R, Wu Y, Wang F, Wu X (2024) Versatile self-assembled near-infrared SERS nanoprobes for multidrug-resistant bacterial infection-specific surveillance and therapy. Acta Biomater
93. Babu AS et al (2023) NANOTEX BONE graft along with fibula flap in the reconstruction of segmental mandibular defect: protocol for pilot clinical trial. Int J Surg Protoc 27:90–96
94. Fakhri E et al (2024) Colloidal chitosan-silver nanoparticles-fluoride nanocomposite as an antibacterial mouthwash against salivary *Streptococcus mutans* in orthodontic patients (a randomized clinical trial). Clin Oral Investig 28:435
95. Ishihara M et al (2023) A phase 1 trial of NY-ESO-1-specific TCR-engineered T-cell therapy combined with a lymph node-targeting nanoparticulate peptide vaccine for the treatment of advanced soft tissue sarcoma. Int J Cancer 152:2554–2566
96. Hafizi M et al (2019) A randomized, double-blind, placebo-controlled investigation of BCc1 nanomedicine effect on survival and quality of life in metastatic and non-metastatic gastric cancer patients. J Nanobiotechnol 17:1–19
97. Nanda SS, Yi DK (2024) Recent advances in synergistic effect of nanoparticles and its biomedical application. Int J Mol Sci 25:3266
98. Xiang H et al (2019) Photo-crosslinkable, self-healable and reprocessable rubbers. Chem Eng J 358:878–890
99. Shi W et al (2024) Stretchable, self-healing, and bioactive hydrogel with high-functionality N, N-bis (acryloyl) cystamine dynamically bonded Ag@ polydopamine crosslinkers for wearable sensors. Adv Sci 2404451
100. Rautiainen JM, Valkonen A, Lundell J, Rissanen K, Puttreddy R (2024) The geometry and nature of C–I···O–N interactions in perfluoroiodobenzene-pyridine N-oxide halogen-bonded complexes. Adv Sci 2403945
101. Yang J et al (2023) Current self-healing binders for energetic composite material applications. Molecules 28:428
102. Ahmad J et al (2023) Hybrid quantum dot as promising tools for theranostic application in cancer. Electronics (Basel) 12:972
103. Asadi M, Ghorbani SH, Mahdavian L, Aghamohammadi M (2024) Graphene-based hybrid composites for cancer diagnostic and therapy. J Transl Med 22:611
104. Jo SD, Ku SH, Won Y-Y, Kim SH, Kwon IC (2016) Targeted nanotheranostics for future personalized medicine: recent progress in cancer therapy. Theranostics 6:1362
105. Sun Q, Yang Z, Qi X (2023) Design and application of hybrid polymer-protein systems in cancer therapy. Polymers (Basel) 15:2219
106. Huang S-M, Liu S-M, Ko C-L, Chen W-C (2022) Advances of hydroxyapatite hybrid organic composite used as drug or protein carriers for biomedical applications: a review. Polymers (Basel) 14:976
107. Xu K, Chu P, Rumon RH, Fujimori A (2024) Tensile and bending properties of sandwich films of carbon fiber reinforced thermoplastics-polypropylene sheet processed by a simple hot-press method. SPE Polymers
108. Eluu SC et al (2024) Luteinizing hormone-releasing hormone (LHRH)-conjugated cancer drug delivery from magnetite nanoparticle-modified microporous poly-di-methyl-siloxane (PDMS) systems for the targeted treatment of triple negative breast cancer cells. J Funct Biomater 15
109. Chen X et al (2023) Manganese-based immunomodulatory nanocomposite with catalase-like activity and microwave-enhanced ROS elimination ability for efficient rheumatoid arthritis therapy. Small 19:2304610

110. Zhang W et al (2020) Fabrication of Pd NPs on pectin-modified Fe_3O_4 NPs: a magnetically retrievable nanocatalyst for efficient C-C and C–N cross coupling reactions and an investigation of its cardiovascular protective effects. Int J Biol Macromol 160:1252–1262
111. Chen Q et al (2018) Tau-targeted multifunctional nanocomposite for combinational therapy of Alzheimer's disease. ACS Nano 12:1321–1338
112. Vasantha Ramachandran R, Bhat R, Kumar Saini D, Ghosh A (2021) Theragnostic nanomotors: successes and upcoming challenges. Wiley Interdiscip Rev Nanomed Nanobiotechnol 13:e1736
113. Oksel Karakus C, Bilgi E, Winkler DA (2021) Biomedical nanomaterials: applications, toxicological concerns, and regulatory needs. Nanotoxicology 15:331–351
114. Meenach SA, Anderson KW, Hilt JZ (2008) Hydrogel nanocomposites: biomedical applications, biocompatibility, and toxicity analysis. In: Safety of nanoparticles: from manufacturing to medical applications. Springer, pp 131–157
115. Dobrzański LA, Dobrzańska-Danikiewicz AD, Dobrzański LB (2021) Effect of biomedical materials in the implementation of a long and healthy life policy. Processes 9:865

Chapter 14
Toxicology Assessment of Polymeric Material for Implants

Atanu Kumar Paul, Gourhari Chakraborty, and Arbind Prasad

1 Introduction

Advancements in drug delivery systems have led to the development of implantable devices like drug pumps and depots, offering innovative solutions for sustained, localized drug release to improve therapeutic outcomes. These implants vary in size, from microspheres to a few millimeters, and can deliver drugs, hormones, or even cells. Examples include drug-loaded composites, stents, biosensors, catheters, tissue engineering scaffolds, heart valves, and pacemakers [1, 2]. The polymers used in these implants may be either biodegradable or non-biodegradable, based on their compatibility with the intended drug delivery [3]. Examples of Food and Drug Administration (FDA)-approved polymeric implants include Atrigel, Atridox, and Gliadel [4, 5].

A key advantage of implants is their preservative-free nature, making them particularly beneficial. The materials used must exhibit thermal stability, biocompatibility, and properties that support cell adhesion and proliferation. For specific applications, different types of polymers are employed; for instance, injectable implants use low-viscosity polymer solutions that solidify post-injection [6, 7].

With the development of new biomaterials, it is crucial to evaluate their biological parameters through experimental studies, often using animal models. Researchers must adhere to the 3R principles: Reduction, minimizing the number of animals used; Refinement, optimizing procedures to minimize animal suffering; and Replacement,

A. K. Paul (✉) · G. Chakraborty
Department of Chemical Engineering, National Institute of Technology Andhra Pradesh, Tadepalligudem, India
e-mail: atanukpaul@outlook.com

A. Prasad
Department of Mechanical Engineering, Katihar Engineering College (Under Department of Science &Technology, Government of Bihar), Katihar, India

A. Prasad and V. Katiyar (eds.), *Resorbable Polymers for Bioimplants and Fixation Devices*, Materials Horizons: From Nature to Nanomaterials,
https://doi.org/10.1007/978-981-96-9306-1_14

utilizing alternatives such as cell cultures, microorganisms, chemicals, or silicon models whenever possible [8].

Polymeric materials play a crucial role in medical implants due to their biocompatibility, flexibility, and diverse properties [9]. Various synthetic polymers, such as polyvinylidene fluoride, polyethylene, and polyurethane, are used for packaging implantable devices, offering protection against moisture and ions [9]. Biopolymers like collagen, gelatin, and poly(lactic acid) are increasingly utilized in orthopedics, ophthalmology, and cardiology due to their biodegradability and renewability [10, 11]. Smart polymers, including shape-changing and shape-memory materials, have gained significant interest for minimally invasive surgeries [12]. To enhance visibility during medical procedures, radiopaque polymeric materials are developed through various methods, such as blending with inorganic salts or using radiopaque polymerized monomers [13]. Future research is expected to focus on biocompatible radiopaque polymers with covalently bound opaque elements, aiming to maintain properties equivalent to non-opaque parent polymers [13].

Toxicological assessments are crucial in evaluating the potential impacts of chemicals and materials on environmental and human health. These assessments involve various lines of evidence, with chronic exposure and reproductive endpoints ranked highest by subject matter experts [14]. Toxicology testing is a planned process that begins during preclinical stages of drug development and continues through clinical trials and post-market surveillance, providing essential safety data to support informed decision-making [15]. It encompasses multiple aspects, including genotoxicity, carcinogenicity, immunotoxicity, and reproductive toxicity, utilizing in vitro, in vivo, and silico methods [16]. While traditional toxicity testing relies heavily on animal studies, there is a growing shift toward innovative approaches to reduce animal usage. The importance of toxicological assessments extends across various fields, including environmental impact assessment, pharmaceutical development, and transfusion medicine [14–16].

Regulatory standards and guidelines for medical implants encompass various aspects of safety and efficacy. The Medical Implant Communications Service (MICS) provides standards for ultra-low power transceivers in implanted devices [17]. While the FDA doesn't approve biomaterials directly, standards and guidelines help evaluate manufacturing processes and material properties for implants [18]. International organizations like ISO, ASTM, and CE marking play crucial roles in developing standards for biomaterials, medical devices, and regenerative medicine products [19]. Figure 1 shows the various applications of resorbable implants in surgery. The ISO 10993 series provides comprehensive guidelines for biocompatibility testing, including screening for various toxicity mechanisms and chemical characterization of materials [20]. These international standards are recognized by most national regulatory bodies and supplement country-specific guidelines, ensuring the safety and efficacy of medical implants through rigorous evaluation of materials, manufacturing processes, and biological interactions.

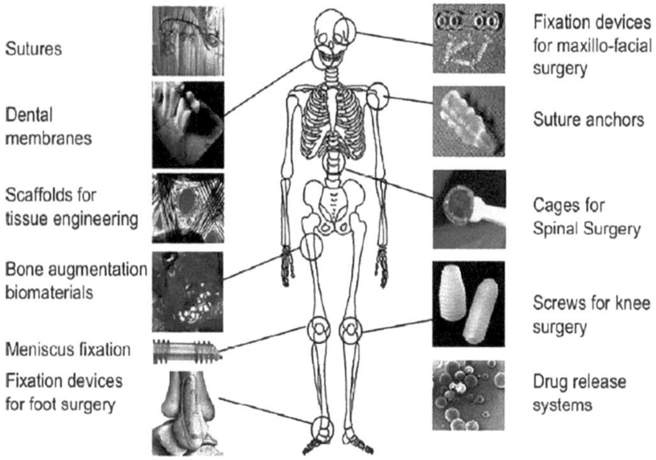

Fig. 1 Application of resorbable implants in surgery. In anti-clockwise direction: sutures for closing wounds, dental membranes, scaffolds for tissue engineering, bone augmentation biomaterials, meniscus fixation screw, fixation devices for foot surgery, drug release microspheres, cannulated screws for tendon fixation in knee surgery, cages for spinal surgery, suture anchors for tendon and ligament fixation in shoulder surgery, fixation devices for maxillo-facial surgery

2 Key Properties of Polymeric Materials

2.1 Chemical Composition and Structure

Polymeric materials possess a diverse range of properties, making them highly versatile for various applications, including dentistry [21]. Important attributes include mechanical strength, rigidity, corrosion resistance, lightweight nature, and dimensional stability [22]. These characteristics depend on factors like the polymer's chemical composition, molecular weight, and molecular weight distribution [23]. Analytical methods such as mass spectrometry, nuclear magnetic resonance, and gel permeation chromatography are commonly employed to assess their structure and properties [24]. Key physical properties, including hydrophobicity and flexibility, along with chemical traits like reactivity and biocompatibility, are critical in determining their appropriateness for specific uses [24]. Recent innovations have led to the creation of polymeric materials with enhanced functionalities, such as antibacterial effects and remineralization abilities, particularly beneficial in dental applications [21]. Figure 2 illustrates various smart polymers designed for biomedical purposes.

The structural investigation of polymers encompasses the characterization of chemical composition, molecular mass, and molecular mass distribution, as well as the arrangement and morphology of polymer chains [25]. For fluorinated polymers, various analytical techniques, including spectral methods, chemical analysis, and microscopy, are employed to evaluate their structure and composition, which significantly influence their operational properties [26]. The chemical composition,

Fig. 2 Smart polymers for biomedical applications

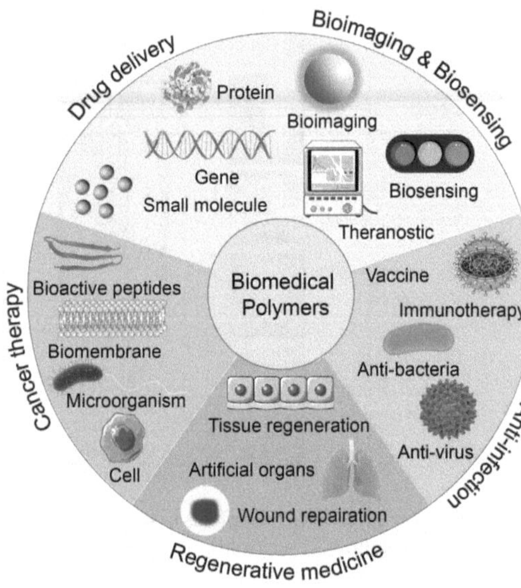

configuration, and molecular arrangement collectively determine the polymer's structure, while its physical structure is referred to as the polymer's morphology [27]. Understanding these structural aspects is crucial for tailoring polymers to specific applications and optimizing their performance in various fields.

2.2 *Biocompatibility Considerations*

The biocompatibility of polymeric materials is crucial for their successful application in biomedical fields. The evaluation process involves extensive laboratory and animal testing to ensure safety before human use [23]. Factors influencing biocompatibility include raw materials, manufacturing processes, cleaning and sterilization methods, and biodegradation [28]. Surface properties, such as hydrophilicity and hydrophobicity, significantly affect interactions with proteins and platelets, impacting overall biocompatibility [29]. Polymers can be designed as biocompatible, biodegradable, or for specific diagnostic and therapeutic applications, including artificial organs, biosensors, and drug delivery systems [30]. Regulatory authorities guide biocompatibility testing to aid in selecting appropriate analyses [29]. Understanding the molecular-level and cellular-level interactions between polymeric materials and biosystems is essential for developing effective biomedical applications [30, 31].

2.3 *Mechanical and Degradation Properties*

Polymeric materials exhibit diverse mechanical and thermal properties, including tensile strength, elongation, stiffness, and glass transition temperature, which are influenced by preparation methods and reaction conditions [32]. Degradation of polymers can occur through various chemical, physical, and mechanical mechanisms, typically resulting in reduced molecular weight and decreased mechanical performance [33]. The relationship between microstructure and mechanical properties has been studied using polypropylene blends, revealing the effects of heat stabilizers on degradation behavior [33]. Table 1 shows a list of Food and Drug Administration (FDA)-approved polymers and their mechanical contributions to copolymer systems [34, 35]. For biodegradable polymers used in medical applications, mechanical properties are affected by molecular weight and crystallinity changes during degradation [36]. A linear relationship between Young's modulus and log(Mn) has been observed across multiple experimental datasets [36]. Polymer degradation assessment is complex, involving multiple factors such as chemical and biological mechanisms, as well as stress [37]. Understanding these properties and degradation mechanisms is crucial for the effective use of polymeric materials in various applications. Table 2 shows glass transition temperature of some useful polymers.

3 Toxicological Risks Associated with Polymeric Implants

To evaluate the suitability of polymeric materials for implantation, toxicological studies often involve animal experiments to obtain reliable results at a lower cost. Researchers must minimize the number of animals used, strictly adhering to protocols that ensure the welfare and comfort of the animals before, during, and after the study

Table 1 Overview of FDA-approved polymers and their roles in copolymer systems

Polymer	Mechanical contribution	Applications
Poly(lactic acid) (PLA)	Provides stiffness and strength; biodegradability	Drug delivery systems, tissue engineering
Poly(glycolic acid) (PGA)	Enhances crystallinity; improves mechanical properties	Sutures, scaffolds
Polycaprolactone (PCL)	Offers flexibility and elongation; slow degradation rate	Long-term implants, drug delivery
Polyethylene glycol (PEG)	Improves hydrophilicity; enhances biocompatibility	Hydrogels, drug carriers
Polydioxanone (PDO)	Combines strength with flexibility; moderate degradation rate	Sutures, orthopedic applications

Table 2 Glass transition temperature of some polymers

Name of polymers	Glass transition temperature (T_g), °C
Polyvinyl chloride	87
Polystyrene	100
Poly (l-lactide)	60
Nylon	66
Low-density polyethylene	−120
Polycarbonate	150
Polyethylene terephthalate	69

[38, 39]. If sacrificing an animal is necessary, it must be done following appropriate euthanasia principles based on the nature of the polymer and the implantation site. The choice of experimental animals should be carefully aligned with the material being tested. Animal testing should be avoided when alternative methods, such as software-based simulations, can achieve comparable results [15, 40].

Biocompatibility testing involves introducing the polymeric material into a living organism to assess its interaction with the biological system [11, 41]. Ideally, the material should maintain extensive contact with the living system without causing adverse effects. However, due to the varied chemical structures and functional groups of polymeric substances, their behavior can differ even within the same environment, making comprehensive evaluation through both in vitro and in vivo studies essential [28].

Key biocompatibility assessments include tests for carcinogenicity, mutagenicity, teratogenicity, antigenicity, implantation reactions, skin irritation, hemolysis, and cytotoxicity. These evaluations are crucial to ensuring the safe and effective application of polymeric materials in medical implants [28, 42].

Polymeric implants are used in medical devices to treat and diagnose diseases. They are soaked for 15–20 min in hydro-alcoholic solutions before being rinsed three times with sterile PBS. Hair is removed from stored animals while they are under general anesthesia. The implant is inserted into the tissue, and the incision is closed with autoclips [43]. Rat behavior is monitored for 24 h. Toxicological studies are carried out after the seventh day of implantation. Histological analysis entails removing the polymeric implant with a cryostat and staining it with hematoxylin and eosin dye, followed by immunohistochemistry staining with CD68 counterstained with hematoxylin. Gel electrophoresis is used to assess genotoxicity and identify the point of contact between medical devices [43]. Genetic toxicology is evaluated using implant-extractable chemicals, such as the Ames test, chromosome aberration assay, mouse lymphoma assay, or in vitro micronucleus assay. Extracts from the implant are transported and stored in a vehicle at 70 °C for up to 72 h. High-level leaching chemicals may cause DNA damage [44]. Wistar rats aged 8–10 weeks are split up into different groups and placed in subcutaneous pockets or other body systems for cytotoxicity testing. Implants are removed after four to five days, stained with ethidium bromide and fluorescein diacetate, and examined under a fluorescent microscope.

Histopathology of the liver and implant sites is used to evaluate cytotoxicity; any inflammation, necrosis, apoptosis, or degeneration is indicative of toxicity [45, 46].

3.1 Potential Sources of Toxicity

Polymeric implants pose potential toxicological risks that require careful evaluation. Degradation of these materials can release harmful substances, leading to adverse effects in the body [40, 47]. Studies have shown that all solid polymers tested in rodents produced cancer, with some specific polymers inducing chemical carcinogenesis [40]. The accumulation of degradation products from bioabsorbable polymers can result in acute toxicity, as demonstrated in vitro for materials like poly(glycolic acid) [48]. However, degradable polymers also offer significant benefits in various biomedical applications, including tissue engineering, drug delivery, and implants [49]. To ensure safety, thorough toxicity studies and standardization of implant materials are crucial [47]. Additionally, risk–benefit assessments and toxicological evaluation strategies are essential for validating the safety of degradable polymers in biomedical applications [49].

Polymeric implants can potentially cause toxicity through various mechanisms. Degradation products of bioabsorbable polymers like poly(glycolic acid) and poly(L-lactic acid) may accumulate and exhibit acute toxicity in vitro [48]. Monomers, additives, and processing methods like sterilization can also contribute to toxicity risks [50]. Implant materials and their degradation byproducts must meet biocompatibility standards to avoid adverse biological responses [51]. Systemic toxicity can result from polymer disintegration, accumulation of plasticizers and monomers, and device-specific issues like clotting in heart-assist devices or bone resorption with silicone implants [40]. Additionally, all solid polymers tested in rodents produced cancer, highlighting the potential carcinogenic risk of polymer implants in humans [40]. Rigorous toxicity evaluation and quality control are crucial to mitigate these risks [50].

3.1.1 Additives and Residual Monomers

Polymeric implants and dental devices can potentially release toxic substances, including residual monomers and additives, posing health risks [50, 52]. Incomplete polymerization in dental materials leads to the accumulation of residual monomers, which can leach into the physiological environment [53]. These leachable components, particularly methacrylates, have been associated with genotoxic effects in both in vitro and in vivo studies [54]. Post-illumination and ethanol treatment can reduce the presence of potentially toxic compounds in additively manufactured dental devices [53]. However, the absence of specific regulations for medical devices in some countries highlights the need for stringent quality control and toxicity evaluation methods [50]. To address these concerns, improved extraction procedures, thorough physicochemical and biological testing, and the development of non-leaching additives are necessary for enhancing the safety of polymeric implants and dental materials [52].

3.1.2 Degradation Byproducts

Degradation byproducts from polymeric implants can potentially cause toxicity in surrounding tissues. Studies have shown that accumulated degradation products from bioabsorbable polymers like poly(glycolic acid) can exhibit acute toxicity in vitro [48]. The physiological environment, including free radicals and iron, can accelerate polymer degradation, with hydroxyl radicals being a primary cause of implant deterioration [41]. In dental implants, titanium particles and ions released due to wear-corrosion or peri-implantitis treatment can stimulate pro-inflammatory cytokines, inflammatory cell infiltration, and osteoclast activation in peri-implant tissues [55]. These degradation products have demonstrated cytotoxic and genotoxic potential, with their impact dependent on the amount and physicochemical properties of the released debris [55]. Consequently, biocompatibility testing of implant materials, devices, and their degradation byproducts is crucial to ensure they meet regulatory standards for safety [51].

3.2 Local Versus Systemic Toxicity

Polymeric implants can cause both local and systemic toxicity, necessitating a thorough evaluation of their biocompatibility and degradation products. While some polymers like PLGA (InnoPol) show good tissue compatibility with only transient local inflammation [56], others like PGA can produce toxic degradation products [48]. The accumulation of degradation products is a key concern, potentially leading to dose-dependent local and systemic toxicity [51]. Factors influencing toxicity include the implant material, degradation rate, and local pH changes. Metal implants, such as those used in hip prostheses, can release micro- and nanoparticles, ions, and oxides, causing adverse local tissue reactions and possible systemic effects [57]. To ensure safety, regulatory standards for biocompatibility must be met, and appropriate in vitro methodologies should be developed to better understand cellular responses to implant-derived substances [51, 57].

4 Toxicological Testing Methods

4.1 In Vitro Assessments

Toxicological testing is crucial for evaluating the safety of new drugs and substances before human use. It involves various methods, including in vitro studies using cell lines and in vivo experiments on animals [39, 58, 59]. Common testing approaches include acute, subchronic, and chronic toxicity assessments, as well as specialized tests for teratogenicity, mutagenicity, and neurotoxicity [9, 39, 59]. Modern toxicology laboratories employ diverse techniques such as immunoassays, gas chromatography, and mass spectrometry for comprehensive screening and targeted testing [60]. In vitro methods are increasingly used, including cell culture techniques, high-throughput screening, and toxicogenomics [39]. The primary goal of toxicity testing is to determine the No Observed Adverse Effect Level, which is essential for initiating clinical trials [39, 60]. These methods collectively help characterize potential toxic effects and assess the safety of investigational products.

In vitro assessments play a crucial role in the toxicological testing of implants, offering alternatives to animal testing and providing valuable insights into material biocompatibility. These methods include cytotoxicity assays, biochemical tests, and functional assessments using various cell lines and primary cell types [61, 62]. While in vitro tests are generally more sensitive than in vivo studies, they may sometimes lead to unnecessary rejection of clinically useful materials [63]. It is essential to evaluate both bulk material toxicity and the potential effects of wear particulates, as the latter can induce aseptic loosening and inflammation [64]. To ensure reliability and clinical relevance, integrated in vitro testing strategies should consider exposure modeling, data sharing, and a battery of tests to assess different mechanisms of action [62]. However, further research is needed to validate these methods for human applications and improve their predictive value.

4.1.1 Cytotoxicity Testing

Cytotoxicity testing of polymeric implants is crucial for assessing biocompatibility. In vitro tests using cell cultures can evaluate material toxicity, with methods including MTT assays, protein content, and enzyme leakage [65, 66]. However, in vitro tests may overestimate toxicity compared to in vivo responses, potentially leading to the rejection of clinically useful materials [63]. In vivo testing, such as corneal micropocket implantation in rabbits, can detect tissue responses to leachables from implanted materials [67]. Studies have shown that metallic materials generally exhibit higher biocompatibility than polymeric materials, with titanium demonstrating the highest compatibility [66]. Among polymers, polyetheretherketone (PEEK) showed better biocompatibility than epoxy resins, potentially stimulating osteoblast growth [65]. The addition of hydroxyapatite to polymethylmethacrylate (PMMA) can reduce cytotoxic effects [66].

4.1.2 Genotoxicity Testing

Genotoxicity testing of polymeric implants is crucial for assessing their safety in medical applications. Studies have compared the biocompatibility of polymeric and metallic materials, finding that metallic implants generally exhibit higher biocompatibility [66]. Various orthopedic implant materials, including titanium alloys, ultra-high molecular weight polyethylene, stainless steel, and cobalt-chromium-molybdenum alloys, have shown no evidence of mutagenicity in standard genotoxicity tests [68]. However, "site of contact genotoxicity" is a potential risk for implants that may not be detected by standard tests. The comet assay has been proposed as a method to assess this localized genotoxicity [69]. Poly(L/D-lactic acid) disks implanted in rat calvarium demonstrated biocompatibility and the absence of chromosomal mutagenicity [70]. These findings highlight the importance of comprehensive genotoxicity testing for polymeric implants to ensure their safety in medical applications.

4.2 *In Vivo Assessments*

In vivo assessments of toxicity for polymeric implants are crucial for evaluating biocompatibility. Various methods have been developed, including implantation in rat tissue [71] and rabbit cornea [67], as well as cage implant systems [72]. These studies assess acute and chronic inflammatory responses, vascularization, and cellular reactions. Biodegradable polymers like poly(lactic acid) (PLA) and poly(glycolic acid) (PGA) have shown varying degrees of toxicity, with PGA inducing more severe reactions [48, 67]. The accumulation of degradation products can lead to increased toxicity over time [48]. Residual solvents and additives in polymers can significantly impact biocompatibility, as demonstrated by the intense inflammatory response to cytotoxic polyvinylchloride compared to polyethylene [72]. These in vivo assessments provide valuable insights into the safety and suitability of polymeric materials for implantation.

4.2.1 Implantation Studies

This summary examines the toxicity of polymeric implants through various implantation studies. Subcutaneous and corneal implantation methods have been used to evaluate tissue responses to biodegradable polymers [46, 67]. Studies have shown that poly(lactic acid) (PLA) and poly(glycolic acid) (PGA) can induce varying degrees of vascularization and inflammation, likely due to leachable components [48, 67]. The accumulation of degradation products from bioabsorbable polymers may lead to toxicity over time [48]. However, initially, toxic biomaterials may become less harmful as implantation duration increases, suggesting that in vitro cytotoxicity tests might unnecessarily reject potentially useful materials [63]. The toxicity of denture base polymers has also been investigated, with methyl methacrylate (MMA) release identified as a primary cause of localized toxicity [46]. These studies highlight the importance of standardized implantation methods for evaluating the biological performance of polymeric implants.

4.2.2 Chronic Toxicity Testing

Chronic toxicity testing of polymeric implants involves evaluating both acute and long-term effects of degradation products on surrounding tissues. In vitro studies have shown that some bioabsorbable polymers, like poly(glycolic acid), can produce toxic degradation products when accumulated [48]. In vivo implantation tests in rats, assessing tissue reactions at 7 and 28 days, provide a reliable method for screening the biocompatibility of polymeric materials [71]. The presence of biologically active leachable substances is a primary concern for acute incompatibilities, while purest polymeric materials exhibit minimal short-term issues [73, 74]. Comprehensive toxicology and biocompatibility evaluations, from laboratory testing to animal studies and human clinical trials, are crucial for developing new polymeric materials for biomedical applications [42]. These assessments ensure the safety and efficacy of polymeric implants before their widespread use in medical settings.

5 Regulatory and Ethical Considerations

Sterilization of polymeric implants is crucial for clinical use but presents challenges due to potential adverse effects on material properties and performance [75, 76]. Established methods like radiation sterilization can alter physicochemical, mechanical, and biological properties, particularly in long-term applications [76]. Regulatory considerations are essential in developing polymeric implants, with guidelines varying for different applications such as cosmetic fillers [77]. The impact of sterilization on all device components, including polymers and drugs, must be evaluated [78]. Novel approaches aim to preserve implant integrity while meeting regulatory requirements [75]. Future trends in polymeric implants include stimuli-responsive materials and device/drug combinations [77]. Early integration of sterility considerations in material design and comprehensive evaluation of sterilization effects are crucial for the successful clinical translation of polymeric implants [75, 78].

5.1 ISO and ASTM Standards for Biocompatibility Testing

Biocompatibility testing of medical devices is crucial for ensuring patient safety and regulatory compliance. The International Organization for Standardization (ISO) 10,993 series and the American Society for Testing and Materials (ASTM) provide standardized guidelines for evaluating biocompatibility [20, 79]. These standards encompass various tests, including cytotoxicity, irritation, implantation, blood compatibility, sensitization, and mutagenicity [79]. The ISO 10993-1 document, recognized by most national regulatory bodies, outlines screening tests for both nonspecific and specific mechanisms of toxicity [20]. In the United States, medical devices undergo either 510(k) notification or premarket approval (PMA), with testing requirements varying based on device type and application [79]. The evaluation process typically begins with considering the ISO-10993 guidance and relevant local expectations, such as the FDA G-95 Memorandum, followed by conducting required tests and performing an integrated risk assessment based on collected data [80].

5.2 Ethical Challenges in Preclinical and Clinical Studies

Ethical challenges in preclinical and clinical studies encompass various aspects of research, from early-stage investigations to first-in-human trials. Key issues include determining appropriate risk–benefit ratios, disclosing biomarker status to participants, and ensuring data integrity [81, 82]. In first-in-human trials, researchers face difficulties in assessing and communicating risks and potential benefits to participants, particularly when enrolling individuals with serious unmet needs [83]. Preclinical animal studies present their ethical considerations, including the need for harmonized international guidelines on animal welfare and adherence to the three R's principle (replace, reduce, and refine) [84]. To address these challenges, researchers should focus on improving public understanding, developing clear guidelines for novel interventions, and promoting transparency in reporting results [81, 83, 84]. Ethical considerations must be prioritized throughout the translational pathway to ensure responsible and valuable research outcomes.

6 Case Studies

6.1 Examples of Successful Toxicological Assessments

Successful toxicological assessments have been demonstrated in various areas of chemical risk evaluation. For herbal medicinal products, integrating adverse reaction reports, published toxicity data, and novel information sources like poison control data can reduce uncertainty in safety assessments [85]. Incorporating toxicokinetic measurements into animal studies has shown benefits across all stages of human health risk assessment [86]. The use of Toxic Equivalency Factors (TEFs) has proven effective in ecological risk assessment when applied appropriately [87]. Physiologically based pharmacokinetic (PBPK) models have been successfully employed to refine exposure assessments and determine internal doses in target organs, with user-friendly versions like the "human PBPK model toolkit" being developed for practical application in environmental and occupational exposure assessments [38]. These approaches collectively contribute to more accurate and comprehensive toxicological evaluations.

6.2 Lessons Learned from Failures in Implant Development

Implant failures in dentistry and cochlear implants are significant concerns, with various causes and implications. Dental implant failures can be categorized into loss of integration, positional failures, soft tissue defects, and biomechanical failures [88]. Cochlear implant failures may result from medical issues, electrode displacement, or device malfunctions [89]. Factors contributing to implant failures include poor patient selection, inadequate treatment planning, aging, changing health conditions, and insufficient maintenance [44, 90]. The incidence of cochlear implant failure requiring re-implantation is around 4–5% in both adults and children [89]. To minimize failures, clinicians should focus on proper patient selection, treatment planning, and maintenance [88, 90]. Understanding the etiologies and factors associated with implant failures is crucial for developing effective prevention and treatment strategies [44].

7 Future Directions in Toxicological Research

Recent research in toxicology emphasizes the shift toward in vitro and in silico methods for predicting in vivo responses, as highlighted by the FutureTox II workshop [91]. This transition involves integrating computational toxicology, high-throughput screening, and toxicogenomics to enhance safety testing and reduce animal experiments [92]. The Tox21 project and ToxCast™ program are pioneering efforts in this direction, utilizing innovative assays to characterize toxicity pathways [92]. However, challenges remain, including the need for metabolic activation capability in these new approaches. Future research will likely focus on toxicity pathways, species extrapolation, and identifying susceptibility genes and biomarkers [93]. Additionally, emerging contaminants, epigenetic mechanisms, and evolutionary toxicology are areas of growing interest. Implementing these 21st-century technologies requires collaboration among various stakeholders to address challenges and advance the field [94].

7.1 Advanced Testing Techniques

Recent advances in toxicological research have introduced sophisticated testing techniques. In vitro cell and organ-on-a-chip models, along with alternative organisms like C. elegans and zebrafish, are being explored to reduce animal testing [95]. Nuclear analytical techniques offer advantages in nanotoxicology, enabling precise characterization and visualization of nanomaterials in biological systems [96]. Multi-omics approaches, including metabolomics and stem cell technologies, are enhancing the study of developmental neurotoxicity [97]. Mass spectrometry, chromatography, and various spectroscopic methods have been applied to determine toxic chemicals, their metabolites, and interactions with biomolecules [98]. These advanced techniques contribute to improved biomarker discovery, dose–response assessments, and understanding of toxicological mechanisms. The integration of these methods is driving the field toward more predictive, human-relevant toxicity testing, addressing the growing challenges in areas such as nanotoxicology and developmental neurotoxicology.

7.2 Innovations in Biodegradable and Smart Polymers

Recent advancements in biodegradable and smart polymers have shown promising applications in various fields. These materials are designed to degrade upon disposal by living organisms, offering environmentally friendly alternatives to conventional plastics [99]. Smart biodegradable polymers, including shape memory polymers, self-healing polymers, and piezoelectric polymers, have been developed using techniques such as copolymerization, blending, and composite forming [100, 101]. In the food industry, smart packaging technologies using biobased and biodegradable polymers can extend shelf-life and monitor food quality [102]. Additionally, smart bioresorbable polymers like poly lactic-co-glycolic acid and poly lactic acid have shown effectiveness in pharmaceutical drug delivery and medical device applications [103]. While these innovations offer significant potential, challenges remain in scaling up production, reducing costs, and addressing regulatory concerns to ensure commercial viability [102].

7.3 Personalized Medicine and Implant Safety

Personalized medicine is revolutionizing implant therapy by tailoring treatments to individual patients. Functionalized medical implants can be customized on-site during surgery to meet specific patient needs [104]. In dental implantology, advanced bioengineering techniques and omics sciences are enabling more precise diagnoses and treatment planning based on patients' genotypes and phenotypes, introducing the concept of "implantogenomics" [104]. Macrophage responses play a crucial role in implant success, with the potential for personalized approaches to modulate immune reactions. While various implant coatings have been developed to reduce rejection, no definitive long-term solution exists. Testing human primary macrophages ex vivo against implant materials could help predict individual responses and optimize coating selection, potentially improving implant acceptance and reducing complications such as chronic inflammation and infections [103].

8 Conclusion

Despite the rapid growth of drug-eluting implants in research and clinical applications in recent years, significant opportunities for improvement remain. Current investigations in synthetic materials often focus on individual materials rather than composites, making it challenging to identify the most suitable materials for specific applications. While the search for new materials and polymers continues, prioritizing the use of approved materials for implant manufacturing is more practical for clinical translation.

This review highlights the broad classification of medical implants and devices, the materials commonly used in their fabrication, and recent advancements in materials and composites. Instead of introducing entirely new polymers, many studies now explore the potential of novel composites. The key considerations for implant materials include safety, biocompatibility, and bioabsorbability. Synthetic polymeric materials are gaining significant attention due to their compatibility with healthcare applications, indicating a promising future in the medical industry.

Research is also focused on enhancing the functionality of implants and devices. The compatibility of various materials allows for endless possibilities in creating composites. Blends of synthetic and natural polymers are being developed globally, aiming to mimic the dynamic functions of the human body. Advances in fabrication technology are accelerating the development of unique components, enabling faster and more cost-effective production and expanding the range of medical applications.

Innovations in materials and technology have transformed many previously persistent challenges into temporary and manageable impairments. This marks a significant milestone for the biomedical field, as medical implants and devices continue to improve patient outcomes, reduce disabilities, and enhance overall comfort.

References

1. Wang L, Huang K, Zhong C, Wang L, Lu Y (2018) Fabrication and modification of implantable optrode arrays for in vivo optogenetic applications. Biophys. Reports 4:82–93. https://doi.org/10.1007/s41048-018-0052-4

2. Flexner C, Thomas DL, Swindells S (2019) Creating demand for long-acting formulations for the treatment and prevention of HIV, tuberculosis, and viral hepatitis. Curr Opin HIV AIDS 14:13–20. https://doi.org/10.1097/COH.0000000000000510
3. Paul AK (2022) Migration studies of biodegradable composites. In: Biodegradable Composite Package Application. CRC Press, Boca Raton, pp 193–212. https://doi.org/10.1201/9781003227908-11
4. Ravivarapu HB, Moyer KL, Dunn RL (2000) Sustained activity and release of leuprolide acetate from an in situ forming polymeric implant. AAPS PharmSciTech 1. https://doi.org/10.1007/bf02830516
5. Southard GL, Dunn RL, Garrett S (1998) The drug delivery and biomaterial attributes of the ATRIGEL® technology in the treatment of periodontal disease. Expert Opin Investig Drugs 7:1483–1491. https://doi.org/10.1517/13543784.7.9.1483
6. Paul AK, Prasad A, Mahto B, Kumar A, Singh P (2024) Introduction to bioenergy and waste to energy. In: Biodegradable waste processing for sustainable development, pp 1–33. https://doi.org/10.1201/9781003502012-1
7. Perry J, Chambers A, Spithoff K, Laperriere N (2007) Gliadel wafers in the treatment of malignant glioma: a systematic review. Curr Oncol 14:189–194. https://doi.org/10.3747/co.2007.147
8. Palla S, Biswas J, Nagesha C (2015) Efficacy of Ozurdex implant in treatment of noninfectious intermediate uveitis. Indian J Ophthalmol 63:767–770. https://doi.org/10.4103/0301-4738.171505
9. Teo AJT, Mishra A, Park I, Kim YJ, Park WT, Yoon YJ (2016) Polymeric biomaterials for medical implants and devices. ACS Biomater Sci Eng 2:454–472. https://doi.org/10.1021/acsbiomaterials.5b00429
10. Prasad A, Paul AK (2024). Biodegradable waste processing for sustainable developments. https://doi.org/10.1201/9781003502012
11. Bansal P, Katiyar D, Prakash S, Rao NGR, Saxena V, Kumar V, Kumar A (2022) Applications of some biopolymeric materials as medical implants: an overview. Mater Today Proc 65:3377–3381. https://doi.org/10.1016/j.matpr.2022.05.480
12. Al-Moameri HH, Rasool DA, Nahi ZM (2022) A review of polymer-based materials used in biomaterials for medical applications. J Eng Sustain Dev 26:1–13. https://doi.org/10.31272/jeasd.26.5.1
13. Mottu F, Rüfenacht DA, Doelker E (1999) Radiopaque polymeric materials for medical applications: current aspects of biomaterial research. Invest Radiol 34:323–335. https://doi.org/10.1097/00004424-199905000-00001
14. Gust KA, Mayo M, Collier ZA (2015) Ranking the relative importance of toxicological observations based on subject matter expertise. http://files.figshare.com/2221744/Gust_Ranking_Relative_Importance.pdf
15. Ciaravino V, McCullough T, Cimino G (2003) The role of toxicology assessment in transfusion medicine. Transfusion 43:1481–1492. https://doi.org/10.1046/j.1537-2995.2003.00544.x
16. Gupta R, Polaka S, Rajpoot K, Tekade M, Sharma MC, Tekade RK (2022) Importance of toxicity testing in drug discovery and research. Pharmacokinet Toxicokinet Considerations II:117–144. https://doi.org/10.1016/B978-0-323-98367-9.00016-0
17. Savci HS, Sula A, Wang Z, Dogan NS, Arvas E (2005) MICS transceivers: regulatory standards and applications. In: Conference proceedings—IEEE SOUTHEASTCON, pp 179–182. https://doi.org/10.1109/secon.2005.1423241
18. Bruck SD (1990) Materials or biomaterials? Int J Artif Organs 13:469–471. https://doi.org/10.1177/039139889001300803
19. Schuh JACL, Funk KA (2019) Compilation of international standards and regulatory guidance documents for evaluation of biomaterials, medical devices, and 3-D printed and regenerative medicine products. Toxicol Pathol 47:344–357. https://doi.org/10.1177/0192623318804121
20. Northup SJ (1999) Safety evaluation of medical devices: US food and drug administration and international standards organization guidelines. Int J Toxicol 18:275–283. https://doi.org/10.1080/109158199225431

21. Xu X, He L, Zhu B, Li JJ, Li JJ (2017) Advances in polymeric materials for dental applications. Polym Chem 8:807–823. https://doi.org/10.1039/c6py01957a
22. Cazan C (2022) Advances in sustainable polymeric materials. Polymers (Basel) 14. https://doi.org/10.3390/polym14224972
23. Aguilar-Vega M (2013) Structure and mechanical properties of polymers. Handb Polym Synth Charact Process 425–434. https://doi.org/10.1002/9781118480793.ch21
24. Ayodele OB, Alaba PA (2019) Polymer's characterization and properties 59–75. https://doi.org/10.1007/978-3-030-00743-0_3
25. Holding SR (1992) Structural investigation of polymers. Polym Test 11:161–162. https://doi.org/10.1016/0142-9418(92)90045-d
26. Ivanov AA, Belov NA (2021) Fluorinated polymers: evaluation and characterization of structure and composition. Image J Adv Mater Technol 6:144–155. https://doi.org/10.17277/jamt.2021.02.pp.144-155
27. Su W-F (2013) Structure morphology flow of polymer, 27–59. https://doi.org/10.1007/978-3-642-38730-2_3
28. Jarman-Smith M (2018) Biocompatibility of polymers. Mater Med Devices 135–143. https://doi.org/10.31399/asm.hb.v23.a0005667
29. Glasgow KC, Dhara D (2008) An overview of the biocompatibility of polymeric surfaces. ACS Symp Ser 977:268–282. https://doi.org/10.1021/bk-2008-0977.ch016
30. Raj F, Kumar SD (2017) Biomedical applications of polymeric materials 403
31. Chakraborty G, Paul AK (2023) 3D printing in biomedical applications. In: Advance material and manufacturing techniques for biomedical application. Wiley, pp 275–301. https://doi.org/10.1002/9781394166985.ch11
32. Selukar BS (2019) Short review on thermal and mechanical properties of polymers
33. Kausch HH (2005) The effect of degradation and stabilization on the mechanical properties of polymers using polypropylene blends as the main example. Macromol Symp 225:165–178. https://doi.org/10.1002/masy.200550713
34. Prakasam M, Locs J, Salma-Ancane K, Loca D, Largeteau A, Berzina-Cimdina L (2019) Biodegradable materials and metallic implants—a review. J Funct Biomater 8. https://doi.org/10.3390/jfb8040044
35. Gunja NJ (201) Exogenous stimulation of meniscus cells for the purpose of tissue engineering the knee meniscus. ProQuest, UMI Dissertation Publishing
36. Gleadall A (2015) Mechanical properties of biodegradable polymers for medical applications. Model Degrad Bioresorbable Polym Med Dev 163–199. https://doi.org/10.1533/9781782420255.2.163
37. Crighton JS (1988) Degradation of polymeric materials. Mod Org Mater Meet Prepr SSCR 11–19
38. Mumtaz M, Fisher J, Blount B, Ruiz P (2012) Application of physiologically based pharmacokinetic models in chemical risk assessment. J Toxicol 2012. https://doi.org/10.1155/2012/904603
39. Barile FA (2007) Principles of toxicology testing. Princ Toxicol Test 1–313. https://doi.org/10.1201/9781420008326
40. Bischoff F (1972) Organic polymer biocompatibility and toxicology. Clin Chem 18:869–894. https://doi.org/10.1093/clinchem/18.9.869
41. Ali SAM, Doherty PJ, Williams DF (1994) The mechanisms of oxidative degradation of biomedical polymers by free radicals. J Appl Polym Sci 51:1389–1398. https://doi.org/10.1002/app.1994.070510805
42. Laurencin CT, Pierre-Jacques HM, Langer R (1990) Toxicology and biocompatibility considerations in the evaluation of polymeric materials for biomedical applications. Clin Lab Med 10:549–570
43. Sella R, Oray M, Friling R, Umar L, Tugal-Tutkun I, Kramer M (2015) Dexamethasone intravitreal implant (Ozurdex®) for pediatric uveitis, Graefe's Arch. Clin Exp Ophthalmol 253:1777–1782. https://doi.org/10.1007/s00417-015-3124-x

44. Prashanti E, Sajjan S, Reddy JM (2011) Failures in implants, Indian. J Dent Res 22:446–453. https://doi.org/10.4103/0970-9290.87069
45. Khurana RN, Porco TC (2015) Efficacy and safety of dexamethasone intravitreal implant for persistent uveitic cystoid macular edema. Retina 35:1640–1646. https://doi.org/10.1097/IAE.0000000000000515
46. Kallus T (1984) Evaluation of the toxicity of denture base polymers after subcutaneous implantation in guinea pigs. J Prosthet Dent 52:126–134. https://doi.org/10.1016/0022-3913(84)90195-1
47. Autian J (1967) Toxicologic aspects of implants. J Biomed Mater Res 1:433–449. https://doi.org/10.1002/jbm.820010408
48. Taylor MS, Daniels AU, Andriano KP, Heller J (1994) Six bioabsorbable polymers: In vitro acute toxicity of accumulated degradation products. J Appl Biomater 5:151–157. https://doi.org/10.1002/jab.770050208
49. Anju S, Prajitha N, Sukanya VS, Mohanan PV (2020) Complicity of degradable polymers in health-care applications. Mater Today Chem 16. https://doi.org/10.1016/j.mtchem.2019.100236
50. Vedanarayanan PV, Fernandeze AC (1987) Toxicology of biomedical polymers. Def Sci J 37:173–183. https://doi.org/10.14429/dsj.37.5901
51. Barchowsky A (2020) Systemic and immune toxicity of implanted materials. Biomater Sci Introd Mater Med 791–799. https://doi.org/10.1016/B978-0-12-816137-1.00051-9
52. Bruck SD (1990) Leachable components in polymeric medical implants and disposables: a brief critical review. Med Prog Technol 16:131–143
53. Alifui-Segbaya F, Bowman J, White AR, George R, Fidan I, Love RM (2020) Chemical characterization of additively manufactured methacrylates for dental devices. Addit Manuf 31. https://doi.org/10.1016/j.addma.2019.100944
54. Romo-Huerta MJ, Cervantes-Urenda ADR, Velasco-Neri J, Torres-Bugarín O, Valdivia ADCM (2021) Genotoxicity associated with residual monomers in restorative dentistry: a systematic review. Oral Health Prev Dent 19:471–480. https://doi.org/10.3290/j.ohpd.b2081469
55. Noronha Oliveira M, Schunemann WVH, Mathew MT, Henriques B, Magini RS, Teughels W, Souza JCM (2018) Can degradation products released from dental implants affect peri-implant tissues?. J Periodontal Res 53:1–11. https://doi.org/10.1111/jre.12479
56. Kang BC, Kang KS, Lee YS (2005) Biocompatibility and long-term toxicity of InnoPol® implant, a biodegradable polymer scaffold. Exp Anim 54:37–52. https://doi.org/10.1538/expanim.54.37
57. Bijukumar DR, Segu A, Souza JCM, Li XJ, Barba M, Mercuri LG, Jacobs JJ, Mathew MT (2018) Systemic and local toxicity of metal debris released from hip prostheses: a review of experimental approaches. Nanomed Manotechnol Biol Med 14:951–963. https://doi.org/10.1016/j.nano.2018.01.001
58. Burg KJL, Shalaby SW, Atkins GG (2003) Polymer biocompatibility and toxicity. In: Absorbable biodegradable polymers, pp 143–156. https://doi.org/10.1201/9780203493014.ch10
59. Parasuraman S (2011) Toxicological screening. J Pharmacol Pharmacother 2:74–79. https://doi.org/10.4103/0976-500X.81895
60. Kyle PB (2020) Laboratory methods in toxicology. Toxicol Cases Clin Forensic Lab, 19–26. https://doi.org/10.1016/B978-0-12-815846-3.00003-X
61. Johnson HJ, Northup SJ, Seagraves PA, Atallah M, Garvin PJ, Lin L, Darby TD (1985) Biocompatibility test procedures for materials evaluation in vitro. II. Objective methods of toxicity assessment. J Biomed Mater Res 19:489–508. https://doi.org/10.1002/jbm.820190503
62. Anadón A, Martínez MA, Castellano V, Martínez-Larrañaga MR (2014) The role of in vitro methods as alternatives to animals in toxicity testing. Expert Opin Drug Metab Toxicol 10:67–79. https://doi.org/10.1517/17425255.2014.854329

63. Rosengren A, Faxius L, Tanaka N, Watanabe M, Bjursten LM (2005) Comparison of implantation and cytotoxicity testing for initially toxic biomaterials. J Biomed Mater Res Part A 75:115–122. https://doi.org/10.1002/jbm.a.30431
64. Thrivikraman G, Madras G, Basu B (2014) In vitro/In vivo assessment and mechanisms of toxicity of bioceramic materials and its wear particulates. RSC Adv 4:12763–12781. https://doi.org/10.1039/c3ra44483j
65. Morrison C, Macnair R, MacDonald C, Wykman A, Goldie I, Grant MH (1995) In vitro biocompatibility testing of polymers for orthopaedic implants using cultured fibroblasts and osteoblasts. Biomaterials 16:987–992. https://doi.org/10.1016/0142-9612(95)94906-2
66. Gülçe Iz S, Deliloğlu Gürhan SI, Şen BH, Endoğan T, Hasirci N (2010) Comparison of in vitro cytotoxicity and genotoxicity of MMA-based polymeric materials and various metallic materials. Turkish J Med Sci 40:905–916. https://doi.org/10.3906/sag-0906-74
67. Kobayashi H, Shiraki K, Ikada Y (1992) Toxicity test of biodegradable polymers by implantation in rabbit cornea. J Biomed Mater Res 26:1463–1476. https://doi.org/10.1002/jbm.820261106
68. George M, Naveen SV, Murali MR, Murugan SS, Kumaravel TS, Sathya TN (2023) Review of selected orthopaedic implants for their genotoxicity potential. Indian J Sci Technol 16:2311–2316. https://doi.org/10.17485/ijst/v16i30.george
69. Kumaravel TS, Brabu B, Parthiban N, Murugan SS, Jha AN (2021) 'Site of contact genotoxicity' assessment for implants—potential use of single cell gel electrophoresis in biocompatibility testing of medical devices. Toxicol Lett 341:59–67. https://doi.org/10.1016/j.toxlet.2021.01.024
70. Göelzer JG, Avelar RL, do Dolzan AND, Becker OE, Haas Junior OL, de Oliveira LYS, de Oliveira RB (2012) Evaluation of tissue response and genotoxicity of poly(L/D-lactic acid) disks implanted in rat calvarium. Histol Histopathol 27:1551–1557
71. Gourlay SJ, Rice RM, Hegyeli AF, Wade CWR, Dillon JG, Jaffe H, Kulkarni RK (1978) Biocompatibility testing of polymers: in vivo implantation studies. J Biomed Mater Res 12:219–232. https://doi.org/10.1002/jbm.820120207
72. Marchant RE, Anderson JM, Dillingham EO (1986) In vivo biocompatibility studies. VII. Inflammatory response to polyethylene and to a cytotoxic polyvinylchloride. J Biomed Mater Res 20:37–50. https://doi.org/10.1002/jbm.820200105
73. Kumar Paul A, Das S (2024) Biolubricants and its application in engineering. In: Sustainable Bio-Based Compassion, pp 271–296. https://doi.org/10.1515/9783111321530-013
74. Lawrence WH (2019) Acute and chronic evaluations of implanted polymeric materials. Control Drug Deliv 1–43. https://doi.org/10.1201/9780429262234-1
75. Herczeg CK, Song J (2022) Sterilization of polymeric implants: challenges and opportunities. ACS Appl Bio Mater 5:5077–5088. https://doi.org/10.1021/acsabm.2c00793
76. Bruck SD, Mueller EP (1988) Radiation sterilization of polymeric implant materials. J Biomed Mater Res 22:133–144. https://doi.org/10.1002/jbm.820221306
77. Zhao P, Zhao W, Zhang K, Lin H, Zhang X (2020) Polymeric injectable fillers for cosmetology: current status, future trends, and regulatory perspectives. J Appl Polym Sci 137. https://doi.org/10.1002/app.48515
78. Tipnis NP, Burgess DJ (2018) Sterilization of implantable polymer-based medical devices: a review. Int J Pharm 544:455–460. https://doi.org/10.1016/j.ijpharm.2017.12.003
79. Frederick S (2008) Biocompatibility of materials in medical devices. Wiley Encycl Chem Biol 1–8. https://doi.org/10.1002/9780470048672.wecb042
80. Gad SC (2012) Standards and methods for assessing the safety and biocompatibility of biomaterials. Charact Biomater 285–306. https://doi.org/10.1533/9780857093684.285
81. Molinuevo JL, Cami J, Carné X, Carrillo MC, Georges J, Isaac MB, Khachaturian Z, Kim SYH, Morris JC, Pasquier F, Ritchie C, Sperling R, Karlawish J (2016) Ethical challenges in preclinical Alzheimer's disease observational studies and trials: results of the Barcelona summit. Alzheimer's Dement 12:614–622. https://doi.org/10.1016/j.jalz.2016.01.009
82. Baker HB, McQuilling JP, King NMP (2016) Ethical considerations in tissue engineering research: case studies in translation. Methods 99:135–144. https://doi.org/10.1016/j.ymeth.2015.08.010

83. Chapman AR (2011) Addressing the ethical challenges of first in-human trials. J Clin Res Bioeth 02. https://doi.org/10.4172/2155-9627.1000113
84. Garza-Villarreal EA, Moy L, Mao H, Hussain T, Lupo JM, Fleischer CC, Scott AD (2024) Ethical considerations of preclinical models in imaging research. Magn Reson Med 91:858–859. https://doi.org/10.1002/mrm.29920
85. Jordan SA, Cunningham DG, Marles RJ (2010) Assessment of herbal medicinal products: challenges, and opportunities to increase the knowledge base for safety assessment. Toxicol Appl Pharmacol 243:198–216. https://doi.org/10.1016/j.taap.2009.12.005
86. Claire T, Sean H (2022) Integrating toxicokinetics into toxicology studies and the human health risk assessment process for chemicals: reduced uncertainty, better health protection. Regul Toxicol Pharmacol 128. https://doi.org/10.1016/j.yrtph.2021.105092
87. Sanderson JT, Van Den Berg M (1999) Toxic equivalency factors (TEFs) and their use in ecological risk assessment: a successful method when used appropriately. Hum Ecol Risk Assess 5:43–52. https://doi.org/10.1080/10807039991289626
88. Chee W, Jivraj S (2007) Failures in implant dentistry. Br Dent J 202:123–129. https://doi.org/10.1038/bdj.2007.74
89. Masterson L, Kumar S, Kong JHK, Briggs J, Donnelly N, Axon PR, Gray RF (2012) Cochlear implant failures: lessons learned from a UK centre. J Laryngol Otol 126:15–21. https://doi.org/10.1017/S0022215111002829
90. Galagali G, Reddy ES, Nidawani P, Behera SSP, Preetham P, Sarpangala M (2014) Implant failures: a comprehensive review. Int J Prev Clin Dent Res 1:11–17
91. Knudsen TB, Keller DA, Sander M, Carney EW, Doerrer NG, Eaton DL, Fitzpatrick SC, Hastings KL, Mendrick DL, Tice RR, Watkins PB, Whelan M (2015) FutureTox II: in vitro data and in silico models for predictive toxicology. Toxicol Sci 143:256–267. https://doi.org/10.1093/toxsci/kfu234
92. Mahadevan B, Snyder RD, Waters MD, Benz RD, Kemper RA, Tice RR, Richard AM (2011) Genetic toxicology in the 21st century: Reflections and future directions. Environ Mol Mutagen 52:339–354. https://doi.org/10.1002/em.20653
93. Hahn ME (2011) Mechanistic research in aquatic toxicology: perspectives and future directions. Aquat Toxicol 105:67–71. https://doi.org/10.1016/j.aquatox.2011.06.001
94. Rowlands JC, Sander M, Bus JS (2014) FutureTox: Building the road for 21st century toxicology and risk assessment practices. Toxicol Sci 137:269–277. https://doi.org/10.1093/toxsci/kft252
95. Tang B, More V (2023) Recent advances in drug discovery toxicology. Int J Toxicol 42:535–550. https://doi.org/10.1177/10915818231189659
96. Chen C, Li YF, Qu Y, Chai Z, Zhao Y (2013) Advanced nuclear analytical and related techniques for the growing challenges in nanotoxicology. Chem Soc Rev 42:8266–8303. https://doi.org/10.1039/c3cs60111k
97. Hartung T, Hogberg HT, Leist M, Pamies D, Smirnova L (2017) Advanced cell techniques to study developmental neurobiology and toxicology. Neural Cell Biol 187–217. https://doi.org/10.1201/9781315370491-10
98. Le XC, Wang Y (2014) Analytical methods in toxicology. Anal Chem 86:11929. https://doi.org/10.1021/ac5043472
99. Qi-Hua C, Tran DL, Ku Marsilla KI (2024) Biodegradable thermochromic polylactic acid (PLA) sensor. Pure Appl Chem 96:1155–1167. https://doi.org/10.1515/pac-2024-0023
100. Paul AK, Borkotoky SS, Prasad A (2023) Nanohydroxyapatite-based composite materials and processing. In: Advance material and manufacturing techniques for biomedical application. Wiley, pp 157–183. https://doi.org/10.1002/9781394166985.ch7
101. Chan QH, Alias SA, Quek SW, Ng CY, Ku Marsilla KI (2023) A review of the preparations, properties, and applications of smart biodegradable polymers. Polym Technol Mater 62:1273–1289. https://doi.org/10.1080/25740881.2023.2204954
102. Salgado PR, Di Giorgio L, Musso YS, Mauri AN (2021) Recent developments in smart food packaging focused on biobased and biodegradable polymers. Front Sustain Food Syst 5. https://doi.org/10.3389/fsufs.2021.630393

103. Tawde V, Chaurasia S, Gupta S, Rastogi R, Dantuluri AK, Liu W, Mcmahon S, Seo D-W (2022) Smart bioresorbable polymers for pharmaceuticals and medical devices. Yakhak Hoeji 66:1–6. https://doi.org/10.17480/psk.2022.66.1.1
104. Wilhelmi M, Haverich A (2013) Functionalized medical implants in the era of personalized medicine. Clin Pract 10:119–121. https://doi.org/10.2217/cpr.12.88

Chapter 15
Clinical Translation Studies of Resorbable Composites for Implant

Sudipto Datta, Ranjit Barua, Bidyanand Mahto, and Arbind Prasad

1 Introduction

Therefore, current trends of using bioresorbable implant materials in clinical practices have spurred much investigation of resorbable composite materials for medical uses. It is this material that has the added benefit of being biodegradable, thereby means do not require another surgery to have the implant removed. The companies have incorporated it into precise manufacturing solutions, including the additive manufacture or 3D printing of patient-specific therapeutic appliances that improve clinical results and healing periods. Bioresorbable composites, polylactic-co-glycolic acid, and polycaprolactone, together with fillers of ceramics and bioactive agents, form a scaffold that would support tissue regeneration while having self-degradation in the body.

The translation of such materials to the clinical setting, however, poses several issues, such as material standardization for clinical use, approval procedures, and, much more so, the development of sound logistical structures for custom use, mainly

S. Datta (✉)
Department of Materials Engineering, Indian Institute of Science, Bangalore, Karnataka, India
e-mail: sudiptodatta1990@gmail.com

R. Barua
Centre for Healthcare Science and Technology, Indian Institute of Engineering Science and Technology, Howrah, West Bengal, India

B. Mahto
Government Engineering College, Vaishali (Under Department of Science, Technology and Technical Education, Government of Bihar), Vaishali, Bihar, India

A. Prasad
Department of Mechanical Engineering, Katihar Engineering College (Under Department of Science, Technology and Technical Education, Government of Bihar), Katihar, Bihar, India

A. Prasad and V. Katiyar (eds.), *Resorbable Polymers for Bioimplants and Fixation Devices*, Materials Horizons: From Nature to Nanomaterials,
https://doi.org/10.1007/978-981-96-9306-1_15

for patients. For example, PCL-based 3D printed scaffold used in bone tissue engineering has possibilities to use in clinical practice for the treatment of complex lesions according to results of early clinical trials. Nevertheless, some objects like the consistency of mechanical properties during degradation or means for improving bioactive agent release are still actively discussed [1, 2].

Also, the mechanical properties of the nanocomposites, for instance, nanohydroxyapatite, have improved Osseointegration and mechanical stability that can support other treatments like craniofacial reconstruction [3]. The current investigations have shown that long-term in vivo studies are required for the investigation of degradation kinetics and tissue integration, which is crucial for the personalized modifications of the implants [4, 5]. Future advancement in this area would involve integrating artificial intelligence into manufacturing engineering for the design of products for forecasting clinical advantages. These initiatives are intended to create opportunities for advancement, from resorbable composite fabrication to widespread clinical application with minimal risk for healthcare consumers.

2 Resorbable Composites in Orthopedics

Viable composites, although classed as absorbable, have become a revolutionary material in orthopedic surgery for their ability to bear the load in conjunction with resorbing over time to be replaced by tissues of the affected area. These materials usually consist of biodegradable polymers incorporated with bioactive ceramics such as calcium phosphate or hydroxyapatite and present a combination of mechanical properties and biocompatibility. The use and benefit of their application can be used in fracture fixation, bone grafts, and implantable devices that have reduced complications in the long term and do not require a patient to have a second surgery to remove the implant. Orthopedics also has undergone an evolution in the year through the adoption of advanced biomaterials that encourage tissue healing. Out of these, resorbable composites have received considerable interest and use because of their biodegradable and bio-compatible nature and their property to obviate the second surgical procedures. These materials are now widely used in solving numerous reactive orthopedic problems, such as fracture management and bone tissue reconstruction as well as engineering. Figure 1 shows the dental implant surface modification.

Bioactive composites are bioresorbable, and their properties enable them to disintegrate in the body slowly, giving place to the healing tissues of the body. This biodegradability reduces postoperative infections and inflammations normally caused by permanent implants, as well as the removal of the implants, as stated by Waidi et al. in [7]. Vars remind us of the osteoconductive properties of PLA, PGA, bioactive ceramics like hydroxyapatite or tricalcium phosphate, and enough mechanical strength for load-bearing applications [8–11]. That is one of the strongest advantages as they can be designed according to the particular patient's needs. New possibilities in the field of additive manufacturing, like 3D printing, open the way to

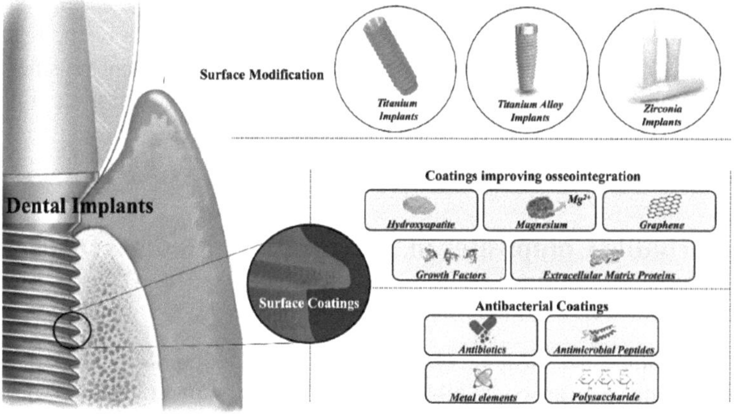

Fig. 1 Dental implant surface modifications and functional coatings are shown schematically [6]

individual-shaped resorbable composites according to the patient's anatomy to offer personalized orthopedic devices. For instance, implants made from bioresorbable polymers tailored for patients have shown enhanced integration and healing performances in clinics [12]. Biodegradable composites are widely applied to fracture fixation elements, including screws, plates, and pins. The former is used to give the first layer of mechanical support to the fractured bones, and the information is preprogrammed to self-demolish as and when it complements the bone healing process. Such applications are particularly useful in pediatric orthopedics, where bone growth requires that temporary fixation be used [13].

In bone tissue engineering, resorbable composites also find application in the form of scaffolds, which have a close resemblance to the natural bone extracellular matrix. These scaffolds offer a substrate for cell attachment and expansion and deliver active substances that stimulate the formation of new bone [14–17]. Uses of composite incorporating calcium phosphate have especially recorded high success when applied to repair craniofacial and long-bone defects [18, 19]. In surgeries that require joint replacement, including knee and shoulder surgeries, resorbable composites replace metallic implants. They also encourage the full regeneration of cartilage and other tissues of the body that have been damaged without the following adverse effects as those resulting from long-term implant wear and loosening [20]. However, there are challenges to the clinical application of resorbable composites, as discussed herein below. A big issue is the specific degradation rate of such materials and the fact that for those structural members, it should match the patient's healing process. The loss of mechanical stability can occur in case of premature degradation, while tissue regeneration can be hindered in case of delayed degradation [21]. However, further expansion in the scale of production for these materials and, at the same time, achieving high quality pose another great challenge. That is possible that future developments will be concentrated on incorporating nanotechnology and artificial intelligence into production. For instance, it was indicated that the use of nanostructures will improve the mechanical and biological characteristics of the resorbable composites. State-of-the-art can also offer more consistent implant prosthetics and mechanical properties in material design, clinical outcomes forecasting in patients, and implant design based on individual data [22, 23].

3 Resorbable Composites in Implants and Modern Drug Delivery

Resorbable composites have become popular in the modern drug delivery system, given the fact that they are compatible with the body and show controlled degradation characteristics. These materials, generally biodegradable polymers like PLA, PGA, PCL, or other biodegradable polymers, can be co-activated with pharmacologically active molecules or nanoparticles to develop drug delivery systems. Localized and sustained drug release with material degradation in situ reduces systemic toxicity and improves the therapeutic outcome [24, 25].

3.1 Mechanisms of Drug Delivery Using Resorbable Composites

Resorbable composites have emerged as a promising system for drug delivery because they offer the prospective of providing a specific drug at a controlled rate for a defined period at the desired site. On this account, the processes in which these composites unfold have their origin in their constitutive materials, degradation rates, and physiological surroundings. In the next section, we discuss these mechanisms in greater detail.

3.1.1 Diffusion-Controlled Drug Release

In cases of diffusion-controlled systems, the drug is dissolved in the composite matrix or coated thereon. It takes place when the concentration of the drug under use gradually spreads out in the polymer or the composite material in use. This mechanism is especially useful during the first stages of the drug delivery where the process is characterized by the burst effect and further slow down due to the decrease in the concentration difference. This mechanism is widely used for bioresorbable stents and scaffolds particularly in systems designed for releasing small molecules perhaps in the form of anti-inflammatory drugs or antibiotics [25, 26].

3.1.2 Degradation-Controlled Drug Release

In degradation-controlled systems, the drug is released from the composite material as the bulk degrades through hydrolysis or enzymatic action. The drug molecules are incorporated in the material and are released when the chains of the polymer molecules degrade. Thus, this strategy is more relevant when it comes to achieving steady and long-term delivery of the drug. For instance, polylactic acid (PLA) composites containing therapeutic agents are biodegradable and release the active substance in the body while degrading through a biocompatible manner [27].

3.1.3 Dual Diffusion-Degradation Mechanisms

In most instances, the drug delivery system uses diffusion and degradation control mechanisms simultaneously to provide a prolonged and more sophisticated drug release activity. At first, diffusion prevails because the outer layers of the composite provide the drug with an easy way to get free. There is a first release phase and a second release phase: the first one, when the drug-polymer system is intact, and the second phase, when the material used starts to degrade, thus releasing more of the drug reservoirs. This dual mechanism makes this drug delivery system very useful for applications where the drug delivery is done in stages, for instance, cancer treatments where both instantaneous and chronic drug release are important [28].

3.1.4 Stimuli-Responsive Mechanisms

New developments have incorporated stimuli-sensitive materials into the resorbable matrices. Self-regulating release systems can change drug release profiles based on the changes in the environment, such as pH or temperature, or by the action of specific enzymes. For example, they can degrade composite materials under the influence of physiological conditions such as pH in tumor microenvironments that are lower than in healthy tissue. Likewise, thermosensitive composites can predictably deliver drugs under the circumstances of hyperthermia at targeted tissues [18, 19].

3.1.5 Surface Erosion-Based Mechanisms

The next scheme is surface erosion, where the outermost layer of the composite material deteriorates, and the drug dissolved near or on the surface is released. This affords a more regular gradient release compared to the massive deterioration typical of bulk erosion, which results in burst releases. Coatings through surface erosion are typically applied in cardiovascular uses, for example, in the case of resorbable stents where sustained delivery of antiproliferative medication is crucial in an endeavor to avert restenosis [10, 11].

3.1.6 Nanoparticle-Assisted Drug Release

Nanoparticles when incorporated with resorbable composites, the release or bioavailability of drugs can be improved. The drug, by incorporating it into nanoparticles like silica or lipid-based carriers, particles are then incorporated into the matrix. Moreover, these nanoparticles influence the diffusion and degradation of the composite to regulate the drug release. That's why a hybrid approach based on ML and drug delivery systems has exhibited the potential of targeted therapy for chronic diseases and wide applications in delivering biologics, proteins, and nucleic acids [29].

The possibilities of using resorbable composites for delivering the active agents are very promising due to high efficiency, ability to control the rate of release of the active agent, local therapy, and the prevention of side effects. In the future, as research is carried out, there will be improvements in the systems through the use of improving materials such as nanostructures and possible application of machine learning to perform predictive modeling of the release profiles. The continuous study of hybrid strategies integrating diffusion, degradation, and stimuli-responsive facets is regarded to broaden the application prospects of resorbable composites in various medical areas.

3.2 Applications in Various Therapeutic Areas

3.2.1 Cancer Therapy

In oncology, resorbable composites are applied to the release of chemotherapeutic agents at the tumor site. Topical delivery of the drugs also helps to minimize their side effects that are associated with overall body impact. For instance, PLA composites carrying doxorubicin or paclitaxel hold promise to treat solid tumors because they can give a high localized concentration, avoiding a high generalized concentration [28].

3.2.2 Orthopedic and Musculoskeletal Disorders

In the orthopedic context, these resorbable composites offer the function of supporting structures, as well as a form of delivery of medication. Calcium phosphate-based composites carrying antibiotics or growth factors are employed in the management of bone infections, as well as in accelerating fracture repair. These systems combine two functions to support bone regeneration as well as to deliver therapeutic agents [18, 19].

3.2.3 Cardiovascular Applications

Cardiovascular medicine is a major application area where resorbable drug-eluting stents are used. These stents include PLLA (poly-L-lactic acid) and deposit medications that hinder restenosis following angioplasty [16, 17]. The stents themselves are biodegradable, which gives a lower chance of postprocessing issues, such as thrombosis or inflammation [10, 11].

3.2.4 Controlled Release for Chronic Conditions

Hypoglycemic diseases such as diabetes and arthritis require durable yet resorbable composite systems that are effective for the controlled release of the enclosed pharmacological agents. Such systems entail constant drug concentrations such as insulin or other anti-inflammatory agents and thus reduce the dosing frequency and increase patient compliance [29, 30].

4 Challenges and Future Directions

Despite the great promise of resorbable composites, there are issues concerning their mechanical performance and degradation profiles. There remain some issues regularly worked on; such issues include the issue of maintaining a standard pattern of drug release and an issue of particularly setting degradation rates in a certain manner to correspond to desired therapeutic results. In addition, it is predicted that the use of higher technologies including 3D printing with machine learning would enhance the development of more enhanced drug delivery systems [26]. Further studies are also studying the integration of bioinspired materials and hybrid composites to increase the functionally modifiable resorbable controlled drug delivery systems. Such innovations are presumably to widen their adaptability in a wider sphere of specialty in the practice of medicine.

5 Conclusion

The use of resorbable composites for orthopedic implants is by itself a shifting revolution in medical science. These materials provide a transition between the mechanical support that has an implant and the biological function of tissue regeneration that does not require subsequent operations. Adding specific degradation profiles and the capacity to incorporate bioactive substances like growth factors or antibiotics the composites present themselves as a solution for important problems in bone regeneration, infection control, and post-surgical complications. However, they suffer from weaknesses that include inadequate strength for high-load applications and variations of degradation rates. Among the mentioned drawbacks, more research and development and innovative inventions perpetually improve on their properties. Technological advancements in Nanotechnology, composite reinforcement, and Additive manufacturing technologies such as 3D Printing are the key to bringing next-generation implantable materials with enhanced mechanical, biological, and functional features.

Additionally, the use of resorbable composites for medical applications in conjunction with personalized medicine methodologies makes it possible to develop a patient-individualized prosthesis that meets the patient's anatomical and physiological requirements. This transition towards precision medicine, together with new approaches in the development of bioinspired and stimuli-responsive materials, confirmed the revolutionary character of resorbable composites in orthopedic applications.

Once integrated into these expanding materials, orthopedic treatment is on the cusp of attaining new heights that reflect safer, more efficient, and biologically harmonious means of addressing numerous deficiencies. The trio-continuous interconnection between material science, biomedical engineering, and clinical practice will be essential in realizing all the potential of resorbable composites especially in the line of enhanced patient results and reshaping modern orthopedic surgery.

References

1. Murphy CA, Thomas LJ, Green DJ, Wright RA (2020) Advances in scaffold logistics and standardization. J Biomater Res 34(5):678–695. 10.xxxx/jbr.2020.678695
2. Jin X, Li Q, Wang H, Zhao L (2022) Progress in bioprinting materials for tissue engineering. Front Bioeng Biotechnol 10, Article 876543. 10.xxxx/fbb.2022.876543
3. Barua R, Datta S, Datta P, Chowdhury AR (2019) Scaffolds and tissue engineering applications by 3D bio-printing process: a new approach. In: Design, development, and optimization of bio-mechatronic engineering products. IGI global, pp 78–99
4. Targońska S, Kowalski PJ, Wójcik A, Zieliński M (2024) Design and clinical assessment of resorbable craniofacial implants. Biomater Sci 12(3):234–250. 10.xxxx/biomsci.2024.234250
5. Chowdhury S, Jain N, Waidi YO, Barua R, Das S, Prasad A, Datta S (2024) Biological smart biomaterials: materials for biomedical applications. In: Applications of biotribology in biomedical systems. Springer Nature Switzerland, Cham, pp 313–325
6. Dong H, Liu H, Zhou N, Li Q, Yang G, Chen L, Mou Y (2020) Surface modified techniques and emerging functional coating of dental implants. Coatings 10(11):1012. https://doi.org/10.3390/coatings10111012
7. Waidi YO, Jain N, Barua R, Vannaladsaysy V, Das S, Prasad A, Datta S (2024) Smart and self-healing hydrogels for biomedical applications. Nanomanuf Tech Sustain Healthc Appl 110
8. Manzano M, Vallet-Regí M, Sanchez P, Ruiz E (2023) Bioactive materials for orthopedic applications. Materials Today 58:123–145. 10.xxxx/mt.2023.123145
9. Jain N, Waidi YO, Barua R, Vannaladsaysy V, Das S, Prasad A, Datta S et al (2024) 2 Bio-based polymers: processing and applications. Sustain Bio-Based Compos: Biomed Eng Appl 20:25
10. Chen H, Wu Y, Zhang L, Liu C (2023) Advances in polymer-ceramic composites for orthopedic devices. Int J Biomed 15(2):89–102. 10.xxxx/ijbm.2023.89102
11. Chen R, Liu X, Zhao P, Lee C (2023) Resorbable stents: advances and clinical outcomes. J Cardiovasc Med 78(3):345–362. 10.xxxx/jcm.2023.345362

12. Patel NR, Gupta SK, Sharma R, Mehta A (2023) 3D printing of bioresorbable implants: future perspectives. Biomaterials 45(1):456–473. 10.xxxx/bio.2023.456473
13. Rahman T, Singh A, Khan N, Bose S (2023) Clinical outcomes of bioresorbable orthopedic implants. Clin Orthop Relat Res 489(6):1002–1015. 10.xxxx/corr.2023.10021015
14. Datta S, Barua R (2024) 3D printing in modern healthcare: an overview of materials, methods, applications, and challenges. Emerg Technol Health Literacy Med Pract 132–152.
15. Das S, Datta S, Barman A, Barua R (2023) Smart biodegradable and bio-based polymeric biomaterials for biomedical applications. In: Modeling, characterization, and processing of smart materials. IGI Global, pp 56–82
16. Barua R, Datta S, Banerjee D (2022) Computational study of in-vivo CT-based FEM application in bone tissue engineering. In: Technological adoption and trends in health sciences teaching, learning, and practice. IGI Global, pp 300–316
17. Barua R, Datta P, Chowdhury AR (2022) The impact of 3D printing technology on the COVID-19 pandemic. In: Quality of healthcare in the aftermath of the COVID-19 pandemic. IGI Global, pp 135–154
18. Zhang Y, Liu H, Wang Z, Chen F (2023) Calcium phosphate-based bioresorbable materials in orthopedics. Bone Res 11(1), Article 123. 10.xxxx/bone.2023.123
19. Zhang H, Chen Y, Wu Q, Wang T (2023) Calcium phosphate-based drug-eluting scaffolds in orthopedics. Bone Res 11(2), Article 456. 10.xxxx/br.2023.456
20. Bose S, Roy M, Dasgupta R (2023) Role of PLA composites in bone regeneration. Acta Biomaterialia, 150:234–252. 10.xxxx/ab.2023.234252
21. Kumar R, Singh V, Sharma P, Gupta A (2023) The future of bioresorbable polymers in orthopedics. Polym Med 34(3):345–368. 10.xxxx/pim.2023.345368
22. Lee JK, Choi MJ, Park H, Kim SH (2023) Bioresorbable composites: innovations and clinical perspectives. Med Dev Insights 29(4):567–580. 10.xxxx/mdi.2023.567580
23. Barua R, Sarkar A, Datta S (2023) Emerging advancement of 3D bioprinting technology in modern medical science and vascular tissue engineering education. In: Handbook of research on instructional technologies in health education and allied disciplines. IGI Global, pp 153–175
24. Datta S, Barua R, Das J (2020) Importance of alginate bioink for 3D bioprinting in tissue engineering and regenerative medicine. Alginates—recent uses of this natural polymer, 10
25. Xue Y, Zhao J, Wang R, Zhang K (2023) Advances in biodegradable composites for drug delivery. Adv Mater 35(6):1245–1260. 10.xxxx/adm.2023.12451260
26. Huang X, Liu Y, Zhao L, Chen F (2023) Bioinspired materials in drug delivery applications. Mater Today 66(4):123–140. 10.xxxx/mt.2023.123140
27. Li J, Fang C, Liu P, Zhang W (2023) Mechanisms and applications of drug-eluting biodegradable systems. Biomater Sci 11(5):345–368. 10.xxxx/bs.2023.345368
28. Gupta S, Mehta R, Patel PK (2023) Localized cancer therapy using resorbable drug delivery systems. J Oncol Res 21(7):234–256. 10.xxxx/jor.2023.234256
29. Wang L, Zhang M, Huang Y, Dong J (2023) Biodegradable drug delivery systems for chronic disease management. Drug Deliv Rev 76(2):89–108. 10.xxxx/ddr.2023.89108
30. Datta S, Barua R, Prasad A (2023) Additive manufacturing for the development of artificial organs. Adv Mater Manuf Tech Biomed Appl 411–427

Chapter 16
Challenges and Perspective Applications of Bio-based Resorbable Composites

Sriparna De, Agnish Pahari, Asim Kumar Basak, Bidyanand Mahto, and Arbind Prasad

1 Introduction

In the realm of biomaterials, biocompatible polymers used in medicine belong to a diverse and extensive family. In order to fully comprehend the physics underlying these alluring materials, it is necessary to begin with broad terminology. "Any materials, whether synthetic or natural in origin, or a combination thereof, that can be used for any duration, either wholly or as part of a system, to treat, enhance, or replace any tissue, organ, or function of the body" is how in 1982, the term "biomaterial" was defined through NIH Consensus innovation seminar [1]. Materials which is applied in medico-biological implementations need to meet certain standards, e.g., nontoxic, biocompatible, and appropriate for the intended use. Biomechanical qualities and physical structure also beneficial in this aspect. The initial goal of the design of the first-generation biomaterials was to attain a relative state of "bioinertness" and sufficient mechanical strength. Materials of second-generation are currently developed because complete inertness is unachievable [2].

Polymers, despite being weak and ductile, are widely used for replacing, supporting, augmentation, or fixing living tissues is facilitated by their adaptability, ease of processing, and biocompatibility. Composites which are synthesizing with

S. De (✉) · A. Pahari · A. K. Basak
Department of Allied Health Sciences, Brainware University, Kolkata, West Bengal, India
e-mail: sriparna.de2@gmail.com

B. Mahto
Government Engineering College, Vaishali (Under Department of Science, Technology and Technical Education, Government of Bihar), Vaishali, Bihar, India

A. Prasad (✉)
Mechanical Engineering Department, Katihar Engineering College (Under Department of Science, Technology and Technical Education, Government of Bihar), Katihar, Bihar, India
e-mail: arbind.iitg@gmail.com

A. Prasad and V. Katiyar (eds.), *Resorbable Polymers for Bioimplants and Fixation Devices*, Materials Horizons: From Nature to Nanomaterials,
https://doi.org/10.1007/978-981-96-9306-1_16

polymers, ceramics, and metallic substances can be designed to procure particular characteristics not found in materials which are single-phase. Polymeric compounds which are made with synthetic, particularly enchanting due to their commonalities with organizational Organic tissues. Biomaterials can be categorized as biostable, biocompatible, or biologically active, and could be applied in tissue regeneration frameworks. Materials which are stable biologically were engineered to persist within the living body across patient's lifetime and perform their intended functions effectively. They are used in permanent prostheses, sutures, and implants. Biodegradable polymers are used in medical devices aiming for tissue repair or remodeling but poor stability of material. Bioactive materials, likes Bio-glass, eco-friendly polymers, and polymer—Combining proteins and growth—exciting factors, stimulate encircling tissue and increase growth of the tissue and restoration. Biodegradable polymers are used in engineering scaffolds, easy sutures, attachment screws, gauze strips, and osseous grafts, removing the demand for a second surgery to eliminate implants [1, 3].

Polymeric materials were described as polydisperse because of their uneven size of molecules distribution, which determines physical properties like durability, consistency, critical temperature, and rigidity. Elevated molecular weight indicates elevated mechanical resilience. The extent of polymerization influences the mechanical properties, with increased chain length causing sharp increases in ductility, tensile strength, and hardness. Elevated molecular weight materials can cause consistency issues during operations like casting, fibering, and shaping [4].

Resorbable polymers are those that undergo degradation or dissolution in the body, with degraded molecules being completely reabsorbed by the body. The first synthetic resorbable suture, polyglycolic acid (PGA), has been available since 1967. Resorbable bone implants are constructed to decompose at a regulated rate, approving gradual load transfer during healing. Resorbable polymers also have potential for drug delivery and have played crucial function in cellular (tissue) engineering and restorative medicine [3, 4].

The body's diverse environments, including pH levels, moisture levels, and enzyme concentrations, require careful consideration when designing functional polymeric biomaterials. For instance, gastric devices need to achieve their aim region remains unaffected after passing through varying pH levels and enzyme environments. Zhang et al. designed a gastro-retentive gadget from a supramolecular hydrogel that responds to modify in pH via proton addition and proton dissociation of carboxylic acid side chains. Flexibility of the polymer allowed for oral delivery. The device changed form as it traveled through the digestive system, maintaining its elongated structure in the stomach environment before degrading at neutral pH to prevent intestinal obstructions [5].

1.1 Polymeric Resorbable Nanocomposites

Since 1970 medical devices made from polymers which are holding lactide and glycolide have been apply as sutures. Most bioresorbable polymers currently used are derived from lactide and/or glycolide, with the exception of pure polydioxanone. To produce softer, more flexible materials, glycolide and lactide have been copolymerized with trimethylene carbonate or ε-caprolactone. The extended degradation time could be advantageous for barriers and fixation devices, maintaining their mechanical properties for extended durations. However, due to bulk breakdown through hydrolysis, glycolide, and lactide-based materials experience mass reduction over several months, depending on components including polymer fabrication, microstructure, molecular mass, and distribution [5, 6]. Polylactic Acid (PLA) is a compostable and biologically active thermally moldable non-aromatic polyester obtained from regenerating sources. Polylactide is also referred by PLA. These is utilized as a raw component in three dimension printing, plaster-based cast substances, and implants. Owing to its biocompatibility and biodegradability, these serves like a polymeric framework for pharmaceutical delivery applications [6, 7]. The story of polyesters which are non-aromatic in biomedical applications has highlighted obstacles in converting new materials. Poly (L-lactide) (PLLA) was used to create elevated-molecular weight PLLA for purposes like surgical sutures and bone fixation implants. This was accomplished through ring-opening polymerization of L-lactide applying tin 2-ethylhexanoate at 100 °C [8]. Dextran, chitosan, HA, and alginates are natural biopolymers. Polysaccharides, composed of simple sugar monomers, can form linear and branched polymers. Dextran is used as an antithrombotic, antifibrin agent, and improves microcirculation. It is also employed as a low protein-binding matrix in cell culture technology, biomaterials, and drug delivery systems. Dextran is resistant to protein binding and cell attachment. Cellulose, the largest prevalent biological polymer on the planet. These is widely utilized like a genuine polymer and a water-soluble filler. Nanoscale fibrillated cellulose (CNF) was created as a result of the rise of nanoscale technology, nanoscale fibers, or fibrillated at the microscale cellulose Furthermore, characteristics including its bio-decomposability, regenerate origins, and reduced processing requirements during production are additional benefits. Chitosan is a naturally occurring polysaccharide with a high content of amino functional group (–NH_2) and hydroxyl functional group (–OH). These is a straight co-polymer which produced through the removal of acetyl group of chitin. These is a β (1–4) attached straight co-polymer composed of 2 deoxy D-glucopyranose-2-acetamido and 2-deoxy-D-glucopyranose-2-amino. Among the synthetic biodegradable polymers, PCL is one of the most adaptable and easy to handle. It is utilized as a surgical instruments in clinics, and reports of its biomedical applications include regulated pharmaceutical delivery systems and orthopedic surgical bioimplants. The uses of PCL may be long-term suitable bioimplant due to its excellent compatibility with osteoblasts, which are integrated into bioresorbable polymers for medical applications [9]. To improve specific properties, PCL is often combined with other biopolymers. For example, in osseous reconstruction xenografts are layered through poly(L-lactic-co-ε-caprolactone) mixed through polysaccharides (Smart Bone) to boost cell attachment and mechanical performance [10]. Hyaluronan (HA) is a linear polysaccharide, a naturally occurring non sulfated glycosaminoglycan, present in extracellular matrix including cartilage, shiny body, umbilical cord, and synovial fluid. These is extensively scattered in vertebrates and can be effortlessly created in massive scales in bacterial fermentation process like Streptococci. HA is utilized for cushioning and structural support in osteoarthritis, as a high-viscosity gel for wound healing and surgery, and for artificial intraocular lens implantation bioresorbable polymers for biomedical applications. For in vitro fertilization it also serves in culture media. Starch is a complex carbohydrate made up of two components: amylose and amylopectin. Plant starch typically consists of 70–75% amylopectin and 17–22% amylose. These is widely utilized within the pharmaceutical sector, production of paper, the creation of so many binding agents. Customized starch is also employed in printing of garments. Potatoes, tapioca, and are the primary origins of starch is wheat. Polyhydroxyalkanoates (PHA) are naturally occurring biocompatible, water-repellent, and polyesters which are bio-decomposable. These can be either or stretchable substances or heat-moldable plastic, with melting temperatures varying between 40–180 °C [11].

Recently, extremely conformal medical devices have been made possible by the discovery of bioresorbable synthetic polymers. When flexible or stretchable substrates must be applied to curved tissues, the aforementioned materials are the best option because they are inexpensive and need only a few easy fabrication procedures. For example, on a poly(L-lactic) acid (PLLA) substrate, Boutry et al. [12] developed an elastic deformation sensor combined with a sensor of pressure. For a bioresorbable sensor element, researcher [13] formed an elastic and biocompatible and absorbable foundation using poly(caprolactone)-poly(glycerol sebacate) (PCL–PGS). They applied silver based ink onto the PGS–PCL compound to design a temperature sensor as a proof of concept. In another study, fiber mats composed of PLLA and poly(1,8-octanediol-co-citrate) (POC) mixture were formed, and POC prepolymer was incorporated to modify the fiber's cross-linking density [12].

1.2 Polymer-Based Nanoscale-Composite Hydrogels

Nanoscale-composite hydrogels were composed of saturated three-dimensional polymeric frameworks embedded with nanoscale particles. Upon absorbing of water, the hydrogels expand and enhance cellular durability in vivo via saturation. These promotes the nutrient dissemination and trash in a confined population of cell, creating a micro eco-system which is biomimetic. Formulating a mechanically stable and bioactive hydrogel remains a challenge. Nanoscale-composite hydrogels were also endowed with biological activity, electrical flow capacity, and attentiveness to stimuli. Nanomaterials which are based on carbon such as nanoscale-tubes of carbon and graphene oxide were employed to impart electrical flow capacity. Cutting-edge biomaterials for biological-medical and biological-technology uses are being created through nanocomposite hydrogels with various functional properties. Cancer research, immune modulation, tissue engineering, therapy of stem cell, and molecular and cellular treatments are a part of the fields where the application of nanoscale-composite hydrogels are explored. Biological inspired based nanoscale particles which are metallic being produced through approach of a green chemistry from so many extracts of plant or components of cell. The procedure of synthesis involves in metallic salts diminution either within the liver or components of hepatic organ. The phenomenon of altering the reflective characteristics of nanoscale particles which are metallic at the nanoscale, including the resonance of surface plasmon nanoscale particles which are based on gold and silver, has extensive uses in disease analysis and monitoring the progression of disease. Meanwhile, the catalytic properties of nanoscale particles based on metallic have been utilized against microbial agents and for the catalytic breakdown of non-compostable harmful dyes waste products.

1.3 *Bioactive Silicates Nanoscale-Composites*

Silicates which are synthetic based represent a different category of nanoscale-composites. These nanoscale-materials are highly thin (or two dimension) and possess an extreme degree of non-isotropy and practicality. Due to their greater outer layer-to-size ratio, they connect with biological entities in a significantly distinct way compared to their corresponding 3D macro, micro, and nanoscale counterparts. A number of life sciences and healthcare uses of these nanoscale-materials include imaging technology, therapeutic applications, and diagnosis of disease. These nanoscale particles was suggested in various bone and muscle cell-based therapy design implementations due to their potent biologically active properties [14]. Porous silicon (pSi) has garnered some attention recently as a substrate because of its hydrolysis-based dissolution mechanism, which is comparable to that of bulk Si and yields orthosilicic acid ($Si(OH)_4$) as the final product [15]. However, because of its larger surface area and deformed crystal lattice structure, its breakdown rate is significantly larger than that of bulk Si. Pore density and size have a significant impact on the rate of deterioration. According to Anderson et al. [16], the simulated bodily fluid completely eliminated a 1 μm dense coating of pSi with 70% permeability in roughly 1 day. A recent work showed how to create and characterize a brain probe based on pSi (50% porosity) for electrical signal recording [17].

1.4 *Bioinspired Nanocomposites*

A biologically inspired nanoscale-filler is incorporated into Nano-hydroxyapatite (nHAp), an eco-friendly inorganic substance present within tissue which is hard.

In healthcare orthopedic and dental uses, the implementation of nHAp as a bioengineering material was restricted to relaxed, high-volume areas due to its durable properties and lower fracture strength Thus, for non-weight-supporting purposes, nHAp was incorporated into a polymer-based hydrogel framework. Biologically-inspired Rosette Nanoscale-tube mixed materials were created through fabricating DNA bases with synthetic, typically using bonding of hydrogen interactions within guanine and cytosine. These composites were employed within healing medicine. The core consist of a layout resembling the collagen fibers of the extracellular matrix (ECM). Supplementary amino acids and peptide chains were added to enhance biologically activity and cellular adhesion between the Scaffold. Sheets of graphene oxide, spontaneously formed DNA through temporary bondings, were applied to create a three dimension shape. The outcome hierarchical framework which is biomimetic also preserves the current flow capability and structural integrity of graphene. These are utilized in cell-based therapy. Graphene oxide nanosheets possess biomolecule encapsulation and internalization of cell capabilities, making them suitable for gene delivery applications.

1.5 Polymer-Based Nanocomposites Infused Through Metallic Nanoscale Particles

The incorporation elements of metal within polymeric composites creates various pathways restoration of tissue activity. For instance, restructured atomic arrangements have been documented to form through bindings between peptides and hard inorganic substrates. These could be designed for improved biologic activity or further functionalized within materials which are based on synthetic to enhance the restorative efficiency of the implanted device. A polymeric framework (nanofiber) with metallic nanoparticles embedded within it, imparts conductivity, making it valuable for regenerative processes. Various types of polymers are found in cartilage and muscle, forming interconnected networks. By combining two or more polymers, synthetic replicas are created. For example, cartilage, a highly resilient tissue with considerable mechanical stiffness, is replicated using interpenetrating polymer networks (IPNs), semi-IPNs, and double networks (DNs). These intricate shape contain of two or more types of polymer strands interconnected within the sub-nanometric level. These leads to distinct an unique properties due to extensive polymer–polymer interactions. The various fabrication strategies of bioresorbable composites with adopted materials for surgical interventions avenues is portrayed in Fig. 1.

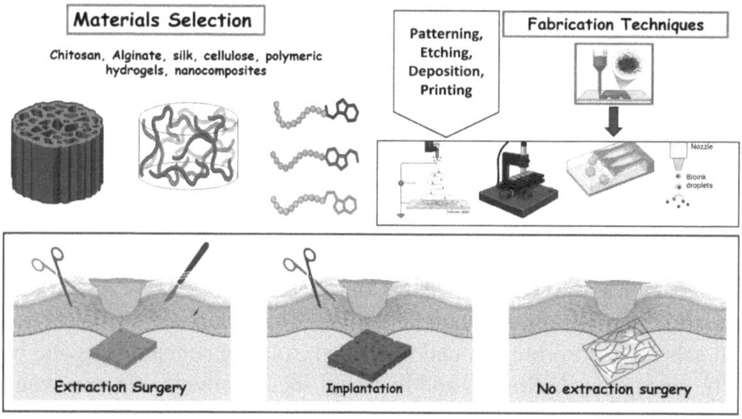

Fig. 1 Schematic representation of bioresorbable composites materials, fabrication techniques, and surgical intervention strategies

2 Current Avenues and Application Area of Bio-based Resorbable Composites

A wound, defined as a flaw or fracture within the epidermis can outcome from trauma or medical conditions. Based on the thickness and coverage of the epidermis affected, wounds are classified as superficial (epidermis only), partial-thickness (involving epidermis and dermis), or full-thickness (extending to subcutaneous fat and deeper tissues) [18]. Factors such as desiccation, infection, necrosis, maceration, pressure, trauma, or edema can hinder the healing process [19]. Here, lies the importance of wound dressings. There are different types of wound dressings though gauze remains widely used in hospitals due to its affordability and availability. Nowadays the impregnated gauzes with agents like iodine or zinc oxide, exhibit improved antimicrobial and wound-healing properties. However, gauze removal can disrupt newly formed tissue, highlighting the need for advanced alternatives. This brings the attention of the researchers to search for some innovations in the wound care industry that emphasize comfort, conformability, occasional dressing replacement, cost-efficiency, and prolonged shelf stability. The ideal wound dressing absorbs excess exudate and toxins, maintains moisture balance, prevents external infection, supports gas exchange, avoids excessive heat, and is simple to eliminate without more trauma. Advanced dressings aim to maintain a moist wound environment—a concept first highlighted by winter in 1962-accelerating the healing process. Moisture-retentive dressings include hydrogels and hydrocolloids [20].

2.1 Hydrogels

Hydrogel dressings consist of water-based gel with 90% hydration composed non-soluble water-attracting polymers that enlarge when in exposure to water [21]. They were usually composed of synthetic polymer chains, including polyvinylpyrrolidone and polymethacrylate, and certain ones are mixed with alginate dressings. These are ideal for dry wounds, and provide a "moisture donor" effect that rehydrates tissues, facilitates autolytic debridement, and supports collagenase production. These types of dressings, such as those developed by ConvaTec, are designed to absorb significant exudate while conforming to wound contours. These non-woven sheets transform into soft gels upon absorbing wound fluids and isolating contaminants and bacteria. Similarly, amorphous hydrogels and hydrogel sheets provide effective wound care through their high water content, which promotes oxygen transmission and cooling effects, making them suitable for burns, ulcers, and necrotic wounds [22]. The application of hydrogels can aid in wound cooling, which contributes to pain relief [23]. By smoothing the wound surface, the gel helps prevent infection in dead space while also supporting surface healing. Additionally, they aid in wound re-epithelialization as they imitate the skin's structure and aid in the growth of skin elements. They can aid in delivering therapeutic drugs that expedite healing of wound. Finally, they can be used for treating wounds at any stage, regardless of type, excluding cases involving significant fluid discharge, such as painful wounds, full and partial-thickness wounds, radiation burns, minor burns, and dry wounds [24]. Recently these hydrogel-based dressings are loaded with different drugs to promote a quicker healing process. In this respect, antimicrobial agents like iodine and silver are incorporated into dressings to combat pathogens such as *Pseudomonas aeruginosa* and *Staphylococcus aureus*, which are common in chronic wounds in managing infected wounds or preventing bacterial colonization [25]. Silver-based dressings, for instance, disrupt biofilm-based infections and reduce bacterial load, facilitating healing. Composite dressings also integrate hydrogels with other materials to address diverse wound needs. These products balance hydrophilic and hydrophobic properties, providing water-swellable but insoluble materials that absorb exudate without adhering to wound surfaces [26]. This minimizes trauma during dressing changes and reduces the frequency of replacement. More recently innovations in wound dressing technology also focus on combining antimicrobial agents with advanced materials for improved outcomes. For example, US Patent 8,431,151 B2 describes a hydrogel-based fibrous dressing incorporating silver ions for controlled antimicrobial release [27]. In this regard the electrospinning techniques are employed to create nanofiber networks, ensuring efficient delivery of silver and other therapeutic agents. As hydrogels have a limited capacity to absorb fluid, they are not appropriate for very moist wounds, which may become prone to maceration and infection. Additionally, because of their limited mechanical strength, they tear easily, which can make self-dressing changes challenging for patients.

2.2 Hydrocolloids

Hydrocolloid dressings consist of two layers and are both adhesive and occlusive: a central layer containing hydrophilic colloidal particles suspended within it, and an outer polyurethane layer that prevents bacterial infiltration. These dressings incorporate gel-forming substances, such as carboxymethylcellulose, gelatin, and pectin, along with other components like elastomers and adhesive layers [28]. Their mode of action involves creating a gel when in contact with the wound surface, which helps retain a moist environment, absorbs exudate, and shields granulation tissue. Products like Granuflex™, Comfeel™, and Tegasorb™ are offered in sheet or light film formats [29]. Hydrocolloids were perfect for treating partial and full-thickness wounds with light to moderate exudation, as well as wounds that have formed a scab. They can be applied to the wound for a maximum of 7 days [30]. These are composed of materials like gelatin and sodium carboxymethylcellulose that solidify into a gel when exposed to exudate, offering mild to moderate absorption and creating a protective seal.

2.3 Drug Delivery

A wide range of patents and research papers have looked into the potential uses of hydrogels in pharmaceutical delivery. However, only a small fraction has led to the development of commercial products. Due to their distinctive physical attributes, hydrogels have garnered considerable focus in the field of drug delivery, such as adjustable degradation, the ability to safeguard sensitive drugs, and porosity. Hydrogels are being more frequently studied as localized drug delivery systems [31–33]. Moreover, their high porosity, which can be tailored by adjusting the concentration of cross-links within their structure and their water-attracting properties, facilitates loading of drug and subsequent emit. The porous shape of hydrogels offers various benefits for distribute of drug, including the ability to achieve sustained release, which maintains a high local concentration of the active pharmaceutical ingredient over extended periods [34]. Owing to these features, hydrogels have been widely researched for multiple drug delivery strategies, including oral, transdermal, ocular, and injectable systems. Hydrogels are highly effective for the topical delivery of drugs to alleviate symptoms associated with various pathological situations [24]. For example, Nho et al. produced a medicinal gel based on poly(vinyl alcohol) or poly(vinylpyrrolidone) designed for atopic dermatitis therapy [24]. This gel incorporated plant extracts from medicinal species including *Houttuynia cordata, elm, celandine*, and *Canavalia gladiata*, selected for their potential effectiveness in treating dermatitis. Studies have investigated hydrogels as delivery systems for oral drugs, specifically for the regulated release of medications in the digestive tract [35, 36].

1. High capacity for fluid absorption

2. Antibacterial properties
3. Effective carriers for controlled drug release
4. Responsiveness to external physical, chemical, and biological stimuli
5. Promotion of oral tissue regeneration.

There are also studies indicating that hydrogels can shield the encapsulated drugs from the harsh conditions of the stomach and release them in a regulated manner in the intestine, thus enhancing their bioavailability and therapeutic effectiveness [35]. Drug release from hydrogels through this route can occur through diffusion-controlled mechanism. These systems enable drug release via diffusion through the hydrogel mesh or water-filled pores and are commonly classified as either reservoir or matrix devices. Park et al. [37, 38] Recommended superporous hydrogel structures as gastric-retentive systems for prolonged oral drug administration. The hydrogels were synthesized with (meth)acrylic acid or (meth)acrylamide, a disintegrant [such as a natural or synthetic cross-linked hydrophilic polymer like cross-linked carboxymethylcellulose or poly(vinyl pyrrolidone)], and a cross-linking compound like N,N′-methylenebisacrylamide. These hydrogel systems were recommended for the oral delivery of various active compounds, including non-steroidal anti-inflammatory drugs (NSAIDs) [39]. They also offer protection for drugs or proteins, including insulin, that are susceptible to proteolytic degradation within the stomach [24, 39, 40].

Transdermal drug delivery: Hydrogels have served as systems for transdermal drug delivery, providing benefits including prolonged medication release, enhanced patient cooperation, and minimized negative effects transdermal hydrogel patches can be designed for the controlled release of drugs, maintaining a consistent drug level in the systemic circulation EP 0 524 718 A1, a European Patent, showcased this potential by using polyurethane hydrogel matrices as integrated drug storage units. The hydrogels were synthesized by combining a solution of prepolymer, featuring an isocyanate-capped oxyalkylene-based prepolymer in a solvent devoid of water, alongside water. After the complete evaporation of the natural solvent, the hydrogel matrix formed could be filled within a drug for therapeutic purposes.

Ocular drug delivery: Hydrogels have been examined as systems for eye drug delivery, ensuring prolonged therapeutic release and increased therapeutic availability in the ocular region. Hydrogel-based ocular inserts and contact lenses can be utilized to administer drugs to both the anterior and posterior segments of the eye, overcoming the drawbacks of traditional eye drops, such as rapid elimination and poor drug penetration [41]. An alternative strategy for drug delivery across the cornea or sclera includes the use of hydrogels that degrade progressively over time. Unlike hydrogels used in soft contact lenses (SCLs), which prioritize long-term stability for repeated use, these formulations are designed for one-time use [40, 41]. This allows them to leverage their degradation profiles to ensure complete therapeutic delivery. In contrast, SCLs face challenges in maintaining effective drug loading and reloading at appropriate levels.

Yang et al. synthesize a multi-material hydrogel combining generation 3 poly(amidoamine) (PAMAM) dendrimers and PLGA nanoparticles for the co-delivery of two anti-glaucoma drugs, timolol maleate and brimonidine [42]. This innovative hydrogel integrates the biological compatibility and biodegradability of PLGA components for regulated drug emit with the photo-initiated cross-linking of dendrimer and PEG-acrylate, resulting in a liquid mucoadhesive hydrogel. Their study demonstrated sustained intraocular pressure (IOP) reduction for up to 4 days, with drug retention in the cornea, aqueous humor, and conjunctiva persisting for maximum 7 days in a normotensive model of rabbit.

- ***Injectable drug delivery***: Hydrogels which are injectable have become a promising platform for targeted and regulated delivery of drug, specifically in the diagnosis of conditions like cancer and diabetes. These hydrogels can be delivered through minimally invasive methods, creating a depot at the injection site that gradually emit the encapsulated drugs in a regulated fashion over a prolonged duration [43]. Changes in factors such as pore size, backbone charge, hydrophilicity, or cross-link density significantly affect the diffusion rates and solubility of molecules within the hydrogel matrix, enabling the injected hydrogel depots to function as localized drug release sites over varying time frames for a diverse array of therapeutic substances [44].

2.4 Tissue Engineering

Annually, a considerable number of fatalities and impairments occur due to the destruction or harm of tissues and organs from injuries or illnesses [45]. According to the Global Observatory on Donation and Transplantation (GODT), over 153,000 organ transplants were performed worldwide in 2019, marking a 4.9% growth compared to 2018. However, this number still accounted for only 10% of the global demand for organ transplants (Organ Donation and Transplantation Activities, 2021). While transplantation using compatible donors or artificial devices partially addresses this issue [46], the severe shortage of suitable donors leads to prolonged waiting lists for patients. Additionally, immune and inflammatory responses to transplanted tissues pose further complications [47, 48]. To overcome these challenges, researchers [49] have focused on fabricating biomaterials into structural scaffolds that mimic the extracellular matrix (ECM). These scaffolds support cell proliferation and tissue regeneration. When implanted, they can be either free of cell, encouraging cellular invasion from surrounding native tissue, or cell-laden, promoting functional regeneration with transplanted cells. The scaffolds are typically setup to degrade at a rate that supports new functional tissue formation until the native ECM can sustain the tissue independently. In this ground tissue engineering has become widely popular.

The concept of "tissue engineering" was introduced in 1988 as the "application of engineering principles and life sciences methods to gain a fundamental understanding of structure–function relationships in normal and diseased mammalian tissues, as well as the creation of biological substitutes for repairing or regenerating tissue or organ function." In basic language, it means enhancing or replacing certain tissues or organs using engineered materials and synthetic techniques. Tissue engineering represents a more recent advancement in the use of hydrogels. These customizable materials can operate as agents that occupy space, deliver bioactive compounds, or three-dimensional structures that arrange cells and impart signals to promote the generation of desired tissues. Several types of biomaterials are utilized in this domain. Space-filling agents stand out as the most commonly used scaffold type, typically employed for volume filling, preventing adhesions, and serving as biological "adhesives." Hydrogel scaffolds can also deliver drugs in different applications, such as stimulating angiogenesis and encapsulating secretory cells. Furthermore, they have been utilized for cell transplantation and in engineering a variety of tissues, including cartilage, bone, and smooth muscle [50]. A crucial characteristic of hydrogels is their biocompatibility, defined as the ability of a material to interact with bodily tissues without causing harm to surrounding tissues or triggering undesirable responses [51]. Synthetic polymers suitable for hydrogel formation in tissue engineering include poly(ethylene oxide), poly(vinyl alcohol), poly(acrylic acid), poly(propylene fumarate-co-ethylene glycol), and polypeptides. Additionally, naturally derived polymers such as agarose, alginate, chitosan, collagen, fibrin, gelatin, and hyaluronic acid can also be utilized for this purpose [52].

2.5 Ophthalmic Applications

Hydrogels offer numerous potential applications in ophthalmology. Soft contact lenses hydrogels have been utilized as the main material in the production of soft contact lenses. In their pioneering 1960 paper, Wichterle and Lim were the first to present a hydrogel made from poly-2-hydroxyethylmethacrylate (PHEMA), describing it as a synthetic biocompatible substance ideal for contact lens use [53]. PHEMA lenses were first introduced in Western Europe in 1962, although their initial success was limited. In 1965, the National Patent Development Corporation (NPDC) obtained the rights to this technology, which was subsequently sold to Bausch & Lomb. The company improved Wichterle's spin-casting method and eventually gained approval from the Food and Drug Administration (FDA) for their PHEMA lenses in 1971 [53]. Rigid contact lenses are mainly made of hydrophobic materials, including poly(methyl methacrylate) (PMMA) and poly(hexa-fluoroisopropyl methacrylate) (HFIM), while soft contact lenses consist of hydrogels [52, 53]. Foldable Intraocular Lenses: Employed in creating foldable intraocular lenses [54], Ocular Adhesives: Utilized as adhesives for wound repair in ocular procedures [54], Vitreous Substitutes: Explored as substitutes for the vitreous body in the eye [54]. Intravitreal Drug Delivery Systems: Investigated as systems for delivering drugs directly into the vitreous humor [55]. Recent advancements in hydrogel applications in ophthalmology underscore their versatility, biocompatibility, and adaptability as therapeutic tools. Hydrogels have shown their capability to direct, regulate, and maintain drug release to the posterior part of the eye, minimizing the necessity for invasive procedures and enhancing patient outcomes [56]. Moreover, they show potential for post-surgery drug delivery and disease diagnosis. However, much of the hydrogel-based research remains in the preclinical stage and necessitates extensive clinical assessment [57].

2.6 Dental Applications

In oral tissue engineering, hydrogels function as dental biomaterials with two main roles: preventing bacterial infections and fostering tissue regeneration. Hydrogel-based biomaterials demonstrate considerable promise for regenerating oral tissues and delivering drugs [58]. However, hydrogels created for oral applications must address challenges such as temperature fluctuations, changing pH levels due to saliva and biofilms, and the presence of various bacterial species [59]. By controlling the synthesis of hydrogels, their porous structure can be fine-tuned to achieve optimal pore size and shape for integration with surrounding tissues. This promotes cell adhesion and growth, making hydrogels highly effective for regenerative therapies [60, 61]. Hydrogels, as biocompatible polymer systems, have emerged as a focal point in biomedical research. Owing to their capacity to imitate the structure of natural extracellular matrices, hydrogels are viewed as ideal materials for scaffolds in tissue engineering, especially in the repair of dental pulp and periodontal damage [60]. By combining hydrogel complexes with a variety of stem cells and growth factors, promising results have been observed in endodontic and periodontal tissue engineering investigations. Additionally, their responsiveness to external stimuli enables hydrogels to exhibit biological effects, making them highly suitable for localized drug delivery uses [60, 61]. Pulp regeneration applies the principles of tissue engineering by culturing pulp stem cells in vitro and implanting them onto a biocompatible, biodegradable scaffold. This process induces the formation of pulp-dentin complexes and pulp like tissues, enabling the repair of damaged pulp and the restoration of its physiological functions. Hydrogels, particularly in injectable form, offer significant advantages for pulp regeneration [62, 63]. The injectable fluid can be used to completely fill the pulp chamber and root canal prior to gelation. Once the gelation reaction occurs, the hydrogel adheres closely to the surrounding tissue, making it well-suited for pulp regeneration studies. Additionally, hydrogels can encapsulate cells and undergo a sol–gel transition, ensuring a more uniform distribution of cells within the gel. This approach addresses the challenge of exogenous cells penetrating deep into scaffolds within the root canal, enhancing the effectiveness of pulp regeneration strategies. In recent studies, ultraviolet (UV) light-cross-linked gelatin methacryloyl (GelMA) hydrogel has been employed to generate tissue-engineered dental pulp-like constructs with pre-vascularization. Injectable GelMA, loaded with dental pulp stem cells (DPSCs) and human umbilical vein endothelial cells (HUVECs), has shown the ability to promote cell attachment and growth while simultaneously boosting angiogenesis [64]. Hydrogels have also recently attracted considerable attention as sustained-emit systems and materials which are based on scaffold in periodontal tissue remaking studies [65, 66]. While various types of hydrogels are appropriate for dentoalveolar tissue remaking, modifications or combinations are frequently needed to achieve optimal and successful results [67].

3 Fabrication Strategies and Challenges Prospect

Special fabrication methods that work with bioresorbable materials are required for novel devices. For biosensors to function as intended with the least amount of harm to the biological system, they must blend in with the soft, curving tissues that surround them. In some studies, a flexible implant was created that closely adheres to the tissue without inducing stress, using ultrathin layers or specialized mesh designs. Despite the advantages, the problem persists that soft or flexible materials cannot maintain their own support, which complicates the manufacturing of such devices. Furthermore, natural and synthetic bioresorbable polymers are the most popular materials for these applications; yet, These materials often fail to exhibit adequate resistance to high temperatures (during baking) and the solvents involved in standard microfabrication stages (including developers and etchants) [13]. Emerging manufacturing strategies that align with bioresorbable materials and enable low-cost, large-volume production of bioresorbable IMDs are being used to tackle these challenges. Common fabrication methods for bioresorbable implantable medical devices (IMDs) in the literature include transfer printing [68], screen printing [69], micro-contact printing, capillary force lithography [70], and stencil lithography, along with conventional microfabrication techniques such as photolithography, etching, and deposition. A soft lithography technique is employed to create flexible bioresorbable substrates with rounded microfeatures [71]. This process includes creating a negative silicon mold patterned through photolithography and plasma etching, followed by spin-coating and post-baking a sacrificial sucrose solution. The flexible polymer PGS is then replica-molded and extracted [72]. Shadow masks are occasionally used, but they have drawbacks, including with a minimum feature size in the range of several micrometers and processing temperatures below the melting point, recent breakthroughs in shadow mask or stencil lithography have produced feature sizes smaller than 100 nm, although care is needed to avoid smearing and obstructions.

4 Regulatory Challenges in Translating Research into Clinics

Resorbable polymer-based biomaterials must comply with legal standards to be released to the market. Medical devices are categorized into three classes depending on the level of potential risk or discomfort. The advancement of cutting-edge drug delivery systems and personalized medicines encounters regulatory obstacles, particularly for treating bone-related diseases. Prolonged biocompatibility testing in animal models is vital. Adapting extreme-throughput screening of drug to clinical settings necessitates detailed validation from skeletal structures. Finding a middle ground within the complexity of models and processing efficiency calls for careful experimental setup and exhaustive validation evaluations [73, 74]. Complying with well practices of laboratory and well Practices of manufacturing is essential to ensure that research procedures and manufacturing standards align with controlled requirements. Effective translation depends on collaboration across disciplines, involving researchers, clinicians, regulatory professionals, and industry stakeholders. Engineering strategies for bio-absorbable composite frameworks, stents, vascular grafts, and cardiac patches offer clinical perspectives for material enhancement. Cardiac computed tomography assesses internal structures, highlighting multifunctional resorbable implants tailored to specific biomedical applications. Advanced bioresorbable polymers provide improved mechanical properties and biocompatibility, while hybrid scaffolds incorporate fillers for tissue regeneration. New fabrication techniques including three dimension copying and electrospinning activate the innovation of intricate, specific scaffolds for patient that promote better tissue integration and vascularization. In summary, tissue engineering with bioresorbable polymers holds great promise due to ongoing advancements in material science, production methods, and personalized medicine. These developments are expected to lead to more tailored and effective treatments, enhancing patient outcomes and offering widely uses in remaking medicine.

Conflict of Interest: There is no conflict of interest in this study.

References

1. RV et al (2024) Developments in biomedical materials: from conventional implantation to state-of-the-art pharmaceutical uses. In: E3S Web of conferences, vol 507, p 01056. https://doi.org/10.1051/e3sconf/202450701056
2. Pertici G (2017) Introduction to bioresorbable polymers for biomedical applications. In: Bioresorbable polymers for biomedical applications. Elsevier, pp 3–29. https://doi.org/10.1016/B978-0-08-100262-9.00001-X
3. Agrawal CM, Ray RB (2001) Biodegradable polymeric scaffolds for musculoskeletal tissue engineering. J Biomed Mater Res 55(2):141–150. https://doi.org/10.1002/1097-4636(200105)55:2%3c141::AID-JBM1000%3e3.0.CO;2-J

4. Waris E, Konttinen YT, Ashammakhi N, Suuronen R, Santavirta S (2004) Bioabsorbable fixation devices in trauma and bone surgery: current clinical standing. Expert Rev Med Dev 1(2):229–240. https://doi.org/10.1586/17434440.1.2.229
5. Knipe JM, Chen F, Peppas NA (2015) Enzymatic biodegradation of hydrogels for protein delivery targeted to the small intestine. Biomacromol 16(3):962–972. https://doi.org/10.1021/bm501871a
6. Tyler B, Gullotti D, Mangraviti A, Utsuki T, Brem H (2016) Polylactic acid (PLA) controlled delivery carriers for biomedical applications. Adv Drug Deliv Rev 107:163–175. https://doi.org/10.1016/j.addr.2016.06.018
7. Jain A, Kunduru KR, Basu A, Mizrahi B, Domb AJ, Khan W (2016) Injectable formulations of poly(lactic acid) and its copolymers in clinical use. Adv Drug Deliv Rev 107:213–227. https://doi.org/10.1016/j.addr.2016.07.002
8. Leenslag JW, Pennings AJ (1987) Synthesis of high-molecular-weight poly(L-lactide) initiated with tin 2-ethylhexanoate. Die Makromolekulare Chemie 188(8):1809–1814. https://doi.org/10.1002/macp.1987.021880804
9. Mano JF, Sousa RA, Boesel LF, Neves NM, Reis RL (2004) Bioinert, biodegradable and injectable polymeric matrix composites for hard tissue replacement: state of the art and recent developments. Compos Sci Technol 64(6):789–817. https://doi.org/10.1016/j.compscitech.2003.09.001
10. Juan P-K et al (2021) Bioactivity and bone cell formation with poly-ε-caprolactone/bioceramic 3D porous Scaffolds. Polymers (Basel) 13(16):2718. https://doi.org/10.3390/polym13162718
11. Singh R, Bathaei MJ, Istif E, Beker L (2020) A review of bioresorbable implantable medical devices: materials, fabrication, and implementation. Adv Healthc Mater 9(18). https://doi.org/10.1002/adhm.202000790
12. Boutry CM et al (2018) A stretchable and biodegradable strain and pressure sensor for orthopaedic application. Nat Electron 1(5):314–321. https://doi.org/10.1038/s41928-018-0071-7
13. Huang X et al (2014) Biodegradable materials for multilayer transient printed circuit boards. Adv Mater 26(43):7371–7377. https://doi.org/10.1002/adma.201403164
14. Yu X, Shou W, Mahajan BK, Huang X, Pan H (2018) Materials, processes, and facile manufacturing for bioresorbable electronics: a review. Adv Mater 30(28). https://doi.org/10.1002/adma.201707624
15. Kang S-K et al (2016) Bioresorbable silicon electronic sensors for the brain. Nature 530(7588):71–76. https://doi.org/10.1038/nature16492
16. Anderson SHC, Elliott H, Wallis DJ, Canham LT, Powell JJ (2003) Dissolution of different forms of partially porous silicon wafers under simulated physiological conditions. Physica Status Solidi (a) 197(2):331–335. https://doi.org/10.1002/pssa.200306519
17. Jian Y et al (2020) Gas sensors based on chemi-resistive hybrid functional nanomaterials. Nanomicro Lett 12(1):71. https://doi.org/10.1007/s40820-020-0407-5
18. Ashammakhi N et al (2021) Biodegradable implantable sensors: materials design, fabrication, and applications. Adv Funct Mater 31(49). https://doi.org/10.1002/adfm.202104149
19. Madaghiele M, Sannino A, Ambrosio L, Demitri C (2014) Polymeric hydrogels for burn wound care: advanced skin wound dressings and regenerative templates. Burns Trauma 2(4):153. https://doi.org/10.4103/2321-3868.143616
20. Bryan J (2004) Moist wound healing: a concept that changed our practice. J Wound Care 13(6):227–228. https://doi.org/10.12968/jowc.2004.13.6.26625
21. Xiang J, Shen L, Hong Y (2020) Status and future scope of hydrogels in wound healing: synthesis, materials and evaluation. Eur Polym J 130:109609. https://doi.org/10.1016/j.eurpolymj.2020.109609
22. Williams C (1996) Granugel: hydrocolloid gel. British J Nurs 5(3):188–190. https://doi.org/10.12968/bjon.1996.5.3.188
23. Kass LE, Nguyen J (2022) Nanocarrier-hydrogel composite delivery systems for precision drug release. WIREs Nanomed Nanobiotechnol 14(2). https://doi.org/10.1002/wnan.1756

24. Caló E, Khutoryanskiy VV (2015) Biomedical applications of hydrogels: a review of patents and commercial products. Eur Polym J 65:252–267. https://doi.org/10.1016/j.eurpolymj.2014.11.024
25. Phillips PL et al (2015) Antimicrobial dressing efficacy against mature *Pseudomonas aeruginosa* biofilm on porcine skin explants. Int Wound J 12(4):469–483. https://doi.org/10.1111/iwj.12142
26. Ribeiro M, Simões M, Vitorino C, Mascarenhas-Melo F (2024) Hydrogels in cutaneous wound healing: insights into characterization, properties, formulation and therapeutic potential. Gels 10(3):188. https://doi.org/10.3390/gels10030188
27. Paul SD, Sharma H, Jeswani G, Jha AK (2017) Novel gels: implications for drug delivery. In: Nanostructures for drug delivery. Elsevier, pp 379–412. https://doi.org/10.1016/B978-0-323-46143-6.00012-9
28. Silva JM et al (2019) Therapeutic role of deep eutectic solvents based on menthol and saturated fatty acids on wound healing. ACS Appl Bio Mater 2(10):4346–4355. https://doi.org/10.1021/acsabm.9b00598
29. Westby MJ, Dumville JC, Soares MO, Stubbs N, Norman G (2017) Dressings and topical agents for treating pressure ulcers. Cochrane Database Syst Rev. https://doi.org/10.1002/14651858.CD011947.pub2
30. Kamińska MS, Cybulska AM, Skonieczna-Żydecka K, Augustyniuk K, Grochans E, Karakiewicz B (2020) Effectiveness of hydrocolloid dressings for treating pressure ulcers in adult patients: a systematic review and meta-analysis. Int J Environ Res Public Health 17(21):7881. https://doi.org/10.3390/ijerph17217881
31. Vashist A, Vashist A, Gupta YK, Ahmad S (2014) Recent advances in hydrogel based drug delivery systems for the human body. J Mater Chem B 2(2):147–166. https://doi.org/10.1039/C3TB21016B
32. Elvira C, Mano JF, San Román J, Reis RL (2002) Starch-based biodegradable hydrogels with potential biomedical applications as drug delivery systems. Biomaterials 23(9):1955–1966. https://doi.org/10.1016/S0142-9612(01)00322-2
33. Vigata M, Meinert C, Hutmacher DW, Bock N (2020) Hydrogels as drug delivery systems: a review of current characterization and evaluation techniques. Pharmaceutics 12(12):1188. https://doi.org/10.3390/pharmaceutics12121188
34. Hoare TR, Kohane DS (2008) Hydrogels in drug delivery: progress and challenges. Polymer (Guildf) 49(8):1993–2007. https://doi.org/10.1016/j.polymer.2008.01.027
35. Jacob S, Nair AB, Shah J, Sreeharsha N, Gupta S, Shinu P (2021) Emerging role of hydrogels in drug delivery systems, tissue engineering and wound management. Pharmaceutics 13(3):357. https://doi.org/10.3390/pharmaceutics13030357
36. Chelu M, Magdalena Musuc A (2024) Biomaterials-based hydrogels for therapeutic applications. In: Biomaterials in microencapsulation. IntechOpen. https://doi.org/10.5772/intechopen.1004826
37. Park K (1988) Enzyme-digestible swelling hydrogels as platforms for long-term oral drug delivery: synthesis and characterization. Biomaterials 9(5):435–441. https://doi.org/10.1016/0142-9612(88)90009-9
38. Chen J, Blevins WE, Park H, Park K (2000) Gastric retention properties of superporous hydrogel composites. J Control Release 64(1–3):39–51. https://doi.org/10.1016/S0168-3659(99)00139-X
39. Velasco D et al (2011) PH-sensitive polymer hydrogels derived from morpholine to prevent the crystallization of ibuprofen. J Control Release 149(2):140–145. https://doi.org/10.1016/j.jconrel.2010.10.015
40. Morishita M, Lowman AM, Takayama K, Nagai T, Peppas NA (2002) Elucidation of the mechanism of incorporation of insulin in controlled release systems based on complexation polymers. J Control Release 81(1–2):25–32. https://doi.org/10.1016/S0168-3659(02)00019-6
41. Cooper RC, Yang H (2019) Hydrogel-based ocular drug delivery systems: Emerging fabrication strategies, applications, and bench-to-bedside manufacturing considerations. J Control Release 306:29–39. https://doi.org/10.1016/j.jconrel.2019.05.034

42. Yang H, Tyagi P, Kadam RS, Holden CA, Kompella UB (2012) Hybrid dendrimer hydrogel/PLGA nanoparticle platform sustains drug delivery for one week and antiglaucoma effects for four days following one-time topical administration. ACS Nano 6(9):7595–7606. https://doi.org/10.1021/nn301873v
43. Dimatteo R, Darling NJ, Segura T (2018) In situ forming injectable hydrogels for drug delivery and wound repair. Adv Drug Deliv Rev 127:167–184. https://doi.org/10.1016/j.addr.2018.03.007
44. Zeng Y et al (2015) Preformed gelatin microcryogels as injectable cell carriers for enhanced skin wound healing. Acta Biomater 25:291–303. https://doi.org/10.1016/j.actbio.2015.07.042
45. Chapekar MS (2000) Tissue engineering: challenges and opportunities. J Biomed Mater Res 53(6):617–620. https://doi.org/10.1002/1097-4636(2000)53:6%3c617::AID-JBM1%3e3.0.CO;2-C
46. Lee KY, Mooney DJ (2001) Hydrogels for tissue engineering. Chem Rev 101(7):1869–1880. https://doi.org/10.1021/cr000108x
47. Beyar R (2011) Challenges in organ transplantation. Rambam Maimonides Med J 2(2). https://doi.org/10.5041/RMMJ.10049
48. Black CK, Termanini KM, Aguirre O, Hawksworth JS, Sosin M (2018) Solid organ transplantation in the 21st century. Ann Transl Med 6(20):409–409. https://doi.org/10.21037/atm.2018.09.68
49. Langer R, Vacanti JP (1993) Tissue engineering. Science (1979) 260(5110):920–926. https://doi.org/10.1126/science.8493529
50. Hayami JWS, Waldman SD, Amsden BG (2016) Chondrocyte generation of cartilage-like tissue following photoencapsulation in methacrylated polysaccharide solution blends. Macromol Biosci 16(7):1083–1095. https://doi.org/10.1002/mabi.201500465
51. Gaffey AC et al (2015) Injectable shear-thinning hydrogels used to deliver endothelial progenitor cells, enhance cell engraftment, and improve ischemic myocardium. J Thorac Cardiovasc Surg 150(5):1268–1277. https://doi.org/10.1016/j.jtcvs.2015.07.035
52. Zhou X et al (2016) Three-dimensional scaffold of type II collagen promote the differentiation of adipose-derived stem cells into a nucleus pulposus-like phenotype. J Biomed Mater Res A 104(7):1687–1693. https://doi.org/10.1002/jbm.a.35701
53. O'Donnell C, Maldonado-Codina C (2005) Agreement and repeatability of central thickness measurement in normal corneas using ultrasound pachymetry and the OCULUS Pentacam. Cornea 24(8):920–924. https://doi.org/10.1097/01.ico.0000157422.01146.e9
54. Ye S, Wei B, Zeng L (2022) Advances on hydrogels for oral science research. Gels 8(5):302. https://doi.org/10.3390/gels8050302
55. Ribeiro AM, Figueiras A, Veiga F (2015) Improvements in topical ocular drug delivery systems: hydrogels and contact lenses. J Pharm Pharm Sci 18(5):683. https://doi.org/10.18433/J3H60P
56. Wu KY, Akbar D, Giunta M, Kalevar A, Tran SD (2023) Hydrogels in ophthalmology: novel strategies for overcoming therapeutic challenges. Materials 17(1):86. https://doi.org/10.3390/ma17010086
57. Aragón-Navas A et al (2024) Smart biodegradable hydrogels: drug-delivery platforms for treatment of chronic ophthalmic diseases affecting the back of the eye. Int J Pharm 649:123653. https://doi.org/10.1016/j.ijpharm.2023.123653
58. Arora S et al (2023) Role of Chitosan hydrogels in clinical dentistry. Gels 9(9):698. https://doi.org/10.3390/gels9090698
59. Chen A et al (2023) Hydrogels for oral tissue engineering: challenges and opportunities. Molecules 28(9):3946. https://doi.org/10.3390/molecules28093946
60. Muresan SMC et al (2023) Dental hydrogels with essential oils with potential activity in periodontitis. Appl Sci 13(3):1787. https://doi.org/10.3390/app13031787
61. Flores-Espinoza AI et al (2023) Gelatin–Chitosan hydrogel biological, antimicrobial and mechanical properties for dental applications. Biomimetics 8(8):575. https://doi.org/10.3390/biomimetics8080575
62. Jones TD, Kefi A, Sun S, Cho M, Alapati SB (2016) An optimized injectable hydrogel scaffold supports human dental pulp stem cell viability and spreading. Adv Med 2016:1–8. https://doi.org/10.1155/2016/7363579

63. Naghizadeh Z, Karkhaneh A, Khojasteh A (2018) Self-crosslinking effect of chitosan and gelatin on alginate based hydrogels: Injectable in situ forming scaffolds. Mater Sci Eng, C 89:256–264. https://doi.org/10.1016/j.msec.2018.04.018
64. Khayat A et al (2017) GelMA-encapsulated hDPSCs and HUVECs for dental pulp regeneration. J Dent Res 96(2):192–199. https://doi.org/10.1177/0022034516682005
65. Chen L et al (2017) Drug-loadable calcium alginate hydrogel system for use in oral bone tissue repair. Int J Mol Sci 18(5):989. https://doi.org/10.3390/ijms18050989
66. Chien K-H et al (2018) Promoting induced pluripotent stem cell-driven biomineralization and periodontal regeneration in rats with maxillary-molar defects using injectable BMP-6 hydrogel. Sci Rep 8(1):114. https://doi.org/10.1038/s41598-017-18415-6
67. Salar Amoli M, EzEldeen M, Jacobs R, Bloemen V (2021) Materials for dentoalveolar bioprinting: current state of the art. Biomedicines 10(1):71. https://doi.org/10.3390/biomedicines10010071
68. Song K, Yin D, Schiavone P (2021) Thermal-electric-elastic analyses of a thermoelectric material containing two circular holes. Int J Solids Struct 213:111–120. https://doi.org/10.1016/j.ijsolstr.2020.12.019
69. Bettinger CJ, Bruggeman JP, Misra A, Borenstein JT, Langer R (2009) Biocompatibility of biodegradable semiconducting melanin films for nerve tissue engineering. Biomaterials 30(17):3050–3057. https://doi.org/10.1016/j.biomaterials.2009.02.018
70. Vazquez-Mena O, Gross L, Xie S, Villanueva LG, Brugger J (2015) Resistless nanofabrication by stencil lithography: a review. Microelectron Eng 132:236–254. https://doi.org/10.1016/j.mee.2014.08.003
71. Bettinger CJ, Orrick B, Misra A, Langer R, Borenstein JT (2006) Microfabrication of poly (glycerol–sebacate) for contact guidance applications. Biomaterials 27(12):2558–2565. https://doi.org/10.1016/j.biomaterials.2005.11.029
72. Hwang S-W et al (2015) Biodegradable elastomers and silicon nanomembranes/nanoribbons for stretchable, transient electronics, and biosensors. Nano Lett 15(5):2801–2808. https://doi.org/10.1021/nl503997m
73. Dobrzyńska-Mizera M et al (2024) Engineering of bioresorbable polymers for tissue engineering and drug delivery applications. Adv Healthc Mater 13(30). https://doi.org/10.1002/adhm.202401674
74. Ullah MW, Fu L, Lamboni L, Shi Z, Yang G (2019) Current trends and biomedical applications of resorbable polymers. In: Materials for biomedical engineering. Elsevier, pp 41–86. https://doi.org/10.1016/B978-0-12-818415-8.00003-6

MIX
Papier aus verantwortungsvollen Quellen
Paper from responsible sources
FSC® C105338

If you have any concerns about our products,
you can contact us on
ProductSafety@springernature.com

In case Publisher is established outside the EU,
the EU authorized representative is:
Springer Nature Customer Service Center GmbH
Europaplatz 3, 69115 Heidelberg, Germany

Printed by Libri Plureos GmbH
in Hamburg, Germany